Education in the Asia-Pacific Region: Issues, Concerns and Prospects

Volume 75

The purpose of this Series is to meet the needs of those interested in an in-depth analysis of current developments in education and schooling in the vast and diverse Asia-Pacific Region. The Series will be invaluable for educational researchers, policy makers and practitioners, who want to better understand the major issues, concerns and prospects regarding educational developments in the Asia-Pacific region.

The Series complements the Handbook of Educational Research in the Asia-Pacific Region, with the elaboration of specific topics, themes and case studies in greater breadth and depth than is possible in the Handbook.

Topics to be covered in the Series include: secondary education reform; reorientation of primary education to achieve education for all; re-engineering education for change; the arts in education; evaluation and assessment; the moral curriculum and values education; technical and vocational education for the world of work; teachers and teaching in society; organisation and management of education; education in rural and remote areas; and, education of the disadvantaged.

Although specifically focusing on major educational innovations for development in the Asia-Pacific region, the Series is directed at an international audience.

The Series Education in the Asia-Pacific Region: Issues, Concerns and Prospects, and the Handbook of Educational Research in the Asia-Pacific Region, are both publications of the Asia-Pacific Educational Research Association.

Those interested in obtaining more information about the Monograph Series, or who wish to explore the possibility of contributing a manuscript, should (in the first instance) contact the publishers.

Please contact Lay Peng, Ang (e-mail: laypeng.ang@springer.com) for submitting book proposals for this series.

Bruno S. Sergi • Elena G. Popkova
Ainura A. Adieva • Nurgul K. Atabekova
Editors

University 4.0 and Educational Technology for Sustainable Development

Best Practices and Perspectives in Russia and Central Asia

Springer

Editors
Bruno S. Sergi
University of Messina
Messina, Italy

Ainura A. Adieva
International University of Kyrgyzstan
Bishkek, Kyrgyzstan

Elena G. Popkova
The Laboratory of Scientometrics and International Rankings
Armenian State University of Economics
Yerevan, Armenia

Nurgul K. Atabekova
Department of Legal Studies
The International University of Kyrgyzstan
Bishkek, Kyrgyzstan

ISSN 1573-5397 ISSN 2214-9791 (electronic)
Education in the Asia-Pacific Region: Issues, Concerns and Prospects
ISBN 978-981-96-7375-9 ISBN 978-981-96-7376-6 (eBook)
https://doi.org/10.1007/978-981-96-7376-6

This Springer imprint is published by the registered company Springer Nature Singapore Pte Ltd.
The registered company address is: 152 Beach Road, #21-01/04 Gateway East, Singapore 189721, Singapore

A Systemic Approach to Sustainable Development of Higher Education in Russia and Central Asia in the Transition to Universities 4.0 (Introduction)

Russia and other countries of Central Asia are distinguished participants in the global economy because the dynamic development of their economies is based on a high level of education. A progressive knowledge society, where higher education and lifelong learning are widely accessible and prevalent, is the cornerstone of outstanding innovative development and rapid economic growth in Russia, along with other countries of Central Asia. Universities play a decisive role in advancing the knowledge society in these countries.

In recent years, under the influence of the Fourth Industrial Revolution, Russia and Central Asian countries have witnessed the emergence of Universities 4.0. They represent new participants in the system of science and higher education, whose activities entail the active integration of educational technologies, commonly referred to as EdTech. The emergence of Universities 4.0 has occurred against the backdrop of the active participation of Russia and other Central Asian countries in global initiatives for sustainable development, particularly encompassing the system of science and higher education.

This book defines a University 4.0 as a scientific and educational institution adapted to the Fourth Industrial Revolution, one that supports and promotes progressive life values (particularly those rooted in the sustainable development goals within the context of sustainable education as established by the UNESCO system), thereby achieving a high degree of inclusiveness in its various manifestations.

First, there is cultural inclusiveness, which embraces all cultures and underpins international cooperation among Universities 4.0. This manifests through transnational academic mobility, the attraction of international students and faculty, and the global dissemination of knowledge and technology. Owing to this cultural inclusiveness, a University 4.0 actively participates in international activity and is featured in global university rankings, including THE, QS, and ARWU.

Second, gender inclusiveness is reflected in the equal opportunities and social and financial conditions afforded to men and women in research and education, whether as students, educators, or university administrators. The conceptual basis for this is laid out in the World Bank report "Overcoming Barriers for Women in

STEM," which calls for a greater engagement of women in science, technology, engineering, and mathematics (STEM) disciplines.

Third, social inclusiveness involves creating favorable conditions for the participation of individuals with disabilities in science and higher education. This includes providing accessible campuses, research laboratories, and classrooms, as well as extensive opportunities for distance learning and remote employment of university educators, alongside a conflict-free and supportive social environment that encourages the involvement of people with disabilities.

Fourth, digital inclusiveness is embodied in the commitment of a University 4.0 to employ advanced information and communication technologies in its scientific, educational (EdTech), and marketing activities, as well as in its administration and the active training of IT professionals for the digital economy.

Fifth, ecological inclusiveness is demonstrated by a University 4.0's responsibility for the environmental impacts of its activities and its commitment to mitigating these effects through the rationalization of resource consumption, the development of eco-friendly technologies, and the training of green professionals who are competitive and in demand in the labor market.

By comprehensively addressing these facets of inclusiveness, this book presents a broad perspective on the University 4.0 model. The highlighted manifestations of inclusiveness define the main directions explored in the book, which are illustrated through the experiences of Universities 4.0 in Russia and Central Asia.

Examples of Universities 4.0 examined in this book include participants of Russia's "Priority 2030" program, such as Peoples' Friendship University of Russia (RUDN), Lomonosov Moscow State University (MSU), St. Petersburg State University (SPBU), National Research University Higher School of Economics (HSE), ITMO University, National University of Science and Technology MISIS, Moscow Institute of Physics and Technology (MIPT), Moscow Engineering Physics Institute (MEPhI), Pirogov Russian National Research Medical University, I. M. Sechenov First Moscow State Medical University (Sechenov University), Voronezh State Medical University named after N. N. Burdenko, Moscow Aviation Institute (MAI), Bauman Moscow State Technical University (BMSTU), Russian Presidential Academy of National Economy and Public Administration (RANEPA), and others.

Examples of Universities 4.0 in Central Asian countries discussed in this book include the leading universities of Kazakhstan—Kazakh State Agrarian Research University, Abai Kazakh National Pedagogical University, Narxoz University, S. Seifullin Kazakh Agro Technical Research University, L. N. Gumilev Eurasian National University, Karagandy State University, South Kazakhstan University named after Mukhtar Auezov (SKU), Karaganda University of Kazpotrebsoyuz, Almaty Technological University, Kazakh National Medical University, and Satbayev University. Universities from Uzbekistan are also considered—Tashkent State University of Economics, Tashkent Institute of Irrigation and Agricultural Mechanization Engineers, Tashkent University of Information Technologies named after Muhammad al-Khwarizmi, Tashkent State University of Law, and the Higher

School of Business and Entrepreneurship under the Government of the Republic of Uzbekistan.

The theoretical understanding of the University 4.0 model developed in this book for Russia and Central Asia allows for a comparative analysis with models from other regions of the world, including the USA. In terms of cultural inclusiveness, the primary challenge for Universities 4.0 in Russia and Central Asian countries lies in embracing the cultures of foreign nations, a task facilitated by their inherently multiethnic social fabric characterized by internal unity and harmonious coexistence of diverse cultures. In contrast, cultural inclusiveness in US Universities 4.0 centers on achieving racial equality within the domestic social environment.

Regarding gender inclusiveness, Universities 4.0 in Russia and Central Asia focus on expanding opportunities for women to access higher education and engage in research, whereas in the USA the emphasis is on enhancing gender diversity in science and higher education beyond the traditional binary distinctions.

In terms of the social inclusiveness, the integration of individuals with disabilities into science and higher education in Russia and Central Asia is primarily supported through state funding. In the USA, it is largely driven by private investments (charitable donations) and tuition-based educational services.

With respect to ecological inclusiveness, Universities 4.0 in Russia and Central Asia prioritize the training of green professionals for industry and energy sectors. In the USA, there is no clear industry-specific focus. Instead, Universities 4.0 cultivate a broad environmental culture that promotes energy conservation, waste reduction, recycling, and the safe disposal of waste.

In the realm of digital inclusiveness, EdTech initiatives in Universities 4.0 in Russia and Central Asia are primarily concentrated on the educational process (e.g., distance learning). In the USA, the focus is on developing digital educational resources, such as scientometric databases and university rankings.

In this context, it is necessary to study Universities 4.0 in connection with sustainable development. Therefore, it is pertinent to determine the current contribution and prospective applications of educational technologies for the sustainable development of the system of science and higher education. Furthermore, it is pertinent to define the unique model of sustainable development of higher education in Russia and Central Asia in the transition to Universities 4.0. References to certain features of this model can be found in existing literature.

One characteristic of the sustainable development of Universities 4.0 is the application of advanced educational technologies to achieve Sustainable Development Goals (SDGs). Among the prospective applications of advanced EdTech is the training of digital professionals in Universities 4.0. While traditional educational technologies aid in developing theoretical knowledge, digital educational technologies are more beneficial for cultivating practical knowledge and skills. Universities 4.0 were initially established to provide personnel and technological support to enterprises in Industry 4.0. Educational technologies assist them in fulfilling this task.

Another characteristic of the sustainable development of Universities 4.0 is the socially responsible utilization of advanced educational technologies aimed at enhancing inclusiveness in the system of science and higher education. There are numerous aspects of inclusiveness within Universities 4.0, including increasing accessibility of higher education for students from different income levels, fostering a tolerant social environment, and providing comfortable learning conditions for individuals with disabilities in Universities 4.0.

Furthermore, it is worth noting the gender neutrality in science and higher education, as well as the learning opportunities for residents of remote and rural areas. EdTech facilitates these by enabling distance learning and automating the educational process. Due to the fact that the mentioned aspects of inclusiveness in Universities 4.0 are individually discussed in existing sources of scientific research literature, a comprehensive understanding of this phenomenon supporting the sustainable development of higher education in Russia and Central Asia has not yet been formed.

The internationalization of activities is another feature of the sustainable development of Universities 4.0. Nowadays, a global scientific and educational space has been formed in which Universities 4.0 from Russia and other Central Asian countries hold strong positions. To flexibly adapt to changes in the global scientific and educational landscape, Universities 4.0 must conduct highly effective marketing activities.

A key manifestation of the sustainable development of Universities 4.0 is the preservation and improvement of their positions in international university rankings. These rankings widely feature universities from Russia and other Central Asian countries. Reputation—a global name—is a criterion for the success of University 4.0. Moreover, attracting international students and faculty and ensuring international academic mobility are of great importance for the sustainable development of Universities 4.0 in the conditions of globalization, significant progress in which has been achieved by Central Asian countries, including Russia.

Another feature of the sustainable development of Universities 4.0 is highly effective university management, flexible personnel training, and close linkage between higher education and the labor market. It is necessary to note the importance of preparing green professionals in Universities 4.0 because they are highly demanded by enterprises in Industry 4.0. The green growth of the digital economy is largely supported by Universities 4.0, which develops environmental competencies among students.

In Russia and other Central Asian countries, educational technologies play an important role in the process of preparing green professionals in Universities 4.0. Thanks to advanced EdTech, the training of green professionals includes the acquisition of practical skills demanded by employers. The linkage between higher education and the labor market in Universities 4.0 is also facilitated by EdTech, including through distance corporate training.

This book aims to bridge the gap in existing scientific research and formulate a systemic approach to the sustainable development of higher education in Russia and Central Asia in the transition to Universities 4.0. The uniqueness of this book lies in the proposal of a new direction for the sustainable development of science and higher education through the establishment of Universities 4.0 and the enhancement of their management. The uniqueness of this book is expressed in that it unveils, for the first time, previously unknown experiences of Russia and other countries of Central Asia in the creation of Universities 4.0 and their management in the interests of sustainable development of science and higher education.

The theoretical significance of the book is determined by its introduction and advancement of a new concept: University 4.0. This concept is construed in the book as a progressive subject of the market of science and higher education, emerging under the influence of the Fourth Industrial Revolution in support of Industry 4.0 and the digital economy. The book systematizes manifestations of the sustainable development of Universities 4.0, including their high-tech nature, inclusiveness, internationality, highly effective management, and close ties with the labor market.

The practical significance of the book lies in its abundance of real-world examples and a series of comprehensive applied solutions aimed at a more complete realization of the potential for sustainable development of Universities 4.0 in Russia alongside other Central Asian countries. The managerial significance of the book is associated with its inclusion of practical recommendations, the implementation of which will improve the practice of managing Universities 4.0 in Russia and Central Asia for the sustainable development of the system of science and higher education.

The book's content is divided into five thematic parts. The first part examines Universities 4.0 as centers for applying advanced educational technologies in Russia and Central Asia. The second part explores the facets of inclusiveness in Universities 4.0 to support the sustainable development of higher education in Russia and Central Asia. The third part is dedicated to communications and the reputation of Universities 4.0 in the context of the internationalization of higher education in Russia and Central Asia. The fourth part investigates the management of Universities 4.0, the development of human resources, and the connection of higher education with the labor market in Russia and Central Asia. The fifth part explains the relationship between EdTech, digital inclusiveness, and university management strategies in Russia and Central Asia.

The book is primarily intended for scholars studying the system of science and higher education. Within its pages, they will find a scholarly interpretation of the new phenomenon within this system—Universities 4.0. Additionally, the book provides them with a scientific vision of the prospects for the sustainable development of the system of science and higher education, with a key focus on the role of Universities 4.0. Detailed descriptions of the unique model of Universities 4.0 in Russia and other Central Asian countries are also included.

The book may also be of interest to direct participants in the system of science and higher education. Representatives of relevant ministries and agencies, acting as subjects of state regulation of the market of science and higher education, will discover innovative applied solutions to current issues in administering this market. Leaders of Universities 4.0 will find authorial scientific-practical recommendations on managing these scientific and educational organizations within the book.

Although the book primarily focuses on the experiences of Russia and Central Asia, it will also be of interest to readers from other countries. The experiences of Russia and Central Asia may prove valuable for state and corporate management in the sustainable development of Universities 4.0 worldwide.

Messina, Italy — Bruno S. Sergi
Yerevan, Armenia — Elena G. Popkova
Bishkek, Kyrgyzstan — Ainura A. Adieva
Bishkek, Kyrgyzstan — Nurgul K. Atabekova

Series Editors Introduction

This engaging and thought-provoking book, *University 4.0 and Educational Technology for Sustainable Development—Best Practices and Perspectives in Russia and Central Asia,* edited by Bruno S. Sergi, Elena G. Popkova, Ainura A. Adieva, and Nurgul K. Atabekova, stands as an important addition to the renowned Springer Book Series ***Education in the Asia-Pacific Region: Issues, Concerns and Prospects***. Having begun over twenty years ago with its first publication in 2002, this volume now stands as the 75th volume in this internationally recognized series.

The rise of Universities 4.0 marks a transformative era in higher education, shaped by the rapid advancements of the Fourth Industrial Revolution. As technological innovation reshapes global and regional landscapes, universities must evolve to meet new challenges. This book explores how universities in Russia and Central Asia are responding to these changes, embracing advanced educational technologies to align with global trends and to contribute to the internationally accepted Sustainable Development Goals (SDGs) set by UNESCO.

At the core of Universities 4.0 is the integration of artificial intelligence, big data, and digital platforms, fostering more inclusive, efficient, and connected learning environments. These technologies not only enhance education but also equip graduates with the skills needed to navigate the digital economy and global workforce. With the world becoming increasingly interconnected, these universities play a critical role in bridging academia, industry, and society. By leveraging innovation and digital transformation, they serve as hubs for research and workforce development, ensuring that education remains relevant in a rapidly evolving economic landscape.

A central theme of this book is inclusiveness, a critical factor in achieving sustainable development. Universities 4.0 embrace inclusiveness across multiple dimensions—cultural, gender, social, digital, and ecological. Cultural inclusiveness fosters international collaboration, attracting students and faculty from around the world and increasing global academic mobility. This international perspective strengthens universities' reputations and enhances their position in world rankings.

Gender inclusiveness ensures equal opportunities for all individuals in research, teaching, and academic leadership, fostering an environment where diverse perspectives contribute to scientific and technological advancements.

Social inclusiveness focuses on accessibility, creating learning environments that accommodate individuals with disabilities. By implementing digital learning tools and offering flexible study options, universities help break down barriers and ensure equal participation in higher education. Digital inclusiveness, another fundamental aspect, democratizes access to knowledge through EdTech, allowing students from diverse backgrounds, including those in remote areas, to benefit from high-quality, relevant education. The growing reliance on digital tools in universities also enhances the efficiency of learning processes, making education more engaging, interactive, and adaptive to students' needs.

Ecological inclusiveness is becoming increasingly relevant as universities take responsibility for sustainable practices. Universities 4.0 are committed to reducing their environmental footprint by optimizing resource consumption and incorporating sustainability into their curricula. By preparing students to work in green industries and develop environmentally responsible solutions, these institutions contribute to broader global efforts to address climate change and promote sustainability. Through a combination of research, policy integration, and education, universities help shape a generation of professionals equipped to lead in the transition toward a greener economy.

Beyond inclusiveness, the book also highlights the growing importance of the internationalization of higher education. As global academic networks expand, universities must engage in cross-border collaborations, joint research initiatives, and the adoption of international educational standards. For Russia and Central Asia, participation in global knowledge exchange has been accelerated by initiatives like the United Nations SDGs and Russia's Priority 2030 program. These efforts enhance universities' ability to attract international students and faculty, fostering cross-cultural dialogue and contributing to global innovation.

Effective university management is another key aspect explored in this book. Universities 4.0 require strong leadership and governance strategies to successfully implement digital transformation and align with global economic demands. The book examines how universities are restructuring their administration, adopting data-driven decision-making, and integrating new educational models that cater to the evolving job market. This includes the shift toward interdisciplinary programs, industry partnerships, and competency-based learning approaches designed to better prepare graduates for professional success.

One of the most significant impacts of Universities 4.0 is their ability to bridge the gap between higher education and the labor market. As industries undergo rapid technological shifts, universities must ensure that their curricula remain relevant to real-world demands. This book explores how institutions are equipping students with both theoretical knowledge and practical skills, fostering innovation-driven economies. By collaborating with industry leaders, offering internship opportunities, and investing in experiential learning, universities enhance students' employability and contribute to national economic growth.

Offering both theoretical insights and real-world applications, this book presents a comprehensive analysis of Universities 4.0 in Russia and Central Asia. It highlights their role in advancing educational technology, inclusiveness, and sustainability, making it a valuable resource for scholars, policymakers, and educators. Through case studies and comparative analysis, the book provides a deeper understanding of how Universities 4.0 operate within different socio-economic and cultural contexts, showcasing best practices that can be adapted globally.

University 4.0 and Educational Technology for Sustainable Development is an essential read for those invested in the future of higher education. It provides invaluable insights into how universities are embracing the Fourth Industrial Revolution and advancing sustainable development goals through educational innovation and global collaboration. The book not only sheds light on the progress already made but also offers forward-thinking perspectives on how universities can continue to evolve in response to emerging challenges.

As institutions navigate this rapidly changing landscape, this book serves as both a guide and a call to action, advocating for a more inclusive, sustainable, and technologically advanced educational future. By integrating cutting-edge technology with inclusive policies and sustainable development goals, Universities 4.0 are shaping the next generation of learners, researchers, and leaders. Their success in adapting to new paradigms will determine the trajectory of higher education in the decades to come, making their transformation a subject of global significance.

This Springer Book Series delves into key topics such as the intersection of technology and education, reforms in primary, secondary, and teacher education, innovative assessment methods, alternative education models, strategies for achieving universal quality education, and fostering active aging through learning. Additionally, the series includes case studies of educational systems from various countries in the region, cross-cultural and cross-country educational studies, and examinations of the sociology of teachers as a professional group, among other related themes. For more information, please visit the Springer Book Series (http://www.springer.com/series/5888).

Each volume is tailored to cater to a broad audience, including researchers, policymakers, educators, higher education students, teachers at all levels, and general readers interested in the latest trends and developments in education across the Asia-Pacific region.

This series is dedicated to exploring the complex and diverse dimensions of education in the Asia-Pacific, shaped by the region's unique challenges and dynamic context. Known for its vast geographical span, cultural diversity, and socio-economic, political, and developmental contrasts, education systems in the Asia-Pacific significantly influence society in various ways, including employment, workforce development, cultural perspectives, and societal norms.

The region exhibits considerable disparities in economic and socio-political development, with both some of the wealthiest nations, such as Japan, and some of the least affluent, like Bangladesh. Asia also has the highest number of individuals living in poverty, with some countries seeing rates as high as 40%. Despite this, many nations in Asia are experiencing impressive economic growth and social progress. However, the challenge of achieving inclusive and sustainable development without sacrificing environmental quality remains pressing. The rise of Asian economies and corporations, coupled with globalization and technological advancements, is transforming trade, business, labor markets, and social structures both within and beyond borders. The Asian Development Bank in Manila has highlighted that the 21st century will be "the Century of Asia-Pacific," signaling a global shift in power toward this region.

Insights from global educational researchers, policymakers, and practitioners confirm that this book series plays a crucial role in fostering knowledge exchange on educational innovations and developments across the Asia-Pacific region.

We invite readers of this volume and others in the series who are interested in writing, co-writing, or editing a volume on any relevant educational topic to reach out to the series editor directly or through Springer. We are dedicated to assisting prospective authors in refining their manuscripts for publication consideration.

School of Education
University of Tasmania
Hobart, TAS, Australia
15 March 2025

Rupert Maclean AO

Faculty of Education
Monash University
Melbourne, VIC, Australia

Philip W. K. Chan

Contents

Higher Education in Universities 4.0 in Russia and Central Asia

1.1 Preface to the First Part of the Book

This first part of the book unveils the scientific essence of Universities 4.0 as a new phenomenon in the realms of science and higher education, emerging under the conditions of the Fourth Industrial Revolution. The main idea of this part is that Universities 4.0 seamlessly intertwine high technology (based on the educational technologies of Industry 4.0) with inclusiveness (through systemic support of sustainable development goals).

The chapter by Elena G. Popkova, "Advanced Experience of Sustainable Development of EdTech Based on Universities 4.0 in Russia and Central Asia for Strategic Academic Leadership," supports the topic of this part by proposing a novel approach to state regulation of sustainable development for EdTech and Universities 4.0 in Russia and Central Asia – a strategy that will strengthen these countries' strategic academic leadership during the Decade of Action, coinciding with the Decade of Science and Technology (until 2030).

The contribution of the chapter by Ayzuura Z. Elchibaeva, Umida F. Sabirova, Konstantin V. Vodenko, and Elena V. Belozerova is explained by the identification of the life values of Universities 4.0 and their significant contribution to social progress.

The chapter titled "Life Values of a University Striving for Digital Inclusion in a Globalizing World" by Dilobar M. Mavlyanova, Sherzod Sh. Axmedov, Meerimai Zh. Karbekova, and Dmitry S. Andreev further develops this topic by proposing a set of life values for Universities 4.0 in Central Asian countries to ensure digital inclusiveness in their higher education systems in a globalized world.

The chapter by Tatiana N. Litvinova, Aktam U. Burkhanov, Barna O. Rakhmankulova, and Elena V. Patsyuk ties into the topic of this part by detailing the evolutionary process from a traditional university to University 4.0 – a digital and inclusive center of higher learning.

The contribution of the chapter by Gulnara K. Tokkulinova, Bibigul Zh. Almukhambetova, Gulmira A. Ayipova, Araily D. Shakirova, and Nurzhanat D. Shakirova, titled "Motivation for Teaching Students in the Pedagogical Field within the Framework of the University 4.0 Education Model," lies in the development of a Model of University 4.0 that embodies a harmonious unity of academic, social, and entrepreneurial potential in higher education.

It is also worth noting the chapter by Yulia N. Zemskaya, Evgeniya A. Kuznetsova, and Anna V. Osmolovskaia, titled "Constructing the Image of the Global University through Its Official Website (Based on About Pages)," in which the authors argue that leading global Universities 4.0 employ a common framework while demonstrating originality in shaping their image on the "About Us" pages of their websites, thus further enriching the topic of this part.

A significant contribution to the discussion in this part is also made by the chapter "The Russian Experience in Identifying Leading Universities" by Svetlana A. Sharonova and Nina V. Trubnikova. From the perspective of this part, the global shifts in the criteria for assessing academic leadership described in this chapter are noteworthy because they set new expectations for Universities 4.0 by the state and society.

In summary, each chapter in this part elucidates the management of Universities 4.0 in its own way and is crucial for uncovering the topic of this part, offering a distinct scientific perspective on the outcomes of higher education in Russia and Central Asia. Together, the chapters in this part form a comprehensive scientific framework that delineates the theoretical criteria distinguishing Universities 4.0 from other institutions and outlines a methodology for evaluating their performance in Russia and Central Asia.

Advanced Experience of Sustainable Development of EdTech Based on Universities 4.0 in Russia and Central Asia for Strategic Academic Leadership

Elena G. Popkova

Abstract The research focuses on developing an approach to state regulation of sustainable development of EdTech based on university 4.0 in Russia and Central Asia for the strategic academic leadership of these countries. To achieve this goal, the research explores the advanced experience of sustainable development of EdTech based on universities 4.0 in Russia and Central Asia in the interests of the strategic academic leadership of these countries using the method of regression analysis. The author compiled economic-mathematical models reflecting the regression dependence of the sustainability of EdTech development on the implementation of measures of state regulation of the activities of universities 4.0 in Russia and Central Asia in 2023. The theoretical significance of the compiled model is associated with the fact that it refined the specificity of sustainable development of EdTech based on universities 4.0 in Russia and Central Asia. In accordance with the model obtained, the author proposes a new approach to state regulation of sustainable development of EdTech and universities 4.0 in Russia and Central Asia. The novelty and essence of the author's approach lie in focusing on the most effective measures of state regulation of sustainable development of EdTech and universities 4.0 in Russia and Central Asia. These measures include stimulating the introduction of advanced technologies, financing science, and creating favorable conditions for academic mobility. The practical significance of the new approach is associated with the fact that its implementation will strengthen the strategic academic leadership of Russia and Central Asia in the Decade of Action, coinciding with the Decade of Science and Technology (until 2030).

Keywords Sustainable development · EdTech · University 4.0 · Russia · Central Asia · Strategic academic leadership

JEL codes H52 · I21 · I23 · I26 · I28

E. G. Popkova (✉)
The Laboratory of Scientometrics and International Rankings,
Armenian State University of Economics, Yerevan, Armenia

B. S. Sergi et al. (eds.), *University 4.0 and Educational Technology for Sustainable Development*, Education in the Asia-Pacific Region: Issues, Concerns and Prospects 75, https://doi.org/10.1007/978-981-96-7376-6_1

1 Introduction

Higher education and science traditionally represent one of the most progressive sectors of contemporary socio-economic systems. Universities are one of the most flexible economic entities. Therefore, in the era of the digital economy, advanced technologies of Industry 4.0 gained widespread adoption in higher education and science among the first sectors.

This phenomenon that took root in the current scientific literature is called EdTech. The economic essence of EdTech lies in its being a scientific-educational system actively incorporating high technologies. Universities 4.0 serve as subjects and sources of EdTech development. Universities 4.0 are understood as scientific-educational organizations applying Industry 4.0 technologies and collaborating with Industry 4.0 businesses.

In the unique socio-economic space of Russia and Central Asia, there exists a distinctive model of the scientific-educational system. The features of this model include the prevalence of state universities, the reliance of the scientific-educational system on state budget financing, and the close interaction and scientific-educational cooperation within the Russian and Central Asian space due to shared geographical proximity and academic cultural traditions.

The specificity of EdTech at the current stage of development of the scientific-educational system in Russia and Central Asia lies in the fact that they face a dual task. In addition to the global task of sustainable development in light of the declared Decade of Action, they also face the task of supporting strategic academic leadership in connection with the Decade of Science and Technology declared in Russia.

Given the noted peculiarities of Russia and Central Asia, these countries require a unique approach to state regulation of EdTech, distinct from approaches implemented in other countries. The development of this approach is a relevant scientific-practical issue addressed in this research. The research aims to develop an approach to state regulation of sustainable development of EdTech based on universities 4.0 in Russia and Central Asia for the strategic academic leadership of these countries.

2 The Concept of EdTech as a High-Tech Scientific and Educational System Driven by Universities 4.0

The theoretical basis of this research is the concept of EdTech as a high-tech scientific and educational system, the drivers of which are universities 4.0 [11, 16]. Relying on this concept, the existing approach to state regulation of sustainable development of EdTech and the activities of universities 4.0 implies the implementation of the following administrative and managerial measures:

- Encouraging the implementation of advanced technologies [1];
- Funding education [5];
- Funding science [4];
- Creating favorable conditions for academic mobility [14];
- Encouraging cooperation between universities and industry, especially Industry 4.0, in R&D [10, 12].

According to the available literature, sustainable development of EdTech implies the following:

- Universal accessibility and wide coverage by higher education [3, 6];
- High citation rates of publications by domestic scientists [17];
- A significant share of revenues from intellectual property in the structure of total foreign trade turnover [9];
- Strong positions of universities in international university rankings, one of the most authoritative of which is the QS ranking [2].

The conducted literature review [7, 8, 13, 15, 20] reveals that the existing approach to the state regulation of sustainable development of EdTech and the activities of universities 4.0 is extensively described, predominantly based on the experiences of developed countries. However, this approach fails to account for the dynamics of developing countries and specific regional characteristics worldwide. This limitation undermines its effectiveness when applied in Russia and Central Asia, where all countries are dynamically evolving and where a distinct specificity exists within the scientific-educational system.

The experience of the functioning and development of the higher education and scientific system, including EdTech, in Russia and Central Asia has been studied in separate publications and is generally undefined. This gap in the literature poses the following research question (RQ): What contributes to achieving strategic academic leadership in Russia and Central Asia, and how can it be strengthened during the Decade of Action, coinciding with the Decade of Science and Technology in Russia? To address this RQ, this research examines the advanced experience of sustainable development of EdTech based on universities 4.0 in Russia and Central Asia in support of their strategic academic leadership.

3 State Regulation of Sustainable EdTech Development in Russia and Central Asia

The author examines the advanced experience of sustainable development of EdTech based on universities 4.0 in Russia and Central Asia in the interests of the strategic academic leadership of these countries using the method of regression analysis. Economic-mathematical models reflecting the regression dependence of the sustainability of EdTech development (denoted as Ed) on the implementation of measures of state regulation of the activities of universities 4.0 (denoted as Reg) are compiled with the application of the method mentioned above.

The data source for the "Frontier Technologies Readiness Index" is UNCTAD [18]. For other indicators, the source is WIPO [19]. The research sample covers 10 out of 12 Central Asian countries because there is insufficient statistical data available for Afghanistan and Turkmenistan to include them in the sample. The research is conducted for 2023. The statistical data of this research is systematized in Table 1.

Based on the results of the regression analysis, the author determined the perspective of strengthening the strategic academic leadership of Russia and Central Asia during the Decade of Action, coinciding with the Decade of Science and Technology in Russia until 2030. For this purpose, the author carried out automated optimization using the simplex method, which makes it possible to identify the prospective combination of state regulatory measures, whereby the full potential of sustainable development of EdTech and the activities of universities 4.0 is most fully disclosed, and the strategic academic leadership of Russia and Central Asia is reinforced.

4 EdTech Model and Perspective on Strengthening Strategic Academic Leadership in Russia and Central Asia until 2030

To systematize the advanced experience of sustainable development of EdTech based on universities 4.0 in Russia and Central Asia in the interests of the strategic academic leadership of these countries, the author conducted a regression analysis of the data from Table 1. This allowed the author to create an econometric model reflecting the cause-and-effect relationships between state regulation of EdTech and universities 4.0 in Russia and Central Asia. The model consists of the following system of multiple linear regression equations:

$$\begin{cases} Ed_1 = 8.58 + 33.90^* \mathrm{Re}g_1 + 2.81^* \mathrm{Re}g_2 + 9.16^* \mathrm{Re}g_3 + 0.03^* \mathrm{Re}g_4 - 0.28^* \mathrm{Re}g_5, \\ Ed_2 = -5.78 + 30.29^* \mathrm{Re}g_1 - 1.00^* \mathrm{Re}g_2 + 15.32^* \mathrm{Re}g_3 + 0.03^* \mathrm{Re}g_4 + 0.22^* \mathrm{Re}g_5, \\ Ed_3 = -7.03 + 8.09^* \mathrm{Re}g_1 + 0.41^* \mathrm{Re}g_2 + 1.37^* \mathrm{Re}g_3 + 0.04^* \mathrm{Re}g_4 + 0.07^* \mathrm{Re}g_5, \\ Ed_4 = -4.39 + 63.51^* \mathrm{Re}g_1 - 3.90^* \mathrm{Re}g_2 + 10.12^* \mathrm{Re}g_3 + 0.13^* \mathrm{Re}g_4 + 0.37^* \mathrm{Re}g_5, \end{cases} \tag{1}$$

Based on model (1), with an increase in the activity of implementing advanced technologies by 1 share of 1, the coverage of higher education in Russia and Central Asia increases by 33.90 points, the citation rate of international publications by scientists from Russia and Central Asia increases by 30.29 points, revenues from international trade in intellectual property objects of Russia and Central Asia increase by 8.09 points, and the position of universities from Russia and Central Asia in the QS university ranking improves by 63.51 points.

Table 1 Factors and outcomes of state regulation of sustainable EdTech development in Russia and Central Asia in 2023

	Tertiary enrolment, % gross, score 0–100	Citable documents H-index, score 0–100	Intellectual property receipts, % total trade, score 0–100	QS university ranking, top 3, score 0–100	Frontier technologies readiness index, share of 1	Expenditure on education, % GDP	Gross expenditure on R&D, % GDP	Tertiary inbound mobility, score 0–100	University-industry R&D collaboration, score 0–100
Country	Ed_1	Ed_2	Ed_3	Ed_4	Reg_1	Reg_2	Reg_3	Reg_4	Reg_5
China	40.4	66.1	8.8	88.8	0.7	3.5	2.4	0.8	86.8
India	18.9	42.8	4.9	48.2	0.7	4.6	0.7	0.1	44.4
Iran, Islamic Rep.	36.8	23.4	0.5	27.0	0.5	3.2	0.8	1.8	12.2
Kazakhstan	45.3	6.2	0.1	39.1	0.6	4.5	0.1	14.4	20.3
Kyrgyz Republic	33.5	4.1	1.0	0.0	0.3	6.6	0.1	61.3	6.0
Mongolia	44.4	4.6	0.5	0.0	0.4	6.5	0.1	2.5	21.7
Pakistan	5.4	19.5	0.5	30.8	0.3	2.1	0.2	0.0	59.2
Russian Federation	56.0	38.1	8.2	49.0	0.8	3.5	1.1	13.0	45.7
Tajikistan	18.4	1.3	0.0	0.0	0.2	5.7	0.1	2.0	31.0
Uzbekistan	11.5	4.1	0.0	0.0	n/a	4.6	0.1	1.7	62.4

Source: Compiled by the author based on UNCTAD [18] and WIPO [19]

With an increase in education funding by 1% of GDP, the coverage of higher education in Russia and Central Asia increases by 2.81 points, and revenues from international trade in intellectual property objects of Russia and Central Asia increase by 0.41 points. With an increase in science funding by 1% of GDP, the coverage of higher education in Russia and Central Asia increases by 9.16 points, the citation rate of international publications by scientists from Russia and Central Asia increases by 15.32 points, revenues from international trade in intellectual property objects of Russia and Central Asia increase by 1.37 points, and the position of universities from Russia and Central Asia in the QS university ranking improves by 10.12 points.

With an increase in academic mobility by 1 point, the coverage of higher education in Russia and Central Asia increases by 0.03 points, the citation rate of international publications by scientists from Russia and Central Asia increases by 0.03 points, revenues from international trade in intellectual property objects of Russia and Central Asia increase by 0.04 points, and the position of universities from Russia and Central Asia in the QS university ranking improves by 0.13 points.

With an increase in the level of collaboration between universities and industry in R&D by 1 point, the citation rate of international publications by scientists from Russia and Central Asia increases by 0.22 points, revenues from international trade in intellectual property objects of Russia and Central Asia increase by 0.07 points, and the position of universities from Russia and Central Asia in the QS university ranking improves by 0.37 points.

Based on model (1), the author determined the prospect of strengthening the strategic academic leadership of Russia and Central Asia during the Decade of Action and the Decade of Science and Technology in Russia. For this purpose, the author conducted automated optimization using the simplex method, allowing the author to identify a prospective combination of state regulatory measures whereby the full potential of sustainable development of EdTech and the activities of 4.0 universities is realized, and the strategic academic leadership of Russia and Central Asia is reinforced (Fig. 1).

According to the perspective demonstrated in Fig. 1, Russia and Central Asia can significantly enhance the sustainability of EdTech and universities 4.0 by 2030, thus substantially reinforcing their strategic academic leadership. To achieve this, the following recommendations are proposed for enhancing the state regulation of EdTech and universities 4.0 in Russia and Central Asia:

- Increase the adoption rate of advanced technologies by 122.22% (from 0.45 in 2023 to a maximum of 1.00 by 2030);
- Raise the funding for science by 46.54 (from 0.57% of GDP in 2023 to 0.84% of GDP by 2030);
- Boost academic mobility by 924.59% (from 9.76 points in 2023 to a maximum of 100.00 points by 2030);
- Enhance collaboration between universities and industry in R&D by 156.61% (from 38.97 points in 2023 to a maximum of 100.00 points by 2030).

These recommendations represent a new approach to state regulation for the sustainable development of EdTech based on universities 4.0 in Russia and Central Asia to ensure the strategic academic leadership of these countries. The distinct

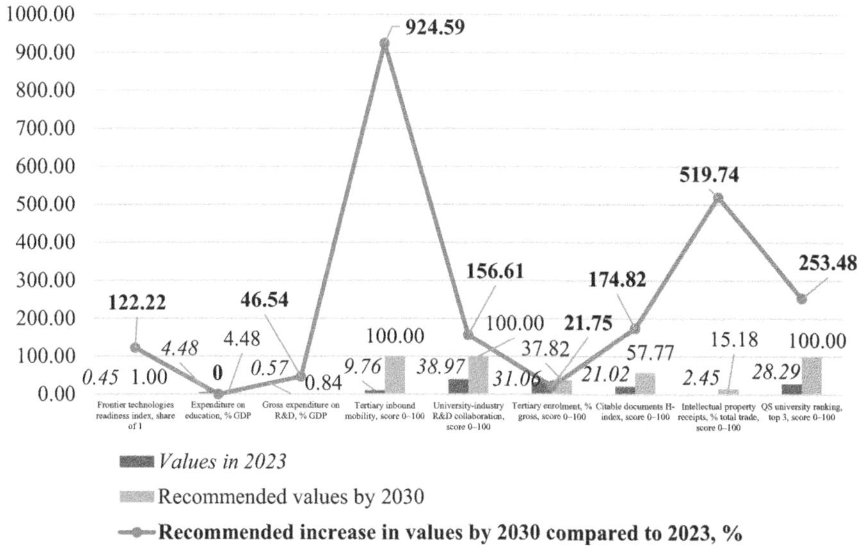

Fig. 1 Perspective on strengthening strategic academic leadership in Russia and Central Asia until 2030. *Source*: Developed and compiled by the author

feature of this approach is the absence of measures to increase educational funding while focusing on other measures.

By implementing the author's recommendations in accordance with the new approach to state regulation of EdTech and universities 4.0 in Russia and Central Asia, the following results can be achieved:

- An increase in higher education coverage by 21.75% (from 31.06 points in 2023 to 37.82 points by 2030);
- A 174.82% increase in the citation of international publications by scientists from Russia and Central Asia (from 21.02 points in 2023 to 57.77 points by 2030)
- A 519.74% increase in revenues from international trade in intellectual property objects in Russia and Central Asia (from 2.45 points in 2023 to 15.18 points by 2030)
- An improvement in the position of universities in Russia and Central Asia in the QS university ranking by 253.48% (from 28.29 points in 2023 to 100.00 points by 2030).

5 Causal Relationships Between State Regulation of EdTech and Universities 4.0 in Russia and Central Asia

The research contributes to the literature related to the EdTech concept [7, 8, 13, 15, 20] by clarifying the causal relationships between state regulation of EdTech and universities 4.0 within the unique scientific and educational system of Russia and

Central Asia. As evidenced by Bogoviz et al. [1], stimulation of the adoption of advanced technologies is established as a key measure of state regulation of EdTech and universities 4.0 in Russia and Central Asia.

In contrast to Fu et al. [5], it is demonstrated that funding education is a contentious measure of state regulation of EdTech and 4.0 universities in Russia and Central Asia, with side effects including a decrease in the citation rates of domestic scientists' publications and a deterioration in the position of domestic universities in the QS ranking.

As supported by Encalada et al. [4], funding for science is identified as an important measure of state regulation of EdTech and universities 4.0 in Russia and Central Asia. Similarly, as confirmed by Przhedetskaya and Borzenko [14], creating favorable conditions for academic mobility is recognized as an important measure of state regulation of EdTech and 4.0 universities in Russia and Central Asia.

In contrast to Kalinina and Chernitsova [10] and Petrenko et al. [12], this research demonstrated that incentivizing collaboration between universities and industry, especially in Industry 4.0 for R&D, is a contentious measure of state regulation of EdTech and universities 4.0 in Russia and Central Asia, with the side effect of reducing accessibility and diminishing the coverage of higher education.

6 Conclusion

Based on the results of the conducted research, the author systematized the advanced experience of sustainable development in EdTech based on universities 4.0 in Russia and Central Asia in the interest of the strategic academic leadership of these countries; its economic-mathematical model has been compiled for the year 2023. The theoretical significance of the compiled model lies in its clarification of the specifics of sustainable development in EdTech based on universities 4.0 in Russia and Central Asia, as well as in revealing previously unknown causal relationships and mathematically describing the patterns of state regulation of EdTech and universities 4.0 in Russia and Central Asia.

In accordance with the obtained model, the author proposed a new approach to the state regulation of sustainable development in EdTech and universities 4.0 in Russia and Central Asia. The novelty and essence of the author's approach lie in focusing on the most effective measures of state regulation of sustainable development in EdTech and universities 4.0 in Russia and Central Asia. These measures include (1) stimulation of the adoption of advanced technologies, (2) funding for science, and (3) creating favorable conditions for academic mobility. The strategic academic leadership of Russia and Central Asia can be achieved through these measures.

Simultaneously, such conflicting measures as funding education and stimulating cooperation between universities and industry in the field of R&D are used to a limited extent and play a secondary role in the developed approach. The practical significance of the new approach is associated with the fact that its implementation will strengthen the strategic academic leadership of Russia and Central Asia in the Decade of Action and the Decade of Science and Technology in Russia (until 2030).

References

1. Bogoviz, A. V., Gimelshteyn, A. V., Shvakov, E. E., Maslova, E. V., & Kolosova, A. A. (2018). Digitalization of the Russian education system: Opportunities and perspectives. *Quality-Access to Success, 19*(S2), 27–32. Retrieved March 5, 2024, from https://case.asu.ru/files/form_312-31738.pdf
2. Bogoviz, A. V., Shvakov, E. E., Tretyakova, O. G., Zakharov, M. Y., & Abramov, A. N. (2020). Globalization of education in the conditions of formation of the global knowledge economy: Regularities and tendencies. In E. Popkova (Ed.), *Growth poles of the global economy: Emergence, changes and future perspectives* (pp. 993–1000). Springer. https://doi.org/10.1007/978-3-030-15160-7_99
3. Denisov, I., Petrenko, Y., Koretskaya, I., & Benčič, S. (2021). The gameover in universities education management during the pandemic COVID-19: Challenges to sustainable development in a digitalized environment. *Sustainability, 13*(13), 7398. https://doi.org/10.3390/su13137398
4. Encalada, D., Bravo-Benavides, D., & Ochoa-Moreno, W.-S. (2023). Expenditures on education, R+D, ICT imports and food security (food production) in Ecuador. In *Proceedings of CISTI 2023: 18th Iberian Conference on Information Systems and Technologies* (pp. 1–5). IEEE. https://doi.org/10.23919/CISTI58278.2023.10211732
5. Fu, C., Ishimaru, S., & Kennan, J. (2024). Government expenditure on the public education system. *International Economic Review, 65*(1), 43–73. https://doi.org/10.1111/iere.12659
6. Galoyan, D. R., Mayilyan, F. N., Bazinyan, Y. A., Movsisyan, M. E., & Torosyan, A. A. (2023). Triple exploitation of human capital in developing countries. In E. G. Popkova & B. S. Sergi (Eds.), *ESG management of the development of the green economy in Central Asia* (pp. 375–383). Springer. https://doi.org/10.1007/978-3-031-46525-3_40
7. Haqyar, Z., Rohila, A. K., Malik, J. S., Kumar, A., & Shubham. (2025). Usage pattern of information and communication technology tools among university faculty members. *Indian Journal of Extension Education, 61*(1), 108–112. https://doi.org/10.48165/IJEE.2025.611RN03
8. Huapaya, E. S. R., Chucos, G. L., Sosa, E. P., & Meza, M. I. (2025). Disruptive technologies in the university curriculum: Use of artificial intelligence. *International Journal of Evaluation and Research in Education, 14*(1), 671–681. https://doi.org/10.11591/ijere.v14i1.30450
9. Ilina, I., Zharova, E., Turginbayeva, A., Agamirova, E., & Kamenskiy, A. (2019). Network platform of commercializing the results of R and D. *International Journal of Civil Engineering and Technology, 10*(1), 2647–2657.
10. Kalinina, I., & Chernitsova, K. (2017). The interaction of universities with high-tech enterprises: Commercialization of innovations. *MATEC Web of Conferences, 121*, 12009. https://doi.org/10.1051/matecconf/201712112009
11. Mamaeva, D. V., Shabaltina, L. V., Garnova, V. Y., Petrenko, E. S., & Borovsky, S. S. (2020). Digital transformation of higher educational system. *Journal of Physics: Conference Series, 1691*, 012081. https://doi.org/10.1088/1742-6596/1691/1/012081
12. Petrenko, E. S., Benčič, S., & Koroleva, A. A. (2019). Digital business in the cyber economy: The organization of production and distribution based on the breakthrough technologies of Industry 4.0. In V. M. Filippov, A. A. Chursin, J. V. Ragulina, & E. G. Popkova (Eds.), *The cyber economy* (pp. 11–17). Springer. https://doi.org/10.1007/978-3-030-31566-5_2
13. Piromalli, L. (2024). Making markets in higher education: State reform, global edtech, and virtual universities in Italy. *Italian Journal of Sociology of Education, 16*(2), 109–130. https://doi.org/10.14658/PUPJ-IJSE-2024-2-6

14. Przhedetskaya, N., & Borzenko, K. (2019). Marketing model of promotion of remote education by modern university. *International Journal of Educational Management, 33*(3), 446–453. https://doi.org/10.1108/IJEM-09-2018-0281
15. Qi, Z. (2025). A modified technology acceptance model for digital learning in Chinese universities. *E-Learning and Digital Media*. https://doi.org/10.1177/20427530251313758
16. Sozinova, A. A., Denisov, I. V., Benčič, S., & Stolyarov, N. O. (2023). Intellectual economic growth in the 'knowledge economy' as the basis for the development of international trade in modern global markets. *International Journal of Trade and Global Markets, 18*(2–3), 230–241. https://doi.org/10.1504/IJTGM.2023.135603
17. Tovmasyan, G., Hakhverdyan, D., Mkrtchyan, T., & Petrosyan, A. (2022). The impact of COVID-19 on education quality and learning efficiency in universities: Evidence from Armenia. *Issues in Educational Research, 32*(4), 1623–2164. Retrieved March 5, 2024, from https://www.iier.org.au/iier32/tovmasyan.pdf
18. UNCTAD. (2023). *Technology and innovation report 2023*. United Nations. Retrieved March 5, 2024, from https://unctad.org/system/files/official-document/tir2023_en.pdf
19. WIPO. (2023). *Global Innovation Index 2023: Innovation in the face of uncertainty*. WIPO. Retrieved March 5, 2024, from https://www.wipo.int/global_innovation_index/en/2023/
20. Zhang, R., & Wang, H.-P. (2024). The impact of teaching beliefs on creative teaching behavior among Chinese university educators: The mediating roles of technology acceptance and organizational innovation climate. *International Journal of Technologies in Learning, 32*(1), 159–188. https://doi.org/10.18848/2327-0144/CGP/v32i01/159-188

Life Values of a Modern University and Their Transformation Under the Influence of Globalization and Digitalization

Aizuura Z. Elchibaeva, Umida F. Sabirova, Konstantin V. Vodenko, and Elena V. Belozerova

Abstract This paper considered the influence of globalization processes on the life values of universities, provided the key definitions of the research, and characterized the process of transformation of traditional given the effect of the factors of the change of the university's existence environment. The design and methodology of this research are based on an interdisciplinary approach, which allows combining the treatment and study of the key provisions of the paper from the position of philosophy, sociology, cultural studies, pedagogics, and institutionalism. Interpretation of the main definitions of this work from the position of the system approach allows considering internal processes of the transformation of the university's traditional values under the influence of external values in the conditions of globalization. This paper presented the definition of the life values of universities, which were considered in the context of their reconsideration given the factors of the external environment, unified by the globalization processes of digitalization, sociocultural exchange, information expansion, etc. According to this, it was determined that universities rethink their mission and transform their life values, which allows them to better adapt to the changing paradigms of social values and to consolidate the social activity of the educational process's participants. The main value of this paper lies in substantiating the value of universities for social development and the role of life values in understanding the place in society's development by all participants of the educational process.

A. Z. Elchibaeva
Kyrgyz State Technical University named after I. Razzakov, Bishkek, Kyrgyzstan

U. F. Sabirova
National University of Uzbekistan named after Mirzo Ulugbek, Tashkent, Uzbekistan

K. V. Vodenko (✉)
Platov South-Russian State Polytechnic University (NPI), Novocherkassk, Russia

E. V. Belozerova
Sebryakovskiy Branch of Volgograd State Technical University, Mikhailovka, Russia

B. S. Sergi et al. (eds.), *University 4.0 and Educational Technology for Sustainable Development*, Education in the Asia-Pacific Region: Issues, Concerns and Prospects 75, https://doi.org/10.1007/978-981-96-7376-6_2

Keywords Values · Life values · Higher education · University · Transformation · Globalization · Sociocultural environment · Society

JEL Codes I23 · I24 · J11 · O15 · P52 · Q01 · R11 · Z13

1 Introduction

In the history of humankind, universities always had an important role in cultural development and evolution. Ensuring the intellectual component of qualitative transformation of society, higher educational establishments influenced these processes directly or indirectly with a time delay, connected with the necessity for the popularization of certain ideas and their achieving a critical mass of followers. Modern universities retain a substantial impact on society, but most often they are treated as "suppliers of knowledge and professions," gradually losing the influence on ideological and value-related processes of cultural environment formation.

The key factor of the change in the influence of educational establishments on the formation of cultural and life values of society is the dynamic development of socio-cultural and technical & economic environment, as a result of which entertainment elements of culture gain more importance, and educational technologies often develop outside of traditional educational establishments. These processes are strengthened by globalization, which erases socio-cultural and, accordingly, value borders, ensuring constant inter-cultural interaction and transformation of traditional value guidelines. In these conditions, the problem of formation and transformation of value guidelines of universities acquires importance and requires thorough rethinking.

As a response to the above problems, the members of Magna Charta Universitatum [9] started (in 2018) the process of reconsideration of the methodology for the development and management of the life values of the higher educational establishments, proposing the tools for assessing the existing and substantiating new life values of educational establishments in agreement with the institutional environment and stakeholders. At that, complying with the principles of saving the fundamental university values of academic freedom, institutional autonomy, and responsibility before society, universities were offered to form their values, which would distinguish them in the global educational environment and conform to the cultural specifics of their countries and regions.

2 Life Values of Universities in the Changing Environment

Socio-philosophical sciences form their methodology around a certain construct, which is aimed at identifying the sense of existence, events, or actions. The issue of life values of universities in the changing environment is the object of research of a wide range of scientific disciplines, and it is most closely connected with sociology and philosophy. These scientific disciplines form a methodological framework and methodological tools for the research. The institutionalization of the process of development of the life values of universities at the global level is, in this context, a tool that offers means and methods to ensure effective management of universities' life values [2, 3, 10]. However, in any case, the methodological tools of this research are of interdisciplinary character and lie in the sphere of philosophical, socio-cultural, and other sciences and cover empirical methods of cognition, analysis, and interpretation of scientific literature, methods of observation and generalization, as well as the graphical method as a tool of reflecting cause-and-effect relationships that are peculiar to dynamic processes.

The key works devoted to the study of universities' life values of universities in the conditions of globalization [15, 16] lie in the sphere of philosophical and socio-cultural disciplines. The problems that were most fully considered in the scientific and expert literature of this profile include works devoted to a rethinking of the role of universities in the development of society [9], the transformation of personal life values in the conditions of globalization [1], the essence of values and their influence on personal judgments [4], the transformation of traditional values under the influence of institutional changes [11], commercialization of higher education and its losing the decisive social role in the development of society [5, 6], the influence of globalization in the development of universities [13], implementation of the concept of Life Values of the Modern University [7, 9], and the issues of saving the values of national education [8, 14].

The main goal of this research was to identify the main life values of modern universities, which form under the influence of different factors, including the globalization of social and educational environment. The main tasks of this research were to determine the essence of the key definitions of the research, characterise the influence of globalization on value landmarks of universities, reflect the mechanism of transformation of traditional values under the influence of different actors, and characterise the role of the concept of Life Values of the Modern University as a tool of development of effective life values and support for their correspondence to the demands of social and cultural development.

3 The General Mechanism of the Transformation of Life Values of Modern Universities

The issue of life values is among relevant world outlook problems, which influence the conscious activities of humans. In this context, the life values of the university go beyond the limits of the personal approach and acquire a wide social status. They accumulate value landmarks of individuals and large social groups, cover regions, state, and, in some cases, global levels and represent the system of aspirations and models of society's behavior.

For the correct achievement of the research goal, it is important to understand and interpret key definitions, which are connected with the problem posed. The main definition is "value." It belongs to different scientific spheres, so it may be treated differently. However, the decisive understanding of this term is set in philosophy, where it is treated as a "certain ideal, collective idea, which is the main driver of human behavior that is supported by real acting forces inherent to the society" [4].

Globalization, as an important factor in the transformation of life values of universities, is, by its nature, the "fundamental changes of the world order, within which development of technologies and active dissemination of mass culture lead to the destruction of such institutional phenomenon as state borders" [13]. These changes influence traditional values and ensure a wide intercultural exchange, which creates conditions for the destruction, rethinking, or enrichment of the cultural environment and related value attitudes of educational establishments.

In this context, traditional values are treated as a certain social experience, which is passed from generation to generation [11]. Values acquire the status of traditional, since they are reproduced—from generation to generation—in almost unchanged form and are basic elements of the social and cultural identity of social groups and are treated in society as life benchmarks and criteria during the making of life decisions.

From the position of universities, life values are treated similarly with a difference regarding the scales and social impact on society in the context of its development. At that, for a long time, universities were assigned a key role in the formation and popularization of actual life values that are necessary for the successful development of mankind. As of now, the role of universities as the main moderators of socio-cultural behavior is diminishing. In this context, B. Readings characterizes modern university as a "post-historical institution that has a small role in the historical development of humanity, because it does not perform the main role of producer and protector of the idea of the national culture" [12].

However, universities, as elements of the global sociocultural environment are also subject to the influence of global processes and transform their value landmarks, internal norms, and conceptual provisions as to their development. Thus, universities, as open systems, actively interact with the surrounding environment; they self-organise and offer certain values that conform to the university's vision. Such values are manifested in communications of the educational institution, its positioning and the activities of its employees and students.

Today, with the penetration of market approaches into higher education, universities are at the stage of rethinking their values. This situation is peculiar to many individuals, social groups, countries, etc. Global acceleration of technological changes, destruction of cultural and information borders, and development of alternative tools for receipt of knowledge and education put into doubt many life values of universities. However, these establishments continue influencing the sociocultural environment, transforming their values according to the changes in the external environment.

The general mechanism of the transformation of life values of modern universities has a similar character, given the influence of different factors of the external environment, including the factor of globalization on the whole. Its main characteristics clearly follow the processes that are peculiar to the transformation of any system and cover the phases of social disintegration, bifurcation, and mobilization of forces. Based on [11], let us present a scheme for the transformation of university values (Fig. 1).

According to the above scheme, traditional values of universities are subject to different influences, and their significance is put into doubt at the stage of social disintegration. The reasons for such a situation might be different. For example, the difference between declared values and the real behavior of the university's management, the mismatch between the university's values and society's cultural values, or the ineffectiveness of the values due to their poor substantiation or absence of the tools of compliance with them.

In any of the above situations, the genuineness of traditional values becomes questionable and interested parties have several variants of behavior, which characterize the phase of bifurcation—compliance with traditional values or refusal from them, or formation of a combination of new and traditional values. These processes are connected with reflections, assessment, or decision-making regarding values. At that, the choice is influenced by different factors, including one's own or someone else's life experience, the presence of advantages or disadvantages with traditional and new values, and their attractiveness in new conditions.

The result of such reflections at the stage of mobilization of forces is the activation of dominating values, at which the ones that are better received by the

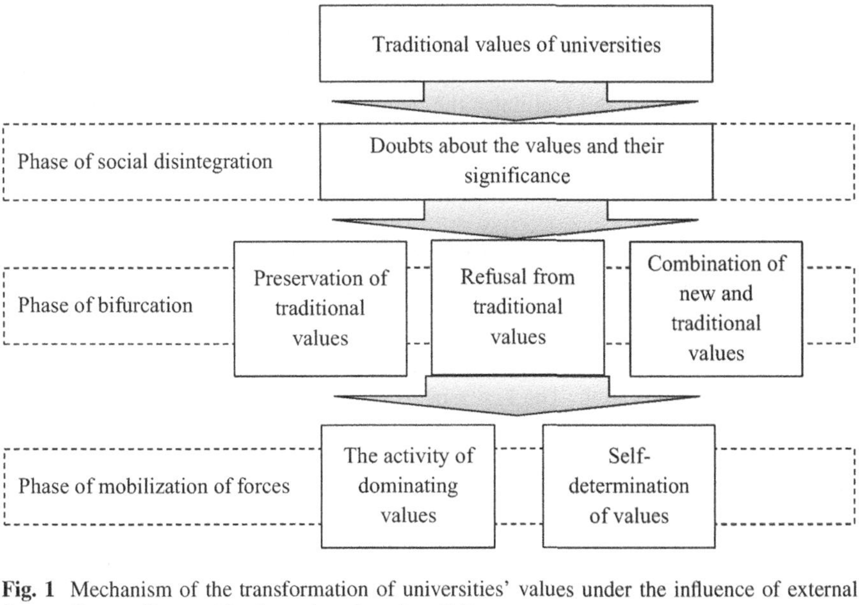

Fig. 1 Mechanism of the transformation of universities' values under the influence of external factors. *Source*: Prepared by the authors based on [11]

interested party and conform to its beliefs dominate over the ones that, due to certain reasons, do not resonate with beliefs of the person who makes decisions. In these conditions, self-determination of values takes place—they may also acquire the status of traditional or widely acknowledged values.

4 Life Values of the RUDN University During Globalization and Digitalization

The described mechanism of the transformation of life values of modern universities poses significant risks for the universities and for the socio-cultural environments within which these universities function. These risks are connected with the possibility of losing important values that are fundamental for the entire mankind or a separate socio-cultural group. That is why the universities of Magna Charta initiated the programs of Living Values [9], which is to reassess and rethink their life values and use efficient tools that can ensure compliance with these values and their treatment in society.

In Russia, this program was joined by RUDN University. As a result of the participation in the program, at the beginning of 2020, the university employees and students were presented with the following life values of their educational establishments: "Professionalism, academic freedom, research culture, and social responsibility" (Fig. 2).

An additional increase in the influence of the above values was set in the motto "from the living values of RUDN to life according to RUDN values" [8]. Thus, against the background of destructive world events, global processes of the constructive character ensure a clear definition, formulation, and translation of the university's life values, which substantially raises the level of coordination within the university's personnel and facilitates the better realization of their role by the university's students.

5 Universities in Russia and Central Asia on their Way of Determination of their Life Values

The obtained results of this research lead to several discussion questions, which require further elaboration. The key aspects that need initial consideration are the complex evaluation of the universities in Russia and Central Asia from the position of determination of their life values and checking the correspondence of the results obtained to the real demands of society and society's acceptance of these values. In case of a mismatch between the university's values and the real demands of society, there emerges a new task of developing and substantiating the program of coordination of these values and their acceptance by interested parties.

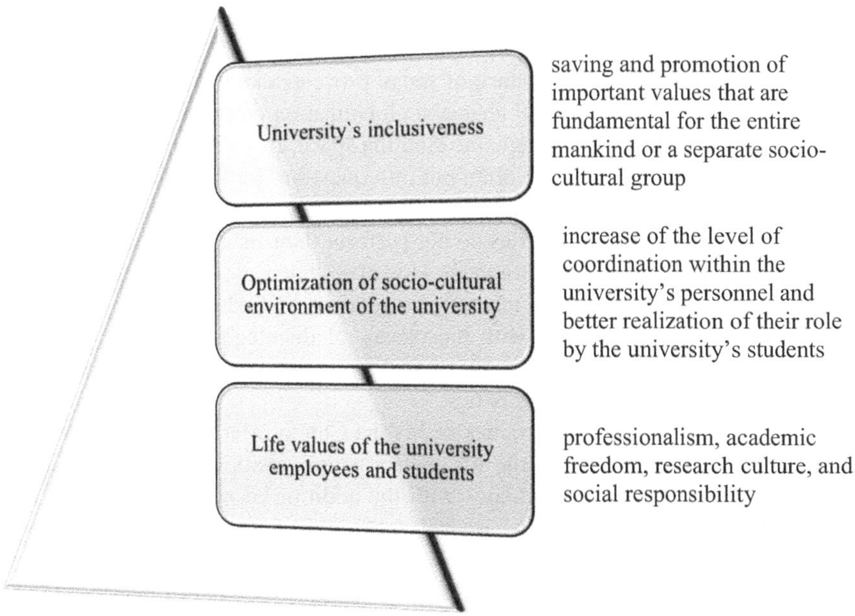

Fig. 2 Life values of the RUDN University during globalization and digitalization. *Source*: Prepared by the authors

Other discussion provisions, revealed in this work, are concerned with the influence of globalization on the formation and rational transformation of modern universities' life values. In this context, the impact of globalization on the studied processes is evaluated ambiguously, which requires additional consideration of positive and negative aspects. In this context, it is also important to identify the reasons for the positive or negative effect of globalization changes on the value-based development of universities and interested parties. This will allow developing tools and mechanisms for the neutralization of the negative effect and rational use of favorable factors.

An equally important task, which requires additional research and disclosure, is an agreement of universities' life values and conditions of their functioning in the market environment with the achievement of financial autonomy and increase in competitiveness. In this case, it is necessary to solve the problem of agreement of commercial and value & cultural provisions of universities' development, which will allow ensuring their financial stability and retaining the role and influence on the sociocultural sphere and general development of society.

Given the above discussion issues, this research can be a theoretical basis and substantiation of the topicality of the above problems, a reservation regarding their importance and limitations, which have to be taken into account during the development and use of the tools for the formation and compliance with rational life values of modern universities in the conditions of globalization.

6 Conclusion

In the conditions of high turbulence of social processes and under the influence of technological and socio-cultural changes, globalization erases the information and cultural borders and forms a single information space. In such conditions, the traditional values of universities are often put into question, for they do not fully allow interpreting of the changing conditions of society's development. In many cases, this allows stating that universities do not perform their main function any more—the formation of a cultural environment. However, this is not so. Educational establishments, together with other institutions, reorient and adapt to new conditions. These changes are connected with the phases of disintegration, bifurcation, and mobilization of forces, and the result of such changes is new combinations of life values and benchmarks for universities.

With this in mind, the universities of Magna Charta offer the tools for developing, checking, and promoting the life values of universities, which are based on fundamental and classical postulates with the addition of new concepts that conform to the topical conditions and are well received in society. RUDN's joining this programme allowed the university to develop a new system of values.

References

1. Aksinjushkin, I. A. (2008). Personal life values in the context of globalization. *Bulletin of the Volga Region Institute of Management, 2*, 134–139. Retrieved October 22, 2023, from https://cyberleninka.ru/article/n/zhiznennye-tsennosti-lichnosti-v-usloviyah-globalizatsii-1
2. Alahmad, A., Al-Abdallat, M., Al-Serhan, A., Al-Tawil, M., Ayyoub, H., Al-Awamleh, A., et al. (2024). Ethics, values, and the promotion of life skills among university students. *Journal of Institutional Research South East Asia, 22*(3), 331–356.
3. Cam Tran, T. T., Thuy Vu, H. T., Van Dam, A. T., & Phan, L. T. (2024). Perceptions of university lecturers and students regarding the manifestation of life values among pedagogical students from Vietnam. *Journal of Educational and Social Research, 14*(2), 93–103. https://doi.org/10.36941/jesr-2024-0029
4. Durkheim, E. (2019). *Sociology. Its subject, method, purpose* (4th ed.). URAIT.
5. Fridman, M. F. (2019). Reinstitutionalization of the university in the context of globalization. *Cennosti i smysly [Values and Meanings], 3*, 156–171. https://doi.org/10.24411/2071-6427-2019-00032
6. Kadirova, M. A. (2022). Globalization and problems of preserving the values of national education. *Endless Light in Science, 5*, 131–135. Retrieved October 22, 2023, from https://cyberleninka.ru/article/n/globalizatsiya-i-problemy-sohraneniya-tsennostey-natsionalnogo-obrazovaniya
7. Korolev, Y. (2021). Life values of a modern university. Modern education: continuity and continuity of the educational system "school – university – enterprise". In *Materials of the XIII international scientific and methodological conference*. GGU named after F. Skoriny.
8. Kulikovskaya, G. A., & Kulikovskaya, N. A. (2020). Student-centered learning and the "life values" of the university: strategic interaction. *Bulletin of the Peoples' Friendship University of Russia Series: State and Municipal Administration, 7*(4), 387–399. Retrieved October 22, 2023, from https://cyberleninka.ru/article/n/studento-orientirovannoe-obuchenie-i-zhiznennye-tsennosti-universiteta-strategicheskoe-vzaimodeystvie

9. MCU. (2023). Living values in higher education institutions. *The Magna Charta Observatory*. Retrieved October 22, 2023, from https://site.unibo.it/magna-charta/en/activities-and-projects/living-values
10. Ortiz-Gómez, M., Melero-Bolaños, R., Muñoz-Ocaña, Y., & de los Ríos-Berjillos, A. (2025). What does sustainable development goals knowledge contribute to? A longitudinal study of values and sustainable behavior in university students. *International Journal of Sustainability in Higher Education*. https://doi.org/10.1108/IJSHE-12-2023-0587
11. Rassadina, T. A. (2006). Transformations of traditional values of Russians in the post-perestroika period. *Sociologicheskie issledovanija [Sociological Research], 9*, 95–102.
12. Readings, B. (2009). *The university in ruins*. BGU.
13. Scott, P. (2000). Globalization and the university. *The Alma Mater, 4*, 3–8.
14. Smetanina, M. D. (2021). University as a cultural center. *Society: Philosophy, History, Culture, 1*(81), 93–95. Retrieved October 22, 2023, from https://cyberleninka.ru/article/n/universitet-kak-tsentr-kultury
15. Taşdemir, N. (2025). Young university students' cultural values, filial obligation norms, and ageism toward older people. *Analyses of Social Issues and Public Policy, 25*(1), e12438. https://doi.org/10.1111/asap.12438
16. Yang, Q., Al Mamun, A., Long, S., Gao, J., & Ali, K. A. M. (2024). The effect of environmental values, beliefs, and norms on social entrepreneurial intentions among Chinese university students. *Humanities and Social Sciences Communications, 11*(1), 978. https://doi.org/10.1057/s41599-024-03501-8

Life Values of a University Striving for Digital Inclusion in a Globalizing World

Dilobar M. Mavlyanova, Sherzod Sh. Axmedov, Meerimai Zh. Karbekova, and Dmitry S. Andreev

Abstract Digitalization of university education in countries of Central Asia influences the satisfaction of the region's citizens' needs for knowledge and skills that are required in the quickly globalizing world. A high level of knowledge and practical skills, their accessibility, the openness of digital universities and teachers, flexibility of the educational process, and, in some cases, partnership with foreign universities are factors that substantiate students' choosing the considered countries of Central Asia (Uzbekistan and Kazakhstan) due to their universities, instead of searching for higher education establishments abroad. The study of the system of life values of digital universities in Uzbekistan and Kazakhstan showed their striving toward the increase in their rankings due to the high level of training and satisfaction of citizens' needs for quality higher education in the context of modern and relevant professions.

This paper aimed to determine the list and characteristics of the life values of universities in the selected countries of Central Asia, which strive towards digital inclusion in the globalizing world. The novelty of this paper consists in the systematization and complex characteristics of the life values of digital universities in Uzbekistan and Kazakhstan, which strive towards digital inclusion in the globalizing world.

Keywords Life values · Universities · Globalizing world · Digitalization · Higher education · Knowledge · Inclusion · Flexibility of learning

JEL Codes I21 · I22 · I23 · I24 · I28 · O33

D. M. Mavlyanova · S. S. Axmedov
Tashkent State University of Economics, Tashkent, Uzbekistan
e-mail: sh.axmedov@tsue.uz

M. Z. Karbekova
Jalal-Abad State University named after B. Osmonov, Jalal-Abad, Kyrgyzstan

D. S. Andreev (✉)
Sebryakovskiy Branch of Volgograd State Technical University, Mikhailovka, Russia

B. S. Sergi et al. (eds.), *University 4.0 and Educational Technology for Sustainable Development*, Education in the Asia-Pacific Region: Issues, Concerns and Prospects 75, https://doi.org/10.1007/978-981-96-7376-6_3

1 Introduction

Universities that strive towards digital inclusion received the advantages of scaling in the globalizing world due to the emergence of new ICT and an increase in the level of coverage, speed, and quality of the Internet. Digital and infrastructural innovations are one aspect of the successful formation of the digital educational environment. Another aspect is the growth of requirements to personnel regarding continuous increase in the level of knowledge and skills within the main professions. This could be implemented in the conditions of digital universities, which allow students to participate in the educational process given their management of the time of learning, forms of education, and financial abilities. Achievement of the effectiveness of digital universities' functioning is connected with their conceptual strategic milestones, which define life values in the modern information society. Implementation of these values allows digital educational platforms to ensure growth in the conditions of the transformations of society and its priorities.

In countries of Central Asia, universities that strive towards digital inclusion have become popular since the start of the COVID-19 pandemic. The formation of such universities facilitates the satisfaction of demand in the sphere of higher education and an increase in the knowledge and skills of students and influences the growth of the level of the system of national education at the global scale.

To reach the main goal of this paper, which is identifying the features of the life values of universities in the selected countries of Central Asia that strive towards digital inclusion in the globalizing world, the following tasks were set: determining the main life values that are traditional for modern digital universities in the world practice and finding the level of implementation and characteristics of these values by the example of the selected countries of Central Asia.

2 Life Values of a University Striving for Digital Inclusion in a Globalizing World

The study of the formation of digital university environments in the selected countries of Central Asia in the context of the implementation of top-priority life values was conducted with the use of a complex analysis of the selected universities' websites. The expert analysis was used to assess the state and level of implementation of the key life values of the studied universities. The experts were five specialists in the sphere of higher education and higher professional education and five specialists in the sphere of digitalization of public and educational platforms. The method of comparison was utilized to reveal the comparative characteristics of implementing life values in inclusive universities.

The issue of life values of a university striving for digital inclusion in a globalizing world was studied in many scientific and analytical materials [11, 12, 14]. To name a few [1, 2, 6, 13], presented theoretical aspects of the formation of life values of digital universities, and [10] elaborated on the features of functioning of digital universities in Uzbekistan [3, 8]. contain data on the formation of two pilot projects of digital universities in Kazakhstan, implemented within national and regional universities of the country. Despite several theoretical and empirical studies on the life values of universities dealing with digital inclusion [15–17], it is still important to determine the modern features of this process, which is done in this paper by the example of countries of Central Asia. In this region, the process of digitalization of the university system of education disseminates and is in demand among students.

3 Life Values of Universities in Kazakhstan and Uzbekistan for Their Digital Inclusion in the Globalizing World

To study the life values that are traditional for modern digital universities in countries of Central Asia, it is necessary to determine their potential list inherent for the activities of such organizations in the world.

Analysis of theoretical materials allows systematizing life values that determine the philosophy of digital universities. They include the following:

- Interactive interaction with the external environment and openness, which involve the existence of channels for incoming queries from students and feedback to them in the form of open systems of knowledge and skills [2];
- Focus on critical pedagogics, connected with the academic course aimed and teaching, exchange of knowledge, and assessment of education efficiency, which involves the development of all participants of the educational process (university, lecturers, students) [1];
- Digital equality and access to knowledge and skills without age, gender, national, or financial discrimination [6];
- Inclusion, which allows attracting to the system of higher education persons who previously did not have opportunities in this sphere (digital universities' services should be accessible to people with special needs, people from remote locations, etc.) [13].

These life values of universities that strive towards digital inclusion in the globalizing world could be integrated, and there may arise new elements depending on the specifics of the university environment and students' demands.

Let us consider the specific features of the formation of life values of universities in Uzbekistan (Table 1).

Analysis of the results from Table 1 allows distinguishing the main characteristics of the activities of the leading digital universities in Uzbekistan. This country has many universities focused on digital inclusion in the globalizing world, but the two selected universities have large advantages compared to other higher education institutions. This is due to public-private partnership and the attraction of a foreign partner, in the case of the Higher School of Business and Entrepreneurship under

Table 1 Characteristics of the activities of Uzbekistan's universities striving for digital inclusion in the globalizing world

	Indicators of development and life values of the universities	Characteristics
1	Higher School of Business and Entrepreneurship under the Government of the Republic of Uzbekistan and Ajou University (South Korea)	
1.1	Web-site	http://e-tinglovchi.gsbe.uz/
1.2	List of university services that are provided in the online form	1. Registration and financial services Filing documents, registration of diploma information, registration of data on work experience, interviews in the online form, testing, payments for studies, and admission of an applicant to the university programme. 2. Participation in the digital university programme (for students of extramural form of study and part-time students). Master's programme and short-term courses (26 directions) with issue of corresponding diplomas (certificates). Provision of online teaching materials. 3. Exams and tests in the online form.
1.3	Requirements for applicants and students	High school diploma, registration and financial procedures. No requirements or restrictions as to the geographical location of applicants or students, age, gender, nationality, religion, etc.
1.4	Life values of the university	Accessibility and digital equality. Access to educational programmes for students from various regions of the country or from abroad. Inclusion: access for people with disabilities. Openness and interactive process: possibility to receive educational materials in the electronic form; and online consultations with teachers; possibility of using three national online platforms and all online platforms of with scientific and methodological materials.
1.5	Key advantage of studies at the university	Dual diplomas (certificates) from the university (Uzbekistan and South Korea), which are acknowledged in two countries.
2	IT-Park University	
2.1	Website	https://itpu.uz/ru/admission/
2.2	List of university services that are provided in the online form	1. Registration for admission. Registration of personal profile and registration for admission exams. Two exams: English language and mathematics. 2. Digital educational programs. The form of studies is largely remote, language—English. Bachelor's and master's programs, as well as short-term courses (26 specialties). 3. Exams and tests in the electronic form.

(continued)

Table 1 (continued)

	Indicators of development and life values of the universities	Characteristics
2.3	Requirements for applicants and students	Apart from a traditional set of documents, it is necessary to provide proof of knowledge of the English language at the level of an IELTS certificate. No requirements as to the location of the applicant or student, since the studies are in the remote form.
2.4	Life values of the university	Accessibility and digital equality: access to educational materials regardless of students' location. Interactive work and openness: students have constant communication with lecturers and IT specialists for the exchange of knowledge, opinions, and approaches. Inclusion: the possibility of providing educational services to talented people who live in remote rural areas and to people with special needs (disabilities, low mobility, etc.). Readiness for new challenges in the labour market. This life value is implemented through the high level of training of students, teaching practical skills by specialists with work experience, and the involvement of the leading IT specialists.
2.5	Key advantage of studying at the university	High level of students' readiness to practical work in their specialties due to the large number of practical lessons and tasks.

Source: Compiled by the authors based on [7, 9, 10]

the Government of the Republic of Uzbekistan and Ajou University (South Korea). Analysis of the activities of IT-Park University revealed such elements as readiness for new challenges in the labor market. Emphasis on this life value is present, to different degrees, in most universities of Uzbekistan that use practical methods in their activities and that strive for digital inclusion.

Let us analyze the life values of digital universities in Kazakhstan (Table 2).

Based on the analysis of the life values of Kazakhstan's universities striving for digital inclusion (Table 2), it should be noted that there are two most developed digital educational platforms, one of which allows for complete receipt of higher education, and another allows supplementing or substituting certain lectures or courses within the traditional full-time training.

4 Life Values of Universities in Central Asia, Which Strive for Digital Inclusion in the Globalizing World on the Example of Uzbekistan and Kazakhstan

In this research, we distinguished the life values of universities in the selected countries of Central Asia, which strive for digital inclusion in the globalizing world. We showed—by the example of Uzbekistan and Kazakhstan—that these conceptual

Table 2 Characteristics of the activities of Kazakhstan's universities that strive for digital inclusion in the globalizing world

	Indicators of activities and life values of digital universities	Characteristics
1	Satbayev University	The structure of digital education in university began functioning in 2004, and in 2020, the digital university began providing services of higher and additional education
1.1	Website	https://satbayev.university/ru/online-learning
1.2	List of university services that are provided in the online form	1. Filing documents in the electronic form, admission exams, etc. 2. Educational services. There are eleven digital educational programs. There are more than 12,000 students in Satbayev University, and more than 600 lecturers. Students have constant access to materials of digital educational programs (through personal accounts). Access to the online database of methodological and practical materials. 3. Exams and tests in the electronic form
1.3	Requirements for applicants and students	Necessity to provide documents stipulated by the regulatory and legal norms. High-speed internet. No requirements as to students' location, social status, religion, gender, race, or nationality.
1.4	Life values of the university	Inclusion: provision of educational services to various categories of population, including people who have difficulties with visiting university (due to work, family responsibilities, special physical needs, financial difficulties). Accessibility: students have access to the materials of digital educational programs of the university in electronic form. Preparedness to new challenges in the labor market: ensuring a high level of training in specialties that are in highest demand in the labor market, the requirement of constant self-development in the professional sphere.
1.5	Key advantage of studying at the university	Flexible schedule of digital educational programs: studied at a convenient time due to the possibility of using video lectures, and the possibility of using consultations of a chatbot and lecturers.
2	D. Serikbayev East Kazakhstan Technical University	
2.1	Website	https://www.ektu.kz/mooc.aspx https://www.ektu.kz/education/videolectures.aspx
2.2	List of university services that are provided in the online form	1. Registration for courses, to participate in video lectures (completing online forms). 2. Educational services. 43 specialties for which there are video lectures, and 11 specialties for which there are popular online courses.

(continued)

Table 2 (continued)

	Indicators of activities and life values of digital universities	Characteristics
2.3	Requirements for applicants and students	No special requirements, apart from an Internet connection for remote education.
2.4	Life values of the university	The flexibility of studies: mass open online courses and video lectures give a choice to students—to visit the university personally or study remotely. Accessibility: students have access to all video lectures and mass open online courses.
2.5	Key advantage of studying at the university	Possibility of saving time due to video lectures and open online courses, which substitute certain lessons.

Source: Compiled by the authors based on [3–5, 18]

milestones of functioning have their specific features depending on capabilities, the resource potential of universities, and digital electronic educational materials in open sources (Fig. 1).

The key digital educational platforms in the sphere of higher education in Uzbekistan are aimed at replacing traditional education with digital education, which allows ensuring the implementation of such value milestones as inclusion. The activities of two digital higher education establishments are a precondition for satisfying the citizens' needs for knowledge, skills, and receipt of relevant specialties that are in demand in the market. The high level of training, which is demonstrated by universities (especially by a digital university created within a partnership in Uzbekistan and South Korea), allows graduates to raise their ranking in the labor market. High openness and interactivity of pedagogics, practiced by the studied digital educational platforms in Uzbekistan, facilitate the constant exchange of knowledge between universities and students, which, in the future, might become tutors for the next generation of young specialists. Significant results of the digitalization of higher education in Uzbekistan are connected with the institutional activities of universities and the government's participation in regulation and support for these initiatives.

It should be stated that the considered leading digital universities in Kazakhstan are focused on the comprehensive provision of higher education services and at the synthesis of digital and full-time training. There is also no doubt that the studied digital universities in Uzbekistan and Kazakhstan are flagships in the formation of the digital economy at the national and regional levels. High digital knowledge and readiness of students and graduates for professional activities in the conditions of implementation and constant upgrade of ICT are ensured due to the realization of life values of these digital educational platforms, which conform to global life values.

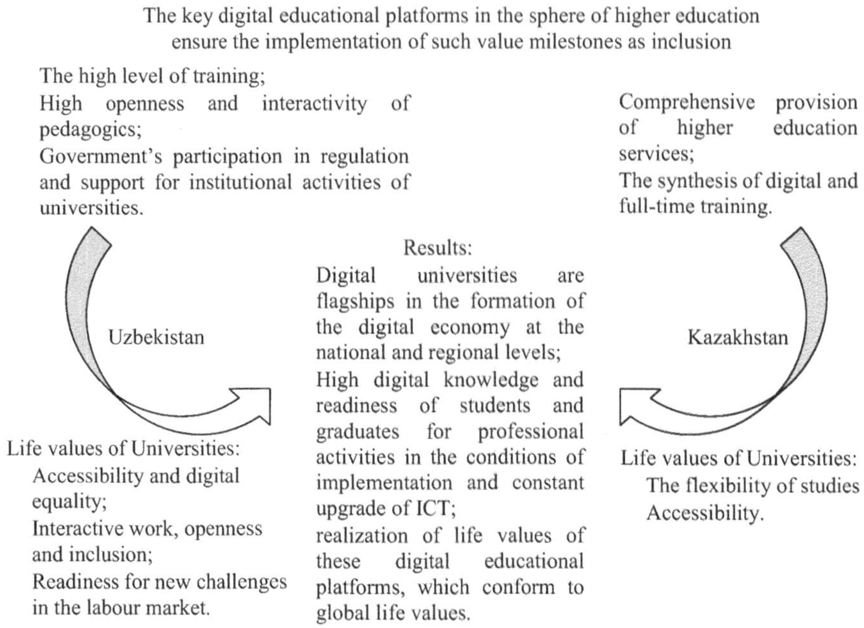

Fig. 1 Life values of universities in Uzbekistan and Kazakhstan, which strive for digital inclusion in the globalizing world. *Source*: Prepared by the authors

5 Conclusion

As a result of this research, we were able to reveal the key life values peculiar for modern digital universities in the world practice and in the practice of Uzbekistan and Kazakhstan. These are interaction with the external environment and openness, focus on critical pedagogics, access to knowledge and skills, digital equality, inclusion, readiness for new challenges in the labor market, and flexibility of training. Based on the materials of the practice of the activities of the analyzed digital educational platforms in countries of Central Asia, these life values can transform and acquire different forms, depending on the scientific & methodological, resource, and personnel potential of universities, government's role, formation of ICT, etc. It is possible to suppose that the emergence of new life values of digital universities that strive for digital inclusion in the globalizing world is also predetermined by changes in the trends in the labor market and the emergence of new professions and requirements to professional training. Given mass digitalization, new professions and spheres of economy can be connected with the requirements as to the ability to work with software and hardware based on AI, big data analysis, machine learning, and interaction with robots of various configurations.

It is possible to state the leading role of digital universities in Kazakhstan and Uzbekistan in attracting talented applicants and providing them with a chance for self-fulfillment in flexible conditions of education. Further reforms in the digitalization of higher education in countries of Central Asia can be created and implemented in view of the existing approach and within the interaction of the government, business environment, and the system of secondary education. The system of secondary education will allow raizing the level of such life value of digital universities as readiness for new challenges in the labour market. We should also emphasize the policy of commitment to the prevention of various forms of discrimination in countries of Central Asia. This policy of the fight against discrimination continued in the mass implementation of inclusive digital education for all categories of the population. This facilitates the integration of the systems of higher education in these countries into the global system of higher education.

References

1. Bannykh, G. A., & Kostina, S. N. (2022). Digital university: Approaches to the concept definition. *The Education and Science Journal, 24*(10), 10–32. Retrieved December 25, 2023, from https://cyberleninka.ru/article/n/tsifrovoy-universitet-podhody-k-kontseptualizatsii-ponyatiya
2. Bineeva, N. (2020). The cognitive model of a digital university: Structural and functional aspects of the analysis. *Don Engineering Bulletin, 12*. Retrieved December 25, 2023, from https://cyberleninka.ru/article/n/kognitivnaya-model-tsifrovogo-universiteta-strukturnye-i-funktsionalnye-aspekty-analiza
3. Dzhilkishinova, M. (2023). Studies in Satbayev University is a golden ticket that not everyone can get. *Liter*. Retrieved December 25, 2023, from https://liter.kz/ucheba-v-satbayev-university-eto-schastlivyi-bilet-kotoryi-vypadaet-ne-kazhdomu-1692272405/
4. Ektu.kz. (2023). *Video collections*. Retrieved December 25, 2023, from https://www.ektu.kz/education/videolectures.aspx
5. Ektu.kz. (2023). *Mass open online course*. Retrieved December 25, 2023, from https://www.ektu.kz/mooc.aspx
6. Ferrante, P., Williams, F., Büchner, F., Kiesewetter, S., Muyambi, G. C., Uleanya, C., et al. (2023). In/equalities in digital education policy – Sociotechnical imaginaries from three world regions. *Learning, Media and Technology, 49*(1), 122–132. https://doi.org/10.1080/17439884.2023.2237870
7. Gsbe.uz. (2023). *Apply modern business education!* Retrieved December 25, 2023, from http://e-tinglovchi.gsbe.uz/
8. Inform.kz. (2023). *The pilot implementation of digital universities in Kazakhstan will start in 2023*. Retrieved December 25, 2023, from https://www.inform.kz/ru/vnedrenie-cifrovyh-universitetov-v-kazahstane-v-pilotnom-rezhime-nachnetsya-v-2023-godu_a4014569
9. It-park.uz. (2023). *Entrance exams in IT Park University are starting*. Retrieved from December 25, 2023, https://it-park.uz/ru/itpark/news/startuyut-vstupitelnye-ekzameny-v-it-park-university
10. Makhmutkhodjaeva, L. S. (2021). Online and digital education at universities of Uzbekistan. *Open Education, 25*(3), 46–61. Retrieved December 25, 2023, from https://cyberleninka.ru/article/n/onlayn-obrazovanie-i-tsifrovye-sredstva-obucheniya-v-vuzah-uzbekistana
11. Najjar, S., & Khawaldah, H. (2024). The emergence and development of the digital educational platforms and their geographical dimensions in Jordan. *Dirasat: Human and Social Sciences, 51*(3), 197–220. https://doi.org/10.35516/hum.v51i3.4560
12. Pacheco-Velazquez, E., Rodes-Paragarino, V., & Marquez-Uribe, A. (2024). Exploring educational simulation platform features for addressing complexity in Industry 4.0: A qualitative analysis of insights from logistics experts. *Frontiers in Education, 9*, 1331911. https://doi.org/10.3389/feduc.2024.1331911

13. Pavlovskaya, S. V. (2023). Specifics of formation of digital universities in certain countries. In M. G. Karaseva (Ed.), *Information technologies in education, science and production: Materials of the scientific and technical Internet conference* (pp. 87–95). BSTU. Retrieved December 25, 2023, from https://rep.bntu.by/handle/data/127969
14. Rahmadi, I. F., Lavicza, Z., Arkün Kocadere, S., Houghton, T., & Spector, J. M. (2024). Investigating the design, participation and experience of teaching and learning facilitated by user-generated microgames on an open educational platform. *Educational Technology Research and Development, 72*(3), 1787–1811. https://doi.org/10.1007/s11423-024-10359-9
15. Rakibi, Y., Alami, A., Selmaoui, S., Bouzit, S., Benjlloun, N., Zaki, M., & Amahmid, O. (2024). The analysis of earth and life sciences educational videos in TelmidTice platform (Morocco): Technical quality, pedagogical and didactic considerations. *International Journal of Information and Education Technology, 14*(5), 732–742. https://doi.org/10.18178/ijiet.2024.14.5.2098
16. Read, R. L., Bechard, N., Suturin, V., Zuiderwijk, A., & Mellenthin, M. (2025). The PolyVent educational platform: An open mechanical ventilation platform for research and education. *HardwareX, 21*, e00615. https://doi.org/10.1016/j.ohx.2024.e00615
17. Said-Hung, E., Arce-García, S., & Mottareale-Calvanese, D. (2024). How Spanish educational researchers used Twitter/X as a platform to promote the dissemination of scientific knowledge: A descriptive study. *Science Editing, 11*(2), 124–133. https://doi.org/10.6087/kcse.336
18. Satbayev University. (2023). *Online learning*. Retrieved December 25, 2023, from https://satbayev.university/en/online-learning

An Algorithm for Achieving Digital Inclusion of a University in a Globalizing World

Tatiana N. Litvinova, Aktam U. Burkhanov, Barna O. Rakhmankulova, and Elena V. Patsyuk

Abstract This work dwells on the problem of developing an algorithm aimed at ensuring the digital inclusion of a university that functions in the conditions of globalisation. To solve this problem, the key theoretical provisions are described and the sequence of evolution of a traditional university to the level of a digital inclusive establishment of higher education is provided. Also, the order of actions for achieving digital inclusion by the university is presented. The considered research topic covers different aspects of technical, pedagogical, social and other disciplines. To reach the set goals, the process approach is used, which reflects the general evolution of inclusive higher education and the achievement of the status of digital inclusive education. The process approach also allows structuring problems and developing the order of their resolution within a certain algorithm. The main methodological tools include analysis and synthesis, the system and process approaches, concrete historical and graphic methods, and the method of grouping. Within the given problems and proposed methodological basis, this work presents the generalisation of conceptual and theoretical provisions on social and digital inclusion, distinguishes their key features and problems, characterises the main groups of people that require inclusion and elaborates on the sequences of the stages of transition from a traditional university to inclusive and digital inclusive university. Based on the obtained conclusions, the stages of the algorithm for achieving the digital inclusion of the university are offered.

T. N. Litvinova
Volgograd State Agricultural University, Volgograd, Russia

A. U. Burkhanov (✉)
Alfraganus University, Tashkent, Uzbekistan

B. O. Rakhmankulova
"Tashkent Institute of Irrigation and Agricultural Mechanization Engineers" National Research University, Tashkent, Uzbekistan

E. V. Patsyuk
Sebryakovskiy Branch of Volgograd State Technical University, Mikhailovka, Russia

B. S. Sergi et al. (eds.), *University 4.0 and Educational Technology for Sustainable Development*, Education in the Asia-Pacific Region: Issues, Concerns and Prospects 75, https://doi.org/10.1007/978-981-96-7376-6_4

Keywords Inclusion · Digital inclusion · Digital technologies · Digitalisation · University · Higher education · Globalisation

JEL Codes I21 · I23 · I24 · I28 · O15 · O33 · O38 · O57

1 Introduction

Due to the gradual evolution of society, we are at the place where processes taking place in one part of our planet can instantly influence the situation on the opposite side of the planet. Apart from this, we have technologies that can connect us with anyone in the world in a matter of seconds, enabling us to work with the entire mankind on the resolution of complex problems or simply communicate on different topics with representatives of different countries. Together with this, we received efficient tools for consolidation and directing the activities of all representatives of the whole of mankind towards understanding the global impact of climate, the use of energy and resources, global conflicts, etc.

Thus, in a globalising world, processes that demand particular understanding and attitude become relevant, accessible, and system-building. These include striving towards the widest possible involvement of all representatives of mankind in public processes. Digital technologies in this segment have a large potential for solving many problems of mankind, including the maximum inclusion of vulnerable groups of the population. They help a lot in the resolution of many problems, which are connected with disabilities (remote accessibility of communications, programmes for people with impaired hearing and vision, etc.), language barriers (translating tools), age restrictions, etc.

Along with this, digital technologies in many cases are barriers that also require an inclusive approach. A quick transition of many analogue processes into the digital sphere, including banking, telecommunications, education, labour market, and other spheres, leads to exclusion from the active social life of a group of people who cannot keep up with changes or do not have sufficient motivation to adapt to new conditions. In such a situation, the notion of inclusion acquires new characteristics, and the achievement of digital inclusion by universities becomes an essential requirement of time.

2 The Notion of Digital Inclusion in Universities

The notion of digital inclusion in universities is at the intersection of several spheres, which include pedagogical, technical & technological, social, and other sciences. According to this, the methodological basis of this research is comprised of different approaches, concepts, and methods, which differ depending on the scientific spheres to which they belong. However, the basic concepts within this research are inclusive universities [2, 6, 12] and digital higher education [1, 7, 10]. According to the process approach, which was used in this work, the digital inclusion of a university is viewed from the position of its evolution from a traditional establishment of higher education to an inclusive digital university. The algorithm of achievement of digital inclusion is presented as a certain sequence of actions and stages, which, in totality, ensure the achievement of the corresponding transition. Given the above, the methodological tools of the research include the process and system approaches, analysis and synthesis, observation and grouping, and concrete historical and graphical methods.

The considered problems belong to a wide range of important scientific tasks, many of which are solved by representatives of different schools and disciplines. Here we used relevant developments that concern the treatment of inclusion as a market phenomenon from the position of its economic potential [4], identification of the problem of inclusion that emerges due to the dissemination of digital technologies [5], determination of the conceptual foundations of the digital age [3] and use of digital technologies in public processes [14], theoretical substantiation of the principles of digital inclusion [11], transition to digital education [8] and application of the appropriate technologies [9, 13], the influence of the digital gap on the development of universities [15] and formation of responsible society [16], etc.

The goal of this work was to substantiate the directions and sequence of actions that would allow achieving digital inclusion in universities. For this, the theoretical essence of inclusion and digital inclusion is disclosed, their specific features are distinguished, the process of transition from traditional to inclusive digital university is described, and the order of actions within the algorithm of achieving digital inclusion by a university is substantiated.

3 Problems and Prospects of Transition from the Traditional to Digital Inclusive University

At the beginning of the mass dissemination of digital technologies, it was clear that this process would be difficult for some countries and groups of the population and would lead to their exclusion from active participation in global or public processes [3]. Such exclusion does not conform to the demands and values of the modern world, so it stimulates the study of the problem of the digital divide and inequality to develop tools for their elimination. In this context, universities are assigned an important role as an intermediate between the digital world and humans who have problems or lack motivation for the digital transition [14].

The term "inclusion" has come a long way in evolution. While recently it was often treated as the "creation of an accessible environment for handicapped people," at present, there is a frequent thought that inclusion is the "inclusion of people with different capabilities and specifics from the position of health into social processes and their adaptation to educational Internet services and apps, including specialised inclusive versions" [4].

Together with the development of the practical basis of inclusion, another concept—"digital inclusion"—formed against the background of digitalisation. It is treated as a certain unfavourable position of a person from the point of view of education, employment, or qualification. Because of this, the person cannot fully use the capabilities of digital technologies and is often used with the absence of access, digital skills, or motivation for the use of technologies to solve everyday, public, or other tasks [4].

In this context, digital inclusion goes beyond the limits of including people with disabilities. It is treated as a "concept according to which digital technologies become accessible for most of users, without the emphasis on their special needs or limitations". From the position of the market approach, the number of people that could be involved in active public and economic life at the global scale is about 1.85 billion, and almost 3.4 billion with their families. In Russia, the number of people with special needs or disabilities exceeds 12%, and with elderly people, who face different limitations, the total number will be above 30% [4].

The main problems that are to be solved by social inclusion are as follows:

- Disability and the related limitations on movement, physical activity, vision, hearing, mental problems, etc.;
- Belonging to a certain culture;
- Belonging to a certain gender or age group;
- Religious beliefs, etc.

Instead, digital inclusion is aimed at the limitation connected with the access and possibility to use technologies. In the conditions of a digital inclusive university, these concepts form additional challenges and opportunities. A general scheme of transition from a traditional university to a digital inclusive university is given in Fig. 1.

The presented scheme reflects the key elements of the transition from a traditional university to a digital inclusive university by passing an intermediary stage—an inclusive university. According to this, a digital inclusive university is formed based on an inclusive one. Problems and barriers given in the left part of Fig. 1 are solved evolutionarily through the formation of the corresponding worldview, technological, or managerial possibilities. Thus, limitations that were determined by different personal limitations and specific features are presented in the "Limitations" block; they were solved primarily due to transforming strategic values and provisions, which was implemented at the global level (block "Transformation of global values"). As a result of this, inclusive universities were formed, which allowed for solving most of the problems of social inclusion.

Some problems that remained from the position of inclusion (block "Barriers") were solved due to digital technologies (block "Digital technologies"). This allowed forming a digital inclusive university, which removed most of the limitations, created additional opportunities for people with special needs, and ensured global involvement. However, against the background of different rates of digitalisation, there appeared new limitations (block "Digital divide"), which led to new challenges, which are being solved thanks to a wide range of measures (block "Digital inclusion").

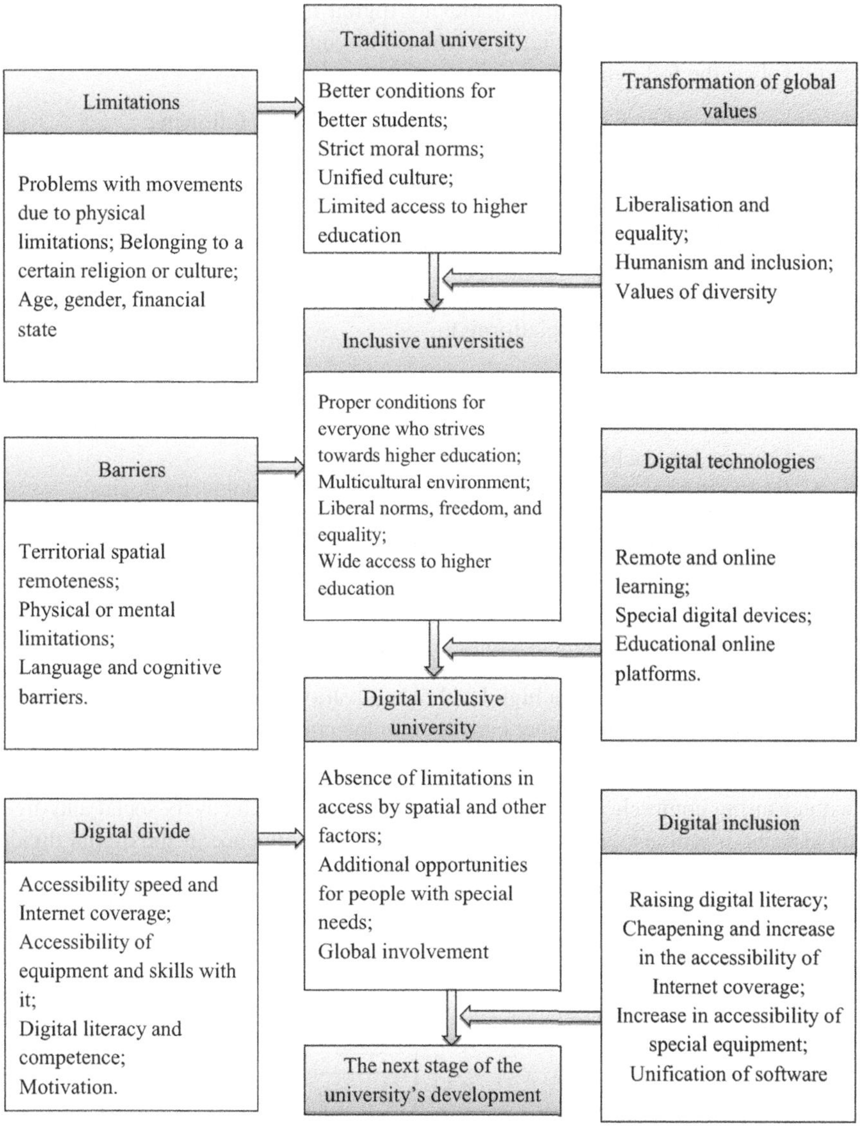

Fig. 1 The scheme of transition from the traditional to digital inclusive university. *Source*: Compiled by the authors based on [9, 17]

It is expected that the resolution of the problem of the digital divide will allow raising the level of digital inclusion of universities. However, based on previous experience, it is also expected that there will be a transition to a new stage of development of inclusion in higher education and the formation of further problems, limitations, and barriers in the development of an inclusive society, which will require the application of a new range of measures and actions.

In any case, the algorithm of achieving digital inclusion of universities in the conditions of globalisation (Fig. 2) is realised through a range of consistent actions and involves the following stages:

1. Creation of a digital educational environment, with the following:
   - Formation of didactic means (online guides, support, inclusive educational materials);
   - Accessibility of hardware and software (technical means and programmes, laboratory equipment);
   - Accessibility of remote education technologies (mobile apps, educational online content, online platforms);
2. Formation of digital competencies:
   - Free resources for mastering basic skills of work with digital equipment;
   - Courses for teaching digital literacy;
   - Promoting values and principles of inclusion of digital technologies;
3. Ensuring accessibility of digital education:
   - Unification of educational online space;
   - Providing accessible high-speed Internet;
   - Providing basic technical equipment according to the needs;
4. Ensuring and supporting a high level of motivation among persons with limited opportunities to obtain higher education in the conditions of digital inclusion.

Implementation of this sequence of actions will allow solving current problems with ensuring comprehensive inclusion in education, which covers social and digital aspects. It will also allow assuaging the negative influence of the digital divide on the process of involvement of all people, including people with special needs, in complete achievement of higher education and, therefore, active participation in social processes.

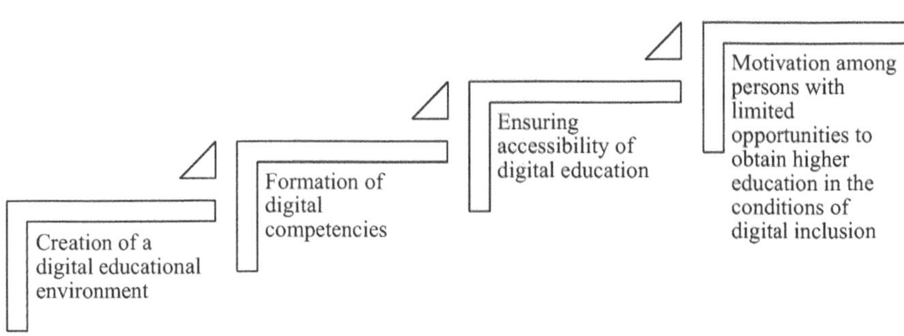

Fig. 2 The algorithm of achieving digital inclusion of universities in the conditions of globalisation. *Source*: Compiled by the authors

4 Topical Issues of the Achievement of Digital Inclusion of Universities in the Conditions of Globalisation

The problem of the achievement of digital inclusion of universities in the conditions of globalisation is among the topics that are actively discussed in expert and scientific circles. It implies the resolution of a range of problems, which are gradually updated and developed together with the development of means for solving them. According to this, digital technologies, as tools of inclusion into higher education lead to the formation of a new type of exclusion in the form of a digital divide. Thus, on the way to the formation of an ideal concept of an inclusive university, there constantly emerge new restrictions and barriers that need resolution and overcoming.

The general algorithm of actions in this context comes from the logical sequence of actions for the creation of the corresponding environment and the formation of conditions for its complete use. However, such a sequence of actions requires the involvement of large resources and the formation of the appropriate culture. These issues require additional discussion to determine the resource bases, assessment of the effectiveness of their implementation, and development of the corresponding mechanism for the implementation of the given plans, actions, and ideas.

Attention should be paid to the development of an algorithm aimed at the achievement of digital inclusion of universities for each university from the position of its solving the main problems aimed at ensuring social and digital inclusion. Such research would involve the development of a special methodological apparatus and its adaptation to local conditions.

Another issue to be discussed is the determination of the problems of the digital divide and its influence on inclusion in higher education. In this context, the main efforts of researchers should be aimed at determining critical digital technologies that influence, to the largest extent, on digital inclusion and the digital divide, as well as the development of measures that facilitate the general accessibility of these technologies and the creation of the corresponding programmes and mechanisms that can ensure maximum involvement of people in higher education and active public life due to studying in a digital inclusive university.

5 Conclusions

The development of digital technologies at the global scale is an important factor in evolution. From the position of public development, digital technologies allowed creating new opportunities for solving many problems, including the involvement of people with special needs in active public life, including obtaining higher education. According to the modern treatment of "digital inclusion", it is seen as a concept according to which digital technologies become more accessible, which allows using them for the resolution of problems of movement restrictions, problems with hearing or vision, and mental or cognitive limitations.

People with special needs and limited capabilities form a large group in society, which actualised the need for their active participation in public life. Thus, processes of social inclusion are set onto the use of digital technologies and facilitate the transition from a traditional university to an inclusive and digital inclusive university. At that, relevant problems are solved, and the transition to the next stage of inclusive university evolution takes place. In general, the algorithm for the achievement of digital inclusion in a university involves a range of successive stages that include the creation of a digital educational environment, formation and support for relevant digital competencies, support for accessibility of digital technologies and education, and support for a high level of motivation for obtaining higher education and application of digital technologies.

References

1. Al-Ghammaz, S. A.-D., Alassaf, H., Alsalti, M., Almomani, H., & Hleiban, T. (2025). The role of e-learning in enhancing digital thinking skills among university students. *International Journal of Innovative Research and Scientific Studies, 8*(1), 451–457. https://doi.org/10.53894/ijirss.v8i1.4178
2. Atanasova, D., & Papen, U. (2025). UK University teachers on inclusive education: conceptualizations, practices, opportunities and challenges. *Studies in Higher Education*. https://doi.org/10.1080/03075079.2025.2455431
3. Castells, M. (2000). *The information age: Economy, society and culture* (Transl. from English; by ed. O. I. Shkaratana). HSE University.
4. Drogichinskaya, V. (2023). Digital accessibility: Why inclusion is a matter of profit. *Forbes*. Retrieved December 5, 2023, from https://www.forbes.ru/mneniya/484771-cifrovaa-dostupnost-pocemu-inkluzia-vopros-pribyli
5. Helsper, E. J. (2017). The social relativity of digital exclusion: Applying relative deprivation theory to digital inequalities. *Communication Theory, 27*(3), 223–242. https://doi.org/10.1111/comt.12110
6. Kohout-Diaz, M. (2024). Promoting inclusive university practices: Fostering diversity and dignity in doctoral supervision. *European Journal of Education, 59*(4), e12710. https://doi.org/10.1111/ejed.12710
7. Majeed, B. H., Abdulsalam, W. H., Ibrahim, Z. H., Ali, R. H., & Mashhadani, S. (2025). Digital intelligence for university students using artificial intelligence techniques. *International Journal of Computing and Digital Systems, 17*(1), 1–10. https://doi.org/10.12785/ijcds/1571029446
8. Malakhova, Y. L. (2022). Digital transformation of inclusive education: Opportunities and risks. *Vestnik magistraturyiu [Bulletin of the Magistracy], 10-1*(133), 95–98. Retrieved December 6, 2023, from https://cyberleninka.ru/article/n/tsifrovaya-transformatsiya-inklyuzivnogo-obrazovaniya-vozmozhnosti-i-riski
9. Muller, O. Y. (2022). The use of digital technologies in inclusive education in a university setting. *Bulletin of the Orthodox St. Tikhon's Humanitarian University. Series 4: Pedagogy. Psychology, 65*, 100–110. Retrieved December 5, 2023, from https://cyberleninka.ru/article/n/primenenie-tsifrovyh-tehnologiy-v-inklyuzivnom-obrazovanii-v-usloviyah-vuza
10. Osiyanova, O. M., Burikova, S. A., & Radyuk, A. V. (2024). Technology for improving students' communicative skills in the university digital educational environment. *Training, Language and Culture, 8*(4), 24–35.
11. Plotichkina, N. V. (2020). Digital inclusion: Theoretical reflection and public policy. *Tomsk State University Journal of Philosophy, Sociology and Political Science,*

58, 216–226. Retrieved December 3, 2023, from https://cyberleninka.ru/article/n/tsifrovaya-inklyuziya-teoreticheskaya-refleksiya-i-publichnaya-politika

12. Ravdansuren, C., Altangerel, A., Ulzii, A., & Spulber, D. (2025). A sustainable and inclusive university on the paradigm of student learning satisfaction, teacher leadership, and professional disposition: The case of Mongolian Universities. *Sustainability, 17*(1), 33. https://doi.org/10.3390/su17010033
13. Ryndak, V. G., Allagulov, A. M., & Chelpachenko, T. V. (2021). Digital technologies as a means of developing inclusive education. *Bulletin of Orenburg State University, 3*(231), 70–78. Retrieved December 6, 2023, from https://cyberleninka.ru/article/n/tsifrovye-tehnologii-kak-sredstvo-razvitiya-inklyuzivnogo-obrazovaniya
14. Vasilenko, L. A., & Zotov, V. V. (2020). Digitalization of public administration in Russia: Risks, incidents, problems. *Tsifrovaia sotsiologiia [Digital Sociology], 3*(2), 4–16. https://doi.org/10.26425/2658-347X-2020-2-4-16
15. Voevodina, E. V. (2022a). Analysis of the "third mission of universities" in the context of educational risks of digital inequality. *Tsifrovaia sotsiologiia [Digital Sociology], 5*(1), 54–63. Retrieved December 4, 2023, from https://cyberleninka.ru/article/n/analiz-tretiey-missii-universitetov-v-razreze-obrazovatelnyh-riskov-tsifrovogo-neravenstva
16. Voevodina, E. V. (2022b). The third mission of universities in the context of sustainable development of the digital society: Gender aspects. *Bulletin of the Russian State University for the Humanities. Series "Philosophy. Sociology. Art History", 1-2*, 182–189. Retrieved December 4, 2023, from https://cyberleninka.ru/article/n/tretya-missiya-universitetov-v-razreze-ustoychivogo-razvitiya-tsifrovogo-obschestva-gendernye-aspekty
17. WAI. (2023). *WCAG 2 overview. Web content accessibility guidelines (WCAG).* World Accessibikity Initiative. Retrieved December 4, 2023, from https://www.w3.org/WAI/standards-guidelines/wcag/

Motivation for Teaching Students in the Pedagogical Field Within the Framework of the University 4.0 Education Model

Gulnara K. Tokkulinova, Bibigul Zh. Almukhambetova, Gulmira A. Ayipova, Araily D. Shakirova, and Nurzhanat D. Shakirova

Abstract The contemporary education system introduces profound changes in the learning and upbringing of the present generation. Each sector of education responds uniquely to the transition to new standards, ranging from early childhood education to university specialization. In doing so, every educational structure must align with the global trend of University 4.0 as a scientific transformation for the sustainable, innovative development of society. The new model of University 4.0 embodies a harmonious unity of scientific-educational, socio-educational, and entrepreneurial potential. The aspect of multiculturalism and inclusivity in education is particularly emphasized. In the realm of university education, some challenges are noted during the transition to the new model, especially in domestic education. On the other hand, the student body keeps pace with the times, thanks to the development of digital technologies. The pedagogical community must adapt to the demands set by society. This research attempts to define the theoretical and methodological aspects, aiming to explore the pedagogical efforts in motivating students of the pedagogical direction within the framework of the University 4.0 education model. The research demonstrates that the university strategy involves orienting towards the end product (graduate) and the client (applicant, student). Therefore, the relevance of this research lies in the importance of motivating students to successful learning and achieving a worthy profession, particularly within the Faculty of Pedagogical Education at the Bishkek State University named after K. Karasaev, the primary

G. K. Tokkulinova (✉)
Kainar Academy, Almaty, Kazakhstan

B. Z. Almukhambetova · A. D. Shakirova
Zhetysu university named after I. Zhansugurov, Taldykorgan, Kazakhstan

G. A. Ayipova
Bishkek Medical College, Bishkek, Kyrgyzstan

N. D. Shakirova
Abai Kazakh National Pedagogical University, Almaty, Kazakhstan

B. S. Sergi et al. (eds.), *University 4.0 and Educational Technology for Sustainable Development*, Education in the Asia-Pacific Region: Issues, Concerns and Prospects 75, https://doi.org/10.1007/978-981-96-7376-6_5

goal of which is to provide high-quality professional education to future educators so that they can realize themselves in a teaching profession.

Keywords University 4.0 · Education · Upbringing · Motivation · Higher education · Pedagogical faculty · Student · Graduate · Teacher · Stimulus · Globalization

JEL Code I2

1 Introduction

The new model of University 4.0 embodies a harmonious unity of scientific-educational, socio-educational, and entrepreneurial potential. The aspect of multiculturalism and inclusivity in education is particularly emphasized. In the system of university education, certain challenges are observed during the transition to the new model, especially within the domestic educational context [10, 11]. On the other hand, the student community keeps pace with the times thanks to the development of digital technologies. The pedagogical collective should transform in response to the demands of today's society [13].

In Kyrgyzstan, as well as in other civilized countries, the reconstruction of all sectors of human life, including education, is underway. The decree of the President of the Kyrgyz Republic, Sadyr Japarov, "On measures to enhance the potential and competitiveness of higher professional education organizations of the Kyrgyz Republic" [4] elevated the potential and competitiveness of university education to the level of the global service market. In this regard, the University 4.0 education model is a necessary implementation of the leading trend in the global world as an interdisciplinary integration within universities [14].

Particularly, it is essential to highlight the educational potential of the pedagogical direction in University 4.0 because it is necessary to modernize the professional training of future teachers through the Concept of the fourth generation as entrepreneurship in science and education [15]. Effective education and upbringing of the youth can be continued thanks to such motivation for students. Therefore, there is a pressing need to transition to the new University 4.0 model in the domestic education system [16].

The Concept of the fourth generation propels global education through the utilization of advanced nanotechnologies and digital tools. However, the education system is currently incapable of meeting the demands set by the rapid development of society. Therefore, within the university education system, it is necessary to reconsider this model as a paradigm for the educational process, starting with motivational efforts directed toward the student community. After all, the students are the clients of any university. Accordingly, the University 4.0 concept prioritizes orientation towards a client.

Consequently, it is crucial to know how to motivate the client for learning and upbringing effectively. If the clients are students of the pedagogical faculty aspiring to become future educators, they must possess the ability to educate the next generation.

2 The University 4.0 Education Model

In exploring this issue, stemming from the objectives and goals of this research, the authors employed the following methods to study pedagogical efforts in motivating students of the pedagogical direction within the framework of the University 4.0 education model:

1. Empirical research: searching for and studying literature related to the research problem;
2. Qualitative research: gaining a detailed understanding of student life through an examination of the university curriculum;
3. Observation: actively observing students during the learning process;
4. Questionnaires: collecting data through formulated questionnaires based on observations of students during university education;
5. Focus group interviews: methods that make it possible to understand students' opinions on the learning process within the university and obtain preliminary information;
6. Analysis: a method used to analyze written and oral materials to gain a factual understanding of the posed problem;
7. Generalization: deciding on subsequent working approaches based on the methods indicated above.

The transition from one model to another in the pedagogical direction follows its regularities associated with complex social changes in the era of digitization. The new model requires a transition from the post-industrial (informational) economy to the knowledge economy [2, p. 17]. J. Dewar called this educational model ecological. Professor O. L. Zhuk referred to it as social-entrepreneurial.

For example, in German education, the University 4.0 model is considered effective in realizing the harmonious development of education through sustainable self-management by students, where the university is governed not only by the administration but also by the student community [6, p. 12]. German researchers (e.g., G. Müller-Christ and B. Gizenbauer) consider such motivation for students to be effective and a key feature of the University 4.0 model because it shapes a personal attitude toward university development through joint decisions and open research.

Nowadays, academic institutions understand that the main factor in developing higher education is the role of human capital in the economic sector. This is achievable through the independence of the student community in implementing startups.

Therefore, any activity emphasizes the importance of human activity, which can only be developed through motivation.

Psychological science regards the motivational sector of the human psyche as one of the main issues in conscious individual activity. "The formation of motivation is one of the most complex and problematic issues faced by a teacher during their professional pedagogical activity" [7, p. 163].

In education, each learner cannot independently show activity in academic or extracurricular activities. This is determined by the consciousness of each individual. Consciousness is the ability of adequate individuals who perceive the surrounding world and events happening around them as a regular phenomenon. Therefore, students' activity manifests in their involvement in the work of the faculty or university. The process of student activity depends on their interest and, most importantly, motivation.

Every person desires to achieve specific goals. The attainment of these goals happens through persistent effort driven by knowledge, skills, and abilities. Thanks to the motivational work of teachers, students exhibit activity and engage in educational and developmental activities. Motivation is deemed successful when students derive satisfaction from academic activities and enjoy the learning process within the chosen professional direction. Therefore, motivation initiates the activity of students as the starting stage of their entire endeavors.

A. N. Leontiev defined motivation as a synthesis consisting of a person's needs and motives with an emotional outlook, representing the personality, and evaluation from the perspective of others [8]. Consequently, to comprehend how to motivate students, it is essential to study the individual formation of each learner during any activity or in the learning process. Hence, the motivation of students in the pedagogical field is necessary. The question of motivating students in the learning process plays a crucial role in developing future educators.

However, it is necessary to understand the essence of the words motive and motivation because these terms do not express the same meaning. I. A. Ghidaryan defines a person's motivation through a motive as their mental state, which can fix an internal human need, consisting of a motive [5].

It can be asserted that a motive and a person's need are equivalent, and motivation is a certain quality of life with a set goal. S. L. Rubinstein argued that a need does not express a specific direction, while motivation is purposeful, especially if it motivates a teacher to educate a student [9].

In the university education system, students' motivation towards learning depends on the professionally set work of the teacher. Nowadays, the professorial and teaching staff understand that the student is a client and the educational program is the product of educators. In the global world, the client chooses a product that can then be used in the contemporary job market. Therefore, the main task of educators is to attract students by motivating and advertising their product (i.e., the educational discipline (program)).

However, the fourth education model talks of a product-oriented approach to student learning, where students are considered "as the product of the university," educational programs are considered as "tools," and the job market is perceived as "the customer" [12]. Therefore, such an approach requires predicting the job market, its requirements, and needs. The university conducts marketing research, identifying the professions demanded by society. On this basis, the university sets tasks for the future and begins to motivate students to learn. Effectively employed graduates become the university's brand, enabling it to have an excellent reputation in the job market.

As is known, recent education models have introduced into the university system elements that do not require applicants to have specific knowledge upon admission because they enter without entrance exams based on ORT [Unified State Exam]. The university learning process adjusts its programs based on the applicants' level of knowledge. Practice shows that the level of school graduates is mostly low. Thus, it is necessary to give lectures at the required level [3]. However, market consumers (e.g., enterprises) do not adapt to the level of knowledge of specialists. Therefore, teachers need to find a golden mean and know how to motivate students.

Recent research shows that strong and weak students differ not in intellectual indicators but in the extent to which their professional motivation is developed [1]. The students' abilities are significant factors in academic work. Professional motivation plays a crucial role concerning the future profession because the motive determines the goal set by each individual. Students of the pedagogical faculty should be motivated to improve academic performance as a crucial factor for a positive attitude toward the future profession. However, even in this case, the future educator must present their profession and their role because this profession is associated with working with children. It would be illogical to deceive them, fail to fulfill promises, and especially to hand them over to another teacher (as a product).

3 Motivating Factors in Choosing a Teaching Profession by Students of the Pedagogical Faculty of Bishkek State University named after K. Karasaev

Based on the above factors that affect students' motivation to learn, the authors conducted a small study among first-, second-, and third-year students of the Pedagogical Faculty of Bishkek State University named after K. Karasaev. The study involved 143 students. Table 1 shows the questions asked to future teachers.

The next questions were posed orally to each student involved in the study. The first question was as follows: "What factor is the most motivating for you when choosing a teaching profession?" The answers to this question are shown in Fig. 1.

The second question was as follows: "What role does your personal motivation play in the process of studying at the pedagogical specialty?" The students' answers are shown in Fig. 2.

According to the respondents' answers, the motivation for students in the pedagogical direction plays a crucial role in their professional preparation and success in their future teaching careers. Therefore, responding to survey questions,

Table 1 Questions asked to future teachers

No.	Questions	Yes	No
1.	Did you enter the Pedagogical Faculty by your choice?	67%	23%
2.	Do your parents motivate you to study?	42%	58%
3.	Do you receive professional motivation from your faculty teachers?	56%	44%
4.	Do you think that you get enough knowledge at the university?	61%	39%
5.	Do you rate your level of motivation to study as high?	41%	59%

Source: Developed by the authors

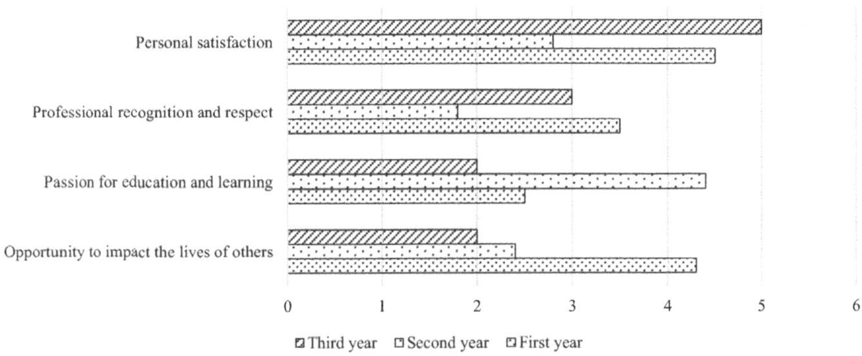

Fig. 1 Motivating factors in choosing a teaching profession. *Source*: Developed by the authors

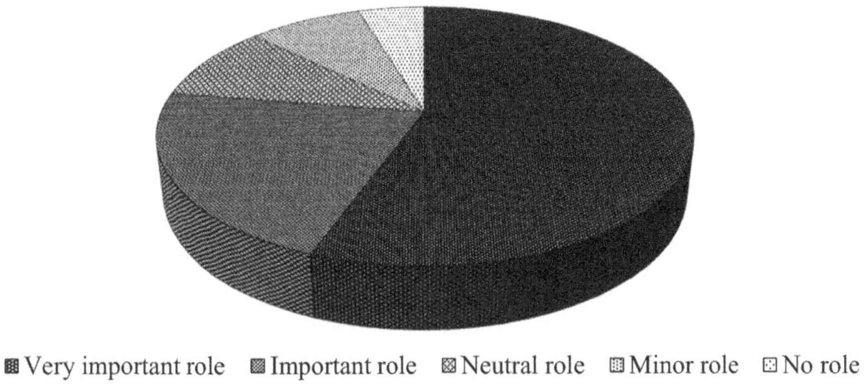

Fig. 2 The role of students' personal motivation in the learning process. *Source*: Developed by the authors

students identified key aspects that can serve as motivating factors, including the following:

1. Recognition of the importance of the profession because teachers play a crucial role in the development of society, shaping the future of the next generation. Awareness of this responsibility and acknowledgment of the importance of the teaching profession can serve as a powerful motivational factor.
2. The opportunity to influence others allows teachers to influence the meaning of their students' lives by helping them to develop, learn new knowledge, and obtain new skills. Realizing that future educators can make a difference in the lives of many children and young people can be an inspiration to learn and develop their professional skills.
3. If students are passionate about education and learning, they may value the process of learning and are always willing to share their knowledge and experience with others. This passion for learning and the ability to enthuse others can be a powerful motivator for professional growth.
4. The ability to creatively apply their methods and approaches during the work is an important indicator for teachers to motivate students. Developing interesting lessons, conducting educational projects, and using innovative teaching methods lead to effective learning for students.

As a result of the conducted research, the authors considered the problems that appear in students' environments during their education. If students initially enroll in a pedagogical university with the intention of receiving education and upbringing, during their studies, some students lose the desire to become a teacher or work in the pedagogical field. In connection with this issue, factors influencing their professional burnout were identified, and methods for motivating students in the pedagogical direction towards learning were developed, which should be employed in the university education system.

4 Conclusion

Therefore, professional recognition and respect for the teaching profession, demonstrated by outstanding results in their work and respect from colleagues, parents, and students, can be a motivating factor for students striving to achieve success in their future pedagogical careers. Thus, continuous development and the opportunity to learn to represent professional growth in the pedagogical field.

A teacher's work brings deep personal satisfaction when students grow and develop; the significant contribution made to their development is undeniable. This can undoubtedly become a powerful source of motivation and meaning in the life of every educator. Therefore, it is important to support and instill confidence in students regarding their abilities and the significance of their future profession. Creating an appropriate educational environment and stimulating interest in the pedagogical process will help students better adapt and successfully cope with the challenges they will face in their future pedagogical careers.

References

1. Clayson, D. E., & Haley, D. A. (2005). Marketing models in education: Students as customers, products, or partners. *Marketing Education Review, 15*(1), 1–10. https://doi.org/10.1080/10528008.2005.11488884
2. Dzhidarian, I. A. (1974). On the place of needs, emotions, and feelings in personality motivation. In E. V. Shorokhova (Ed.), *Theoretical problems of personality psychology* (pp. 145–169). Nauka.
3. Giesenbauer, B., & Müller-Christ, G. (2020). University 4.0: Promoting the transformation of higher education institutions toward sustainable development. *Sustainability, 12*(8), 3371. https://doi.org/10.3390/su12083371
4. Kabar News Agency. (2022, July 18). *The President of Kyrgyzstan has signed a decree granting special status to a number of universities in the country.* Retrieved June 25, 2023, from https://kabar.kg/news/prezident-podpisal-ukaz-o-pridanii-osobogo-statusa-riadu-vuzov/
5. Kaplun, R. N. (2023). Cadets' motivation for training at the university of the strategic missile forces. In A. A. Sukiasyan (Ed.), *Scientific transformation – The basis of sustainable innovative development of society: Collection of articles of the international scientific and practical conference (May 12, 2023, Kazan, Russia)* (pp. 162–165). Omega Science.
6. Koklevsky, A. V. (2021). Educational potential of pedagogical university of the model "University 4.0" in professional training of a teacher. *Bulletin of the Belarusian State Pedagogical University. Series 1: Pedagogic. Psychology. Philology, 3*(109), 17–21. Retrieved June 25, 2023, from https://elib.bspu.by/bitstream/doc/52920/1/%2bВести%203_2021%20Серия%201%20270921_0017-0021.pdf
7. Leontiev, A. N. (1971). *Needs, motives, emotions*. Moscow University Press.
8. Maruhina, V. V. (2014). Motivation of students to study at the pedagogical university (by example of Tuvan State University). *Bulletin of Tuva State University. No. 4 Pedagogical Science, 4*(23), 26–34. Retrieved June 15, 2023, from https://cyberleninka.ru/article/n/motivatsiya-studentov-k-obucheniyu-v-pedagogicheskom-universitete-na-primere-tuvinskogo-gosudarstvennogo-universiteta
9. Popova, E. G., Popov, S. I., & Zagulova, D. V. (2017). Higher education institution – Customer-oriented or product-oriented? *Educational Technology & Society, 20*(3), 430–441. Retrieved June 25, 2023, from https://cyberleninka.ru/article/n/vysshie-uchebnye-zavedeniya-orientiruemsya-na-klienta-ili-na-produkt
10. Roopa, M. (2023). Business engineering 4.0: The transformation of a university course in response to industry 4.0, sustainable development goals, & COVID-19 in South Africa. *International Journal of Engineering Pedagogy, 13*(8), 131–147. https://doi.org/10.3991/ijep.v13i8.38517
11. Rosak-Szyrocka, J. (2024). The era of digitalization in education where do universities 4.0 go? *Management Systems in Production Engineering, 32*(1), 54–66. https://doi.org/10.2478/mspe-2024-0006
12. Rubinstein, S. L. (2000). *Foundations of general psychology*. Peter.
13. Suárez-Brito, P., Vázquez-Parra, J. C., Alonso-Galicia, P. E., & Cruz-Sandoval, M. A. (2024). Creative thinking as a relevant element for industry 4.0 professionals. A study in Mexican university students. *Journal of Latinos and Education, 24*(1), 48–61. https://doi.org/10.1080/15348431.2024.2355146

14. Thuy, L. T. T. (2024). Factors affecting the quality of university accounting education in the context of industry 4.0: An empirical study at universities in Vietnam. *Edelweiss Applied Science and Technology, 8*(6), 8860–8872. https://doi.org/10.55214/25768484.v8i6.3882
15. Ülker, N. (2023). Total quality management in the context of University 4.0: New game new rules. *Frontiers in Education, 8*, 1146965. https://doi.org/10.3389/feduc.2023.1146965
16. Ülker, N., & Otrar, M. (2024). University 4.0 essentials: Influence of industry 4.0 on higher education. *Industry and Higher Education*. https://doi.org/10.1177/09504222241295625

Constructing the Image of the Global University Through Its Official Website (Based on "About" Pages)

Yulia N. Zemskaya, Evgeniya A. Kuznetsova, and Anna V. Osmolovskaia

Abstract The research aims to identify the themes that leading universities worldwide focus on when constructing their image. The authors selected 50 international universities, leaders in the Times Higher Education World University Rankings (THE) for the year 2023, for the research. The research was carried out using content analysis. The material for analysis included the headlines, subheadings, and links posted on the "About" pages of university websites. The authors compiled themes through which universities construct their image, including "Facts and figures," "Leadership and management," "History," "Mission, vision, and values," "People," and others. The research yielded several conclusions. Leading global universities use a unified thematic model for image construction, addressing the same themes on the "About" pages of their websites. This suggests that the image, theoretically created to distinguish an organization from its competitors, can, in practice, be formed uniformly and in a templated manner. Thus, among the fifty universities that are global leaders in higher education, none have developed a unique image through a thematic model on the "About" page using an original set of topics.

Keywords Image · Image communications · Website · University · Thematic model

JEL Codes M31 · M37 · L14 · I 23

Y. N. Zemskaya · A. V. Osmolovskaia
RUDN University, Moscow, Russia
e-mail: zemskaya_yun@pfur.ru; glagoleva_av@pfur.ru

E. A. Kuznetsova (✉)
Higher School of Economics, Moscow, Russia
e-mail: evgeniya.kuznetsova@hse.ru

B. S. Sergi et al. (eds.), *University 4.0 and Educational Technology for Sustainable Development*, Education in the Asia-Pacific Region: Issues, Concerns and Prospects 75, https://doi.org/10.1007/978-981-96-7376-6_6

1 Introduction

Higher education has become highly competitive nowadays [20, 22]. In the struggle for prospective students, universities are compelled to develop marketing communications and engage with their audiences to shape a positively perceived image and reputation [24, 25, 27]. K. Saichaie and C. Morphew [19] indicate that universities are increasingly behaving in the market as commercial entities, actively promoting their services.

The study of university image has become more relevant in the twenty-first century, as highlighted in the literature review on the image and reputation of higher education by A. Lafuente-Ruiz-de-Sabando et al. [12]. In 2008, M. Sung and S.-U. Yang referred to the university image as a new topic receiving increasing attention [20, p. 358].

S. Alessandri et al. defined the university image as the public perception of the university [2, p. 259]. Other authors understand it as the sum of all individual beliefs about the university [6, p. 23]. According to J. Ressler and R. Abratt [18], the image is related to how an organization wants stakeholders to see it (assumed image) and how, in their opinion, they see it (interpreted image).

This research focuses on studying the assumed image in the terminology of J. Ressler and R. Abratt [18], the desired image in the terminology of F. Jefkins [10] (i.e., the representation that is considered accurate from the company's perspective). According to F. Jefkins [10], this type of image can be formed when people lack complete information about the organization or know nothing about it at all. To achieve this, presentation materials are created and placed on various online and offline information carriers (such as corporate websites or company brochures).

R. E. Landrum and R. Torrisi cite J. Weissman's assertion that colleges and universities should remember that people react to the organization's image, not necessarily its reality [13]. This idea becomes particularly relevant in the context of digital communication and the practice of remote university admissions.

The university website has become the primary channel of communication with its target audiences [5]. A website makes it possible to place information of various kinds: image, advertising, and sales-oriented information for diverse audiences (students, prospective students and their parents, alums, staff and faculty, journalists, etc.).

Communication with prospective students is a crucial task for the university because this audience needs to get acquainted with the university. Image information posted on the website can significantly influence the choice of a future student. The university's image can help the prospective student determine which university will best meet their educational needs [16]. The website forms the image not only for prospective students but also for other interested audiences.

Many researchers, such as E. J. Winter [26] and representatives of the National Association for College Admission Counseling (USA) [15], emphasize the importance of the website in getting to know the university and shaping its image.

Studies of the university's image often rely on survey methods, allowing researchers to understand what image is formed for a particular audience. For example, audiences of students and faculty were studied from this perspective by J. P. Pérez and E. M. Torres [17], faculty and staff by T. Luque-Martínez and S. Del Barrio-García [14], and students by P. Alcaide-Pulido and others [1]. However, these methods are not suitable for this research because the authors want to study the image that is desirable for the university (i.e., specifically constructed on the website on the "About" page).

Many researchers, including G. Blanco and A. Metcalfe [3], B. Büyükdoğan and G. Hasan [4], K. Gottschall and S. Saltmarsh [8], K. Saichaie and C. Morphew [19], and Z. Zhang et al. [28] studied university websites from the perspective of content. However, these studies were not dedicated to examining other aspects of Internet communication.

2 "About" Pages of the Leading Global Universities

The research was conducted using content analysis. For examination, the authors selected 50 international universities leading in the Times Higher Education World University Rankings (THE) for 2023 [23].

The material for this research was one of the traditional pages in the website structure—the "About" page. It contains information about an individual or collective subject of communication—a person or an organization. On this page, the university can share information it deems useful and interesting to its audiences, highlighting the unique features that set it apart from other universities. This allows the university to shape the desired image.

For the research, the authors chose leading global universities to understand which specific topics are considered most important from their perspective for shaping the university's image. The research material was collected in July–August 2023.

On the "About" pages, the authors examined headlines and subheadings (if the "About" page contains text) or the topics of links (if the "About" page represents a separate section of the website, it includes a set of links to thematic subsections). The authors identified the main thematic blocks through which the university constructs its desired image. Texts placed on the "About" page or its equivalent section were not analyzed. The authors also did not consider the following:

- Topics indicated on navigation pop-up panels of the "About" section;
- "Cross-cutting" elements present on all pages of the site.

In cases where there was no "About" section on the website, the authors used the materials from pages such as "Meet us," "Portrait," and "Who we are."

3 International Case Study of "About" Pages' Topics Contributing to the Uniqueness of Universities

Leading global universities commonly feature the following topics or blocks on the "About" page:

1. "Facts and figures": 40 out of 50 analyzed websites (80%) include this block;
2. Topics related to university leadership and management.

Specifically, 40 out of 50 analyzed websites (80%) include blocks such as "Leadership," "Management," "Organization," "Administration," and "Governance." The second most common topic is "History" (33 out of 50 (66%)), followed by "Mission, vision, and values" (31 out of 50 (62%)). Examining these leading topics, we can note that they are quite universal for shaping the image of almost any organization, not just a university. Therefore, they serve as the foundation for corporate image formation (Fig. 1).

In addition to general statements about university leadership, 17 websites (34% of those examined) specifically emphasize the position of the president. This may include providing a link to the president's page on the website or the president's

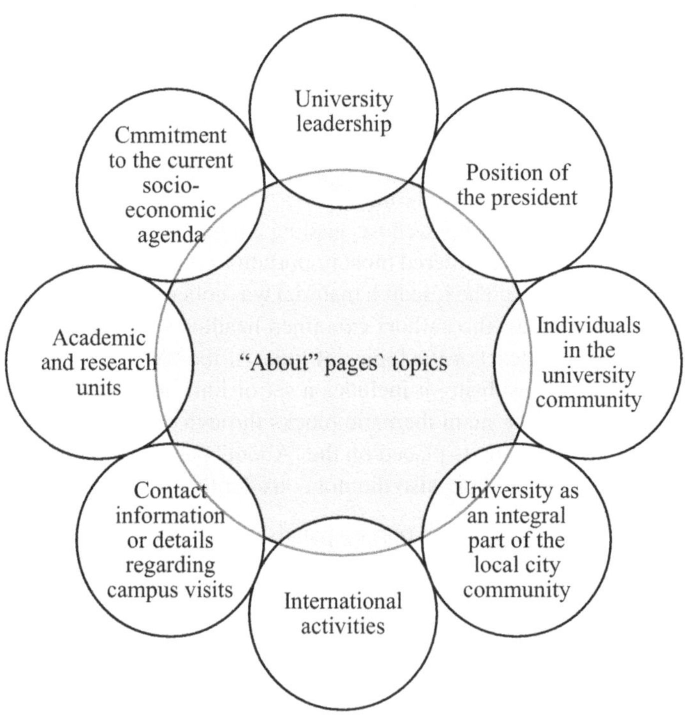

Fig. 1 "About" pages' topics contributing to the uniqueness of universities. *Source* Created by the authors

office page, quoting the president's address, or providing a separate quote from their speech. This emphasis on the president is crucial for the university's image because the president often serves as the public face of the organization, making it possible to personalize the university.

Significant topics related to individuals who are part of or contribute to the university community play a crucial role in shaping the university's image. The "About" pages of the two universities begin with statements emphasizing people: "People are at the heart of Princeton. Princeton is about people" and "Illinois students, scholars, and alumni are a community with the power to change the world," highlighting the importance of this topic. For instance, on the "About" page of Imperial College London, the section about the university's people is titled "Imperial together." On 26% of the websites, there are general formulations such as "Oxford People," "Human resources," or "Our community." One in five universities specifically discusses its alums and one in ten talks about its students or staff.

Furthermore, a quarter of the universities present themselves on their websites as an integral part of the local city community, such as "Harvard in the community." Almost every fifth university discusses the place where it is located and its campuses. This topic might be labeled with a general title, like "Campus location" or "Our locations"; the city's name is often used (e.g., the page "Visitors" on the Chinese University of Hong Kong's website, the page "LSE and London" on the London School of Economics and Political Science's site, the page "Campuses in Abu Dhabi and Shanghai" on New York University's site, etc.). In general, every fourth university among those studied provides information about its campuses.

Nearly half (22 out of 50 universities) highlight their international activities. Headings such as "Global university," "Harvard worldwide," and "Yale and the World" are used for this purpose. This presents the university as a global entity, emphasizing its global status and constructing a corresponding image.

It is a common practice to include contact information or details regarding campus visits on the "About" pages. Sections like "For visitors" and "Visiting" are found on 17 websites (34%), "Campus map" on nine websites (18%), "Contacts" on eight websites (16%), and "Virtual tour" or "Campus excursion" on six websites (12%).

Less than half of the universities (40%) provided information about their academic and research units (departments, faculties, schools, and colleges). This might be attributed to the fact that only the topics or blocks stated on the "About" pages were studied.

Several crucial aspects of constructing the university's image revolve around various achievements, such as competition wins, participation in awards, rankings, collaboration with world-class scholars, and association with Nobel and Pulitzer Prize laureates. Moreover, the university takes pride in its graduates who have received such honors. These pages may have different titles, such as "Awards and honors," "Pride," or "Prizes." Sections on "Nobel Prize Laureates" or "Rankings" are present on every fifth website.

To shape the image of a global university, institutions showcase their involvement in addressing society's significant issues and their commitment to the current socio-economic agenda (similar to how commercial companies often present this in sustainability reports). Diversity topics are found on every third website, while sustainability, inclusion, and affordability are present on every fifth website. Less common are topics related to the global or economic impact of the university, mentioned by 7 out of 50 universities. Nine universities emphasize the preservation of heritage and traditions, while five highlight health (including mental health) and safety.

The cultivation of an image implies the demonstration of the university's uniqueness and distinctive features that set it apart from all others. Various tools associated with the creation of identity can be employed for this purpose.

Let us consider several examples of topics contributing to the uniqueness of universities. For instance, the University of Oxford's website, specifically on the "About" page, contains a link to the "Oxford glossary," elucidating the meanings of certain Oxford-specific terms for the uninitiated. The "Legacies of enslavement" page on the University of Cambridge's site (linked from the "About" page) contains diverse information about the ongoing efforts within the university to combat inequality, including a documentary film. The "Cost of living support" page, linked from the "About" page of Imperial College London, presents information about how the university assists students and staff when the cost of living rises. The "Rethinking public art" page on the website of the University of California contains information about The Stuart Collection, a unique compilation of works by leading contemporary artists oriented toward specific locations (in this case, the university campus and the local community of San Diego). It enriches the cultural, intellectual, and scientific life of the local community by featuring art objects created by contemporary artists specifically for this location.

4 Significance of "About" Pages for the Leading Global Universities' Image

The university's image, in its objectives, aligns with the corporate image of a commercial organization. It is essential to present the university in a favorable light and distinguish it from competitors. According to P. Alcaide-Pulido and others [1], universities should showcase characteristics that make them unique in their environment through their social networks and websites. However, studies by Saichaie and Morphew [19] and Labaree [11] indicate that universities, on their online platforms, often fail to articulate aspects that could shape their identity. The majority of universities portray themselves using the same topics: on every second website of a global university leading in THE rankings, we encounter thematic blocks like "Facts and figures," "Leadership," "History," and "Mission, vision, and values."

M. Hartley and C. Morphew [9] presented a study of the content of college viewbooks (which largely align with the goals and content of "About" pages on university websites). The authors argue that if a university emphasizes specific characteristics in its image, it may alienate some prospective students who do not resonate with these features. Therefore, in the market, it is advantageous to appear more universal. “It seems that many institutions are left in the paradoxical position of wanting to emphasize their special nature while remaining conventional—exciting and enticing but somehow familiar, unique but not weird” [9, p. 685]. If a university portrays itself as excessively unique, it narrows its potential audience, which is not beneficial. Hence, an image is constructed that aligns with the audience’s expectations. This is generally confirmed by the findings of this research: the unique characteristics listed on university websites generally correspond to expectations. Universities have traditions they aim to preserve. They provide support to those in need, combat inequality, and strive to enhance the “living environment”—the locations where their campuses are situated.

The data on leading topics obtained during this research align with the conclusions reached by G. Blanco and A. Metcalfe [3]: figures and facts become preferred evidence in marketing statements. As a result, information about rankings, accreditation, and membership in reputable associations is reflected in the images that higher education institutions carefully craft and present to the public through their websites [3]. On the examined university websites, details about rankings, award winners, and other sources of pride are encountered approximately in every fifth case, and the thematic block “Facts and figures” appears in most cases.

P. Alcaide-Pulido et al. [1] argue that messages conveyed by a university should emphasize values such as commitment to society, ethical actions, and social and environmental responsibility. Based on the results of this research, we can affirm that such topics are relevant for leading global universities because diversity is featured on every third website, and sustainability and inclusivity on every fifth.

The university’s image is linked to personifications [21], with the central figure being the president, presented through addresses or personal pages on a third of the examined websites. However, discussions about the president often fall within the thematic block “Leadership and management.”

Equally important for influencing the university’s image are the individuals who study or have studied at the university and work there. This is evidenced by the data from our study of the corporate culture of the university conducted in 2019. During that time, students and staff at RUDN were surveyed; they identified the people who study and work there as the university’s main strength [7].

5 Conclusion

Based on the results of the conducted research, we can conclude that universities, in constructing their image, actively promote relatively similar representations of themselves. When shaping their image through the “About” page on their websites, universities often limit themselves to audience expectations and narrate the history of their establishment and development, describe their mission and values, present key facts about the people who manage the university, work or study there, and provide information about the campus. The topics used for self-presentation are largely repetitive. Even when they are unique, they remain somewhat predictable.

A paradox arises: the image created in part to distinguish the organization from its competitors ends up being constructed according to a unified thematic model. Among the fifty universities leading in global rankings, not a single one shaped its image through the content of the "About" page using an original set of topics. Many universities prefer to tell their story, offering the audience the same set of topics. Perhaps this pursuit of an ideal is the recipe for successful image communication.

References

1. Alcaide-Pulido, P., O'Sullivan, H., & Chapleo, C. (2021). The application of an innovative model to measure university brand image. Differences between English, Spanish and Portuguese undergraduate students. *Journal of Marketing for Higher Education, 34*(1), 283–300. https://doi.org/10.1080/08841241.2021.1977450
2. Alessandri, S. W., Yang, S.-U., & Kinsey, D. F. (2006). An integrative approach to university visual identity and reputation. *Corporate Reputation Review, 9*(4), 258–270. https://doi.org/10.1057/palgrave.crr.1550033
3. Blanco, G. L., & Metcalfe, A. S. (2020). Visualizing quality: University online identities as organizational performativity in higher education. *The Review of Higher Education, 43*(3), 781–809. https://doi.org/10.1353/rhe.2020.0007
4. Büyükdoğan, B., & Hasan, G. (2014). The using of websites of the universities for prospective students for purpose of institutional promotion. *Researchgate*. Retrieved September 3, 2023, from https://www.researchgate.net/publication/311713779_THE_USING_OF_WEBSITES_OF_THE_UNIVERSITIES_FOR_PROSPECTIVE_STUDENTS_FOR_PURPOSE_OF_INSTITUTIONAL_PROMOTION
5. Das, A., & Gurey, P. (2021). Web content analysis of state university websites of west Bengal: An evaluative study. *Library Philosophy and Practice*. Retrieved September 3, 2023, from https://digitalcommons.unl.edu/libphilprac/6691
6. Duarte, P. O., Alves, H. B., & Raposo, M. B. (2010). Understanding university image: A structural equation model approach. *International Review on Public and Nonprofit Marketing, 7*(1), 21–36. https://doi.org/10.1007/S12208-009-0042-9
7. Glagoleva, A. V., Zemskaya, Y. N., Kachesova, I. Y., & Kuznetsova, E. A. (2021). Academic discourse and corporate culture of a modern university. *Linguistic & Rhetorical Paradigm: Theoretical and Applied Aspects, 26*(1), 81–86.
8. Gottschall, K., & Saltmarsh, S. (2017). 'You're not just learning it, you're living it!' Constructing the 'good life' in Australian university online promotional videos. *Discourse: Studies in the Cultural Politics of Education, 38*(5), 768–781. https://doi.org/10.1080/01596306.2016.1158155
9. Hartley, M., & Morphew, C. (2008). What's being sold and to what end?: A content analysis of College Viewbooks. *Journal of Higher Education, 79*(6), 671–691. https://doi.org/10.1353/jhe.0.0025
10. Jefkins, F. (1998). *Public relations* (5th ed.). Prentice Hall.
11. Labaree, D. F. (1997). Public goods, private goods: The American struggle over educational goals. *American Educational Research Journal, 34*(1), 39–81. https://doi.org/10.2307/1163342
12. Lafuente-Ruiz-de-Sabando, A., Zorrilla, P., & Forcada, J. (2018). A review of higher education image and reputation literature: Knowledge gaps and a research agenda. *European Research on Management and Business Economics, 24*(1), 8–16. https://doi.org/10.1016/j.iedeen.2017.06.005

13. Landrum, E., Torrisi, R., & Harless, C. (1998). University image: The benefits of assessment and modeling. *Journal of Marketing for Higher Education, 9*(1), 53–68. https://doi.org/10.1300/J050v09n01_05
14. Luque-Martínez, T., & Del Barrio-García, S. (2009). Modelling university image: The teaching staff viewpoint. *Public Relations Review, 35*(3), 325–327. https://doi.org/10.1016/j.pubrev.2009.03.004
15. National Association for College Admission Counseling (NACAC). (2019). *2019 State of College Admission Report.* Retrieved August 15, 2023, from https://www.nacacnet.org/2019-state-of-college-admission-report/
16. Paden, N., & Stell, R. (2006). Branding options for distance learning programs: Managing the effect on university image. *International Journal of Instructional Technology and Distance Learning, 3*(8), 45–54.
17. Pérez, J. P., & Torres, E. M. (2017). Evaluation of the organizational image of a university in a higher education institution. *Contaduría y Administración [Accounting and Management], 62*(1), 123–140. https://doi.org/10.1016/j.cya.2016.01.007
18. Ressler, J., & Abratt, R. (2009). Assessing the impact of university reputation on stakeholder intentions. *Journal of General Management, 35*(1), 35–45. https://doi.org/10.1177/030630700903500104
19. Saichaie, K., & Morphew, C. C. (2014). What college and university websites reveal about the purposes of higher education. *The Journal of Higher Education, 85*(4), 499–530. https://doi.org/10.1353/jhe.2014.0024
20. Schlesinger, W., Cervera-Taulet, A., & Wymer, W. (2023). The influence of university brand image, satisfaction, and university identification on alumni WOM intentions. *Journal of Marketing for Higher Education, 33*(1), 1–19. https://doi.org/10.1080/08841241.2021.1874588
21. Sung, M., & Yang, S.-U. (2008). Toward the model of university image: The influence of brand personality, external prestige, and reputation. *Journal of Public Relations Research, 20*(4), 357–376. https://doi.org/10.1080/10627260802153207
22. Suranta, S., & Rahmawati, R. (2024). The role of higher education image and service quality on the effect of university social responsibility (USR) on student loyalty in Indonesia. *Journal of Applied Research in Higher Education, 16*(2), 378–390. https://doi.org/10.1108/JARHE-11-2022-0338
23. Times Higher Education (THE). (2023). *World University Rankings 2023.* Retrieved July 12, 2023, from https://www.timeshighereducation.com/world-university-rankings/2023/world-ranking
24. Tran, Q. B., Nguyen, T. T. Q., Nguyen, H. N., Tran, T. H. M., Duong, T. Q. L., Nguyen, T. H. D., et al. (2023). The impact of brand image on public university links in the context of autonomy: A case study in Vietnam. *International Journal of Data and Network Science, 7*(2), 553–562. https://doi.org/10.5267/j.ijdns.2023.3.016
25. Vartiak, L., Chrancokova, M., Garbarova, M., & Fabus, J. (2024). Responsible email marketing in higher education. *International Review of Management and Marketing, 14*(6), 55–60. https://doi.org/10.32479/irmm.17028
26. Winter, E. J. (2022). *The presentation and role of place in prospective student decision making: A case study of the UK undergraduate market* (Doctoral Thesis … Doctor of Philosophy). University of Portsmouth, Portsmouth.
27. Xia, D., Zhao, P., Wang, J., & Li, Y. (2024). The projection of Chinese University online image and social media engagement based on Bayesian model. *PLoS One, 19*(4), e0300625. https://doi.org/10.1371/journal.pone.0300625
28. Zhang, Z., Tan, S., & O'Halloran, K. L. (2022). Managing higher education and neoliberal marketing discourses on Why Choose webpages for international students on Australian and British university websites. *Discourse & Communication, 16*(4), 462–481. https://doi.org/10.1177/17504813221074076

The Russian Experience in Identifying Leading Universities

Svetlana A. Sharonova and Nina V. Trubnikova

Abstract As the practice of contemporary life shows, the competition for leadership in higher education is increasingly intensifying. At the beginning of the twentieth century, the competition took place at the level of national higher education systems. In turn, the beginning of the twenty-first century, thanks to IT achievements, elevated this competition to a global level. This research aims to analyze the transformation of criteria for assessing university leadership in the era of digital technologies. Additionally, the authors emphasize the impact of international rankings on traditional criteria for determining leading universities in Russia. The methodological foundation is the theory of leadership and the theory of network methodology. Based on the analysis of the values of three models of universities (the Humboldtian, the market-oriented, and the digital), the authors illustrated the shifts occurring in the criteria for assessing university leadership. The authors analyzed the socio-cultural influence on the global system of leadership assessment in international and national ranking databases. In connection with the escalation of international relations related to sanctions imposed by the international community on various areas of life, including education, Russia has decided to build its own higher education system. However, Russia's participation in the Bologna Process has brought about certain transformations in the national higher education system.

Keywords World university rankings · National system of higher education · Digitalization of education · Digital literacy · University research model

JEL Code I23

S. A. Sharonova (✉) · N. V. Trubnikova
RUDN University, Moscow, Russia
e-mail: sharonova-sa@rudn.ru; trubnikova-nv@rudn.ru

B. S. Sergi et al. (eds.), *University 4.0 and Educational Technology for Sustainable Development*, Education in the Asia-Pacific Region: Issues, Concerns and Prospects 75, https://doi.org/10.1007/978-981-96-7376-6_7

1 Introduction

To comprehend the significance of socio-cultural peculiarities in the educational systems when determining leading universities, let us examine the Russian experience. A hierarchical system of leadership among Russian universities took shape over the Soviet period as a result of numerous reforms. This system comprised regional universities, specialized universities, and classical universities. The classical universities were the leaders in the fields of science and higher education. They secured this leading position in the higher education system during the 1930s. On the one hand, they served as the primary suppliers of scientific and government personnel [1]. On the other hand, the leading positions of classical universities were supported by additional funding. As noted by L. V. Ten, the funding norm per student in leading universities, on average, exceeded the corresponding indicators in regular universities by more than two times [2].

Alongside classical universities, the Soviet Union witnessed the establishment of specialized universities to prepare narrowly specialized professionals to address specific challenges of the developing economy. With the rapid growth in the number of such institutions (from 148 to 832 in the 4 years from 1928 to 1932 [3]), there arose a need to identify flagship universities responsible for developing curriculum and regulatory documents for all institutes within each specific sector. Thus, flagship universities determine the policy and directions of development for the corresponding sector of the national economy [4].

With the collapse of the USSR, the established structure of higher education, geared towards swift responsiveness to the production needs of a planned economy, disintegrated. These changes primarily affected specialized universities. The government embarked on a course of forming liberal market relations, marking the 1990s with the creation of a private entrepreneurial sector and a decline in industrial and agricultural production. From 2000 to 2007, Russia experienced economic growth. To sustain this trend, the need arose to restructure the higher education system, aligning it with the new demands of existing realities. The first stage, initiated in 2006 as part of the national project "Education," was primarily directed at establishing new institutions (Siberian Federal University and Southern Federal University) and supporting classical multi-profile universities [5]. The principle of identifying leading universities was based on the necessity to shift the prevailing trend of concentrating educational resources in the metropolitan cities of Moscow and St. Petersburg. In the second stage, under the "5–100" project adopted in 2012, ten federal universities were established in all Federal Districts of the country [2]. Simultaneously, in an attempt to decentralize federal educational universities based on a territorial principle, the government began creating a network of research universities according to the Decree of the President of the Russian Federation dated October 7, 2008, No. 1448 [6]. The new status of a research university was awarded for 10 years and on a competitive basis. In the initial two rounds of competitions, 29 universities were granted the status of research universities, receiving additional funding for developing research infrastructure. Their primary task was to ensure the competitiveness of domestic developments in the global market.

To expand the number of leading universities at the regional level, creating a network of flagship universities was proposed. In 2015, the first competition was announced, with 11 universities emerging as winners. In 2017, the status of a flagship university was conferred upon an additional 22 institutions. The significance of these flagship universities particularly impacted regions where federal and research universities were absent. The government's latest initiative is the "Priority-2023" project, encompassing two tracks, "Territorial Leadership and/or Research Leadership" and "Research Leadership." This project includes 54 universities (30 of them regional and 24 from Moscow and St. Petersburg).

Thus, the traditional structure of leadership in Russian universities, inherent to the national education system, is rejuvenating despite all vicissitudes of economic and political upheavals and the influence of the international community.

2 Criteria for Leadership of Russian Universities

For the Soviet period, the fundamental criterion for designating a university as a leader was its scientific achievements, which gave rise to unique academic schools. These scientific schools emerged in the presence of a leading scientific figure. Among such scholars were L. D. Landau, A. N. Kolmogorov, P. Kapitsa, and others. Even though foreign countries did not recognize higher education diplomas obtained in Russian universities, the achievements of these scientists shaped the image of Russian higher education on the international stage. The government's primary goal was to attain leading positions in developing specific disciplines such as mathematics, physics, chemistry, medicine, and engineering.

In the post-Soviet reforms, criteria for determining leading universities were based on efficiency strategy indicators, relying on metrics from external rankings like QS and THE. In this context, the objective is to enhance the competitiveness of Russian universities on the international stage. Scientific achievements are considered one of the factors contributing to the attractiveness of a university. "However, the actual funding for science in Russia is increasingly shifting towards a competitive model, greatly complicating the planning of scientific activities, as the allocation of funds to research organizations is unstable and inconsistent from year to year, month to month," notes M. A. Polozhikhina [7].

Analyzing the financial support for research activities in Russian universities, G. A. Krasnova points out that "research activities in universities remain secondary compared to educational ones, as evidenced by both the volumes of funds from various sources to higher education institutions and the personnel composition engaged in research and development" [8].

Next, we will examine the specifics of forming criteria for assessing the leading positions of universities in global rankings. During the research, the authors used the analytical research method. International and Russian scientific papers were analyzed to create a certain picture of the impact of world university rankings and digitalization processes on identifying leading Russian universities.

2.1 Alternative Models of University Leadership

The authors focus on two international rankings, QS and THE, which are considered by the government of the Russian Federation in identifying leading universities at the national level. To understand the role of global university rankings in national higher education systems, let us delve into the aspects that these global rankings emphasize and explore how these emphases influence national higher education systems. These rankings have undergone certain changes since their inception [9].

Initially, the QS ranking (Quacquarelli Symonds Ltd) collaborated with THE (Times Higher Education); the first ranking was published in 2004, highlighting four indicators: the quality of scientific research, comprising two indicators (academic expert assessment and citation index), graduate employment, internationalization (the number of international students and faculty), and the quality of teaching (teachers to students ratio). In 2010, THE–QS ranking acquired a new name: QS World University Rankings, which did not impact the evaluation methodology.

The creation of these rankings coincides with the period of implementing economic ideology in higher education [10, 11]. Accordingly, all rankings were aimed at creating a sense of "reputation deficit" on a global scale [12]. The formation of a university's image became a primary goal, on the one hand, to secure greater funding from the government of the state to which a particular university belongs [13, 14]. On the other hand, it encouraged universities to invest in their brands [15]. Furthermore, the market-driven ideology of ranking research conducted by media organizations such as Times Higher Education or The Guardian manifests in the fact that these rankings exist for their own benefit [16]. For instance, the ranking by Times Higher Education is funded by the publication itself and by the company Thomson Reuters. Notably, Thomson Reuters is associated with services within the Web of Science framework, responsible for the global ranking of scientific journals and analyzing the citation of authors' publications.

Market interests of developed countries are also evident in the formulation of criteria and methodology for assessing global rankings. Despite the presence of unique national factors, the prevailing model serving as a criterion for comparisons is the Western research-intensive university with a strong natural science component and a predominantly English-language orientation [17]. The justification for such an approach was the fact that global comparisons of universities could be made based on a single model. The benchmark model chosen was that of a comprehensive research-intensive university since this model "reflects informal cross-border norms influenced by practices in the English-speaking nations" [17 , p. 308]. Consequently, the most influential positions in international rankings are occupied by universities belonging to English-speaking countries.

Currently, there is a growing sense of political influence on ranking indicators. In 2016, the Sustainable Development Goals (SDGs) officially came into effect after the approval by heads of states and governments during the UN Sustainable Development Summit at the 70th session of the UN General Assembly in New York [18]. As a result, QS is set to launch the QS World University Rankings: Sustainability in October 2023 to measure the environmental and social impact of institutions based on all 17 SDGs. The fierce competition among the top ten universities in Europe and America is driven by the fact that many of them received the highest scores of 100 on indicators such as "Academic Reputation," "Employer Reputation," and "Student-Faculty Ratio," making these positions unattainable for other universities. The noble intentions of SDGs to involve universities in addressing social issues of sustainable development allowed other universities from developed countries to take leading positions in the "Sustainability" indicator: the University of Pennsylvania (ranked 12th), the University of Toronto (ranked 21st), and the University of British Columbia (ranked 34th)—all with a score of 99.9 [19].

Thus, rankings compel universities to align with common standards and compete, often overlooking local issues and their social roles [20]. Claire Englund, Anders D. Olofsson, and Linda Price also note that teaching issues are virtually disregarded in institutional and national policies. Goals are expressed in terms of enhancing the competitiveness of the educational institution rather than improving student learning [21]. Many scholars highlight the need to balance teaching and research activities, which poses a serious challenge for staff: on the one hand, succeeding in obtaining the highest possible scores in research assessments, and on the other hand, meeting the requirements for enhancing the quality of educational programs and teaching skills [22–24].

The Russian model of university leadership is an alternative to the European-American model (Fig. 1).

If we relate the proposed criteria for assessing international rankings to Russian practice, rectors of Russian universities note that the international ranking system does not reflect the quality of education [25]. The ranking criteria include many points that Russian universities lack for objective and historical reasons: Russian universities traditionally focus primarily on education, while fundamental research is conducted in the structures of the Russian Academy of Sciences (RAS). Even the previously mentioned examples of the existence of scientific schools in classical Russian universities during the Soviet period would not meet the requirements of a research university, as all research was conducted in the RAS or under the auspices of the RAS. As for the citation index indicator, it is crucial to note that Russia was virtually in political isolation from the global community for almost the entire twentieth century: political and economic blockade by Western countries in 1917–1941, the years of World War II in 1941–1945, and the Cold War in 1946–1980.

A list of scientific journals recommended by the Higher Attestation Commission under the Ministry of Higher Education was created to provide a scientific platform for exchanging ideas and achievements. Publications were in the Russian language. The editorial board of the journal, in collaboration with the publishing house,

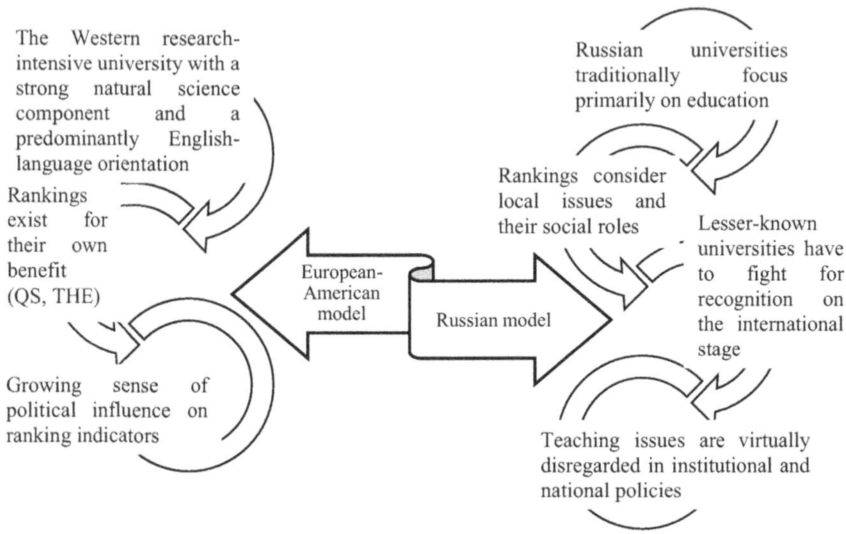

Fig. 1 Alternative models of university leadership. *Source* Compiled by the authors

selected the best scientific articles, translated them into foreign languages, and published them in international journals. Naturally, after 1991, it was impossible to instantly solve the problem of including these journals in international citation systems like the Web of Sciences or Scopus; all authors had to seek opportunities to translate their articles into English, which was rather challenging.

According to Wendy Alexander, "Generative AI will, over time, transform content creation, learning support, and assessment at universities" [26]. In the digital space, media organizations will compile rankings, ensuring competition based on frames [27]. For universities with a high reputation, it will be easy to maintain an international brand of "excellence," while lesser-known universities will have to fight for recognition on the international stage by creating a portfolio: the first place in the national ranking, the title of the best university, a leading position in a professional association, etc. Currently, global rankings do not consider online education. As noted by John J. Cheslock and Jaquette Ozan, even before the COVID-19 pandemic, in 2012–2018, the number of students completing all their courses online increased by approximately 22%, while the number of students attending in-person classes decreased by about 9% [28]. Before the COVID-19 pandemic, this growth was associated with the development of distance education as an independent direction. After the pandemic, almost all universities began to widely use blended learning (a combination of online and offline) in traditional forms of education. Thus, it is necessary to reconsider ranking indicators such as "Internationalization" and "Quality of Teaching."

2.2 New Criteria for Assessing Leadership in the University Environment in Conditions of Globalization and Integration of Digital Innovations

The digitalization of education entails the transformation of the higher education system. According to G. Parmentier and V. Mangematin, the integration of digital innovations signifies the following:

1. The redefinition and reconsideration of organizational boundaries;
2. Ensuring feedback with the community and limiting property rights;
3. Restructuring product and organizational identity [29].

In education, this implies the blurring of conventional education systems, changes in the educational landscape, alterations in communication forms, and the restructuring of educational programs due to rapid transformations in the professional labor market. Such profound and extensive changes require the development of new criteria for assessing leadership in the university environment. In an experiment conducted in the Persian Gulf state, the following criteria were developed by the ministry:

1. The number of students and teachers who underwent training;
2. The use of digital tools;
3. The development of digital culture within the university;
4. The enhancement of overall research knowledge in postgraduate studies;
5. The development of creative abilities in proposed courses;
6. The creation of a strong target group to support the initiative [30].

Russian scientists proposed their indicators for assessing the leadership of an educational organization: educational activities, scientific and teaching activities, management of digital activities, and control and reporting [31]. Given the current trend of homogenization in creating global education rating indicators, identifying a unified set of indicators with their criteria is not challenging. Global rankings have years of experience in developing unified directions and criteria. However, the digitalization of education brings about instability, uncertainty, complexity, and the hybrid nature of reality. Depending on the specific characteristics of the implementation conditions of educational services in different countries and regions, the number of indicators may vary, and the set of these indicators may differ from each other (e.g., in the Persian Gulf state, there are 106 of them, while Russian scientists propose 17). Determining the importance of each variable increases when determining the final assessment [32]. It is challenging to evaluate variables because the university organism is complex; each variable depends on the degree of implementation of others, enhancing the overall quality of the educational process [33]. The circle of stakeholders expands as universities orient themselves toward a knowledge-based economy, forcing them to respond to the demands of various population groups [34]. In the views of Ben Jongbloed, Jurgen Enders, and Carlo Salerno, a formula based on indicators may not be the most suitable approach. Instead, they suggest using a forward-looking approach with leading indicators instead of retrospective ones [34]. In 2000, the European Commission proposed considering not only the indicators of "education" and "research" but also "innovation" when compiling university rankings [35]. The prototype of the research university model in the network space loses clear identification features [36].

Considering these uncertainties in the characteristics of variables, programmers are seeking an adequate approach to developing tools for AI. Some propose using fuzzy methods, which are closer to the way people think [37].

3 Conclusion

Leadership among universities will always be relevant and in demand. Examining three value models that succeeded each other over the past century, the authors state that changes are observed in forms, methods, and the value orientations of universities. Alongside the models, approaches to assessing leadership in higher education have also evolved. The value orientation of the Humboldtian model, focusing on the education of an enlightened citizen of society, was replaced by the provision of educational services commodifying teaching and research. The market competition model among universities marked the substitution of academic standards with a race for prestige.

A new stage in the development of the higher education system is associated with the processes of digitalization, the introduction of artificial intelligence into the analysis of university ranking indicators. The ability of universities to adapt, demonstrate flexibility, and continually adjust their course of action and development trajectory comes to the forefront [29].

Digitalization brings new challenges: the intensification of digital inequality, transparency issues, and security problems, which must be considered when developing variables for creating rankings. However, the capabilities of AI can address the return of academic values from the Humboldtian model to the new "University 4.0" and "University 5.0" models and reduce the level of subjectivity in choosing assessment indicators for universities with digitalization in mind. The basis for this conclusion is the fact that developers of educational technology increasingly focus on ethical and sociocultural aspects. Proposed methods of pedagogical practice using IT aim to follow students' interests and contribute to their comprehensive development. If, as programmers say, they are not ready to create an adequate measurement tool today, it is entirely possible in the near future.

References

1. Leshukov, O. V., & Froumin, I. D. (2017). Flagship universities: From soviet experience to searching for new model. *University Management: Practice and Analysis, 21*(4), 22–29. https://doi.org/10.15826/umpa.2017.04.046
2. Ten, L. V. (2007). The market and management of higher education in the modern period. *Council of Rectors, 1*, 103–108.
3. Kharatyan, G. M. (Ed.). (1971). *Public education, science, and culture in the USSR: Statistical collection*. Publishing house "Statistics".
4. Avrus, A. I. (2001). *History of Russian universities*. Moscow Public Science Foundation.
5. Froumin, I., & Povalko, A. (2014). Top-down push for excellence: Lesson from Russia. In Y. Cheng, Q. Wang, & N. C. Liu (Eds.), *How world-class universities affect global higher education. Influences and responses* (pp. 47–64). Sense Publishers. https://doi.org/10.1007/978-94-6209-821-3_5
6. Presidential Executive Office. (2008). *Decree "on the implementation of the pilot project for the establishment of national research universities"* (October 7, 2008 No. 1448). Accessed June 10, 2023, from http://www.kremlin.ru/acts/bank/28121
7. Polozhikhina, M. A. (2019). Ambiguous overall results in reforming of Russian science. *Economic and Social Problems of Russia, 2*, 100–138. https://doi.org/10.31249/espr/2019.02.05
8. Krasnova, G. A., & Burangulov, E. R. (2022). Analysis of the principles, models and mechanisms of financing of university science in Russia. *Science Governance and Scientometrics, 17*(1), 108–135.
9. Thessin, R. A., Tekleselassie, A. A., Trimmer, L. B., Shepard, S. D., & Clayton, J. K. (2025). Leveraging a candidate assessment system to develop an equity-centered school leadership pipeline through a university–district partnership. *Education Sciences, 14*(12), 1408. https://doi.org/10.3390/educsci14121408
10. Wang, D., Liu, Z., Wang, Y., & Zhang, J. (2025). Antecedents of university students' self-leadership: The roles of teacher developmental feedback and student proactive vitality management. *British Journal of Educational Psychology., 95*, 627–642. https://doi.org/10.1111/bjep.12743
11. Yakob, W. R. W., & Don, Y. (2025). Influence of digital leadership on Malaysian public university branding: Professionals' identity as mediator. *International Journal of Evaluation and Research in Education, 14*(1), 124–132. https://doi.org/10.11591/ijere.v14i1.28286
12. Brankovic, J. (2018). The status games they play: Unpacking the dynamics of organizational status competition in higher education. *Higher Education, 75*(4), 695–709. https://doi.org/10.1007/s10734-017-0169-2
13. Yaseen, U., Idrees, R. N., Shakil, M. H., Haider, S. Z., & Khalil, J. (2025). Influence of academic leadership on organizational commitment of faculty members in private sector universities: Mediating role of work engagement. *Quality Assurance in Education, 33*(1), 17–30. https://doi.org/10.1108/QAE-04-2024-0072
14. Yu, H., Hamid, A. H. A., Abdul Wahab, J. L., & Mahmud, M. I. (2025). The impact of deans' distributed leadership on university teachers' job performance: The mediating role of job satisfaction. *Cogent Education, 12*(1), 2458713. https://doi.org/10.1080/2331186X.2025.2458713
15. Drori, G., Delmestri, G., & Oberg, A. (2016). The iconography of universities as institutional narratives. *Higher Education, 71*, 163–180. https://doi.org/10.1007/s10734-015-9894-6
16. Marques, M., & Powell, J. J. W. (2020). Ratings, rankings, research evaluation: How do schools of education behave strategically within stratified UK higher education? *Higher Education, 79*, 829–846. https://doi.org/10.1007/s10734-019-00440-1
17. Marginson, S., & van der Wende, M. (2007). To rank or to be ranked: The impact of global rankings in higher education. *Journal of Studies in International Education, 11*(3–4), 306–329. https://doi.org/10.1177/1028315307303544

18. United Nations. (2022, July 13). *Sustainable Development Goals can be reached 'despite our grim times': ECOSOC president.* Accessed June 10, 2023, from https://news.un.org/en/story/2022/07/1122422
19. QS Top Universities. (2024). *QS world university rankings 2024: Top global universities.* Accessed June 10, 2023, from https://www.topuniversities.com/world-university-rankings
20. Gadd, E., Holmes, R., & Shearer, J. (2021). Developing a method for evaluating global university rankings. *Scholarly Assessment Reports, 3*(1), 2. https://doi.org/10.29024/sar.31
21. Englund, C., Olofsson, A. D., & Price, L. (2018). The influence of sociocultural and structural contexts in academic change and development in higher education. *Higher Education, 76*, 1051–1069. https://doi.org/10.1007/s10734-018-0254-1
22. Lomas, L. (2004). Embedding quality: The challenges for higher education. *Quality Assurance in Education, 12*(4), 157–165. https://doi.org/10.1108/09684880410561604
23. Osseo-Asare, A. E., Longbottom, D., & Murphy, W. D. (2005). Leadership best practices for sustaining quality in UK higher education from the perspective of the EFQM excellence model. *Quality Assurance in Education, 13*(2), 148–170. https://doi.org/10.1108/09684880510594391
24. Skelton, A. (2012). Colonized by quality? Teacher identities in a research-led institution. *British Journal of Sociology of Education, 33*(6), 793–811. https://doi.org/10.1080/01425692.2012.692047
25. Pugach, V. N. (2011). Ratings of universities as one of the ways to assess the quality of education. *Internet Journal Naukovedenie, 2*(7), 11.
26. Alexander, W. (2023, April 26). *What does AI mean for international recruiters? QS insights magazine.* Accessed June 10, 2023, from https://www.qs.com/what-does-ai-mean-for-international-recruiters/
27. Hasse, R., & Krücken, G. (2013). Competition and actorhood: A further expansion of the neo-institutional agenda. *Sociologia Internationalis, 51*(2), 181–205.
28. Cheslock, J. J., & Shamekhi, Y. (2020). Decomposing financial inequality across U.S. higher education institutions. *Economics of Education Review, 78*, 102035. https://doi.org/10.1016/j.econedurev.2020.102035
29. Parmentier, G., & Mangematin, V. (2014). Orchestrating innovation with user communities in the creative industries. *Technological Forecasting and Social Change, 83*, 40–53. https://doi.org/10.1016/j.techfore.2013.03.007
30. Ghamrawi, N., & Tamim, R. M. (2023). A typology for digital leadership in higher education: The case of a large-scale mobile technology initiative (using tablets). *Education and Information Technologies, 28*, 7089–7110. https://doi.org/10.1007/s10639-022-11483-w
31. Varnavskiy, A. V., Volkova, E. S., Buryakova, A. O., & Klimova, E. A. (2020). Model for assessing digital leadership of organization. *Vestnik universiteta [University Bulletin], 4*, 23–32. https://doi.org/10.26425/1816-4277-2020-4-23-32
32. Guarini, M. R., Battisti, F., & Chiovitti, A. (2021). A methodology for the selection of multi-criteria decision analysis methods in real estate and land management processes. *Sustainability, 10*(2), 507. https://doi.org/10.3390/su10020507
33. Alvarez-Blanco, L., Castro-Lopez, A., & Cervero, A. (2022). Intelligent analysis of the quality of education through teaching practices on virtual campuses. *European Journal of Psychology of Education, 38*, 1111–1128. https://doi.org/10.1007/s10212-022-00649-2
34. Jongbloed, B., Enders, J., & Salerno, C. (2008). Higher education and its communities: Interconnections, interdependencies and a research agenda. *Higher Education, 56*, 303–324. https://doi.org/10.1007/s10734-008-9128-2
35. European Council. (2000). *Presidency conclusions: Lisbon European Council 23 and 24 March 2000.* Accessed June 10, 2023, from https://www.europarl.europa.eu/summits/lis1_en.htm
36. Geuna, A., & Muscio, A. (2009). The governance of university knowledge transfer: A critical review of the literature. *Minerva, 47*(1), 93–114. https://doi.org/10.1007/s11024-009-9118-2

37. Castro-López, A., Cervero, A., Galve-González, C., Puente, J., & Bernardo, A. B. (2021). Evaluating critical success factors in the permanence in higher education using multi-criteria decision-making. *Higher Education Research and Development, 41*(3), 628–646. https://doi.org/10.1080/07294360.2021.1877631

Sustainable Development of Higher Education in Russia and Central Asia Through Inclusive Universities 4.0

Preface to the Second Part of the Book

This second part of the book focuses on a comprehensive exploration of the inclusiveness of Universities 4.0. From the sustainable development perspective, it reconsiders the essence of openness and inclusiveness in higher education. The central scientific topic addressed in this part include cultural and gender inclusiveness, accessibility for individuals with disabilities, and lifelong learning.

The key conclusion of the chapter by Elena V. Susimenko, Elena G. Popkova, Natalia V. Przhedetskaya, and Yury A. Tikhomirov is that the importance of cultural inclusiveness for Universities 4.0 is growing simultaneously with the development of the international higher education market and intercultural communications.

Equally noteworthy in this part are the recommendations provided by Aktam U. Burkhanov, Mavluda Sh. Akhmedova, Ainura A. Adieva, and Elena I. Litvinenko in the chapter titled "Gender Neutrality of the University as the Basis for Its Inclusiveness." This chapter advocates for achieving gender neutrality in Universities 4.0 by ensuring equal rights for all higher education applicants regardless of gender, age, or social background.

The chapter by Gulnora K. Abdurakhmanova, Aidarbek T. Gyiazov, and Yury A. Tikhomirov, "Anti-ageism and Openness to Lifelong Learning, Work-Life Balance in an Inclusive University," further develops the topic by establishing fundamental principles of anti-ageism policies in Central Asian countries. These principles support the promotion of continuous learning in inclusive Universities 4.0 to achieve a work-life balance.

In support of this part's topic, the chapter "The Inclusion of People with Disabilities in Higher Education" by Abrorbek Kozimjonov, Olga V. Dybina, Galina A. Dorofeeva, and Talgat T. Talantbekov offers comprehensive recommendations for more fully integrating people with disabilities into the higher education environment from the standpoint of empathy and as rational actions aimed at the intellectual advancement of society and the better appreciation of contemporary human values.

The contribution of the chapter by Aktam U. Burkhanov, Dilobar M. Mavlyanova, Victoria N. Ostrovskaya, and Svetlana E. Karpushova lies in the formulation of theoretical and methodological foundations for enhancing accessibility.

The chapter by Gulayym A. Dzhumabayeva and Gulnur M. Amanova, "Influence of Science on the Development of the Moral Consciousness of Students," further supports the topic of this part by arguing for the need to establish inclusive higher education that is accessible regardless of financial or material constraints based on Universities 4.0.

The chapter by Shavkat I. Ilkhamov, Timur M. Israilov, Victoria S. Ustenko, Elena V. Makarova, and Natalya A. Levchenko, titled "Macroeconomic Implications and Prospects for Increasing Inclusiveness of the National Science and Higher Education System," contributes to the discussion by demonstrating that creating conditions for inclusiveness in Universities 4.0 supports employment, boosts GDP and per capita income, accelerates technological growth, and attracts foreign investment.

Thus, the chapters in this part highlight various aspects of the inclusiveness of Universities 4.0 in Russia and Central Asia and provide evidence that these inclusive institutions play a central role in ensuring the sustainable development of higher education in the region.

Cultural Inclusiveness of Universities in the Context of the Development of International Higher Education

Elena V. Susimenko, Elena G. Popkova, Natalia V. Przhedetskaya, and Yury A. Tikhomirov

Abstract This research is devoted to the identification of the role of cultural inclusiveness of universities in view of the development of the international higher education market, increase in the problem of intercultural communications, and support for cultural inclusiveness in higher education institutions. Problems of international higher education are presented from the position of a combination of several approaches, including the economic concept of educational services' export, the theory of social interaction from the position of development and use of soft power, and the pedagogical and cultural vision of the problem of universities' cultural inclusiveness under the conditions of an increase in the number of foreign students. The main conclusions are made on the basis of the use of the methods of dialectics and critical analysis, observation and generalisation, and graphical method and modelling. This research dwells on the theoretical aspects of the development of international higher education and determines the role of Russian universities in it, as well as characterises the topicality of the problem of a multicultural environment and the necessity for the development of cultural inclusiveness in higher education institutions in the conditions of an increase in the number of foreign students. To solve the revealed problems, the model of cultural inclusiveness of universities that work in the sphere of international education is proposed.

E. V. Susimenko
Platov South-Russian State Polytechnic University (NPI), Novocherkassk, Russia
e-mail: e.susimenko@npi-tu.ru

E. G. Popkova (✉)
The Laboratory of Scientometrics and International Rankings,
Armenian State University of Economics, Yerevan, Armenia

N. V. Przhedetskaya
Rostov State University of Economics, Rostov-on-Don, Russia

Y. A. Tikhomirov
RUDN University, Moscow, Russia
e-mail: Tikhomirov-yua@rudn.ru

B. S. Sergi et al. (eds.), *University 4.0 and Educational Technology for Sustainable Development*, Education in the Asia-Pacific Region: Issues, Concerns and Prospects 75, https://doi.org/10.1007/978-981-96-7376-6_8

Keywords Cultural inclusiveness · University · Higher education · International higher education · Internationalization of higher education

JEL Codes F16 · F22 · F23 · I23 · I24 · I25 · I28 · O15 · O38 · Z13

1 Introduction

Globalisation and international integration in different spheres became important trends of the early twenty-first century. The creation of global institutions in the sphere of trade and business, labour migration and education, and scientific and technical environment led to important preconditions for the active interaction of different countries and cultures' representatives. According to this, scientific concepts have formed in the sphere of international higher education—they cover the theory of internationalization of education, export of educational services, international migration of students as a factor of soft power, etc. Internationalization of higher education is linked with the opportunities for cooperation and student exchange. The concept of export of educational services has an economic direction and is primarily focused on the balance of payments of the country and the growth of the cost of services export.

Each of the proposed approaches is directly connected with the problem of intercultural communication, which is accompanied by language difficulties, rejection of the manifestations of other cultures, historical preconditions, etc. Given this, universities' ability to involve the representatives of different ethnic, language, or confessional groups in higher education and their ability to organise the adaptation of foreign students and lead the intercultural dialogue form a relevant scientific task, which becomes particularly important in the conditions of the development of international higher education.

In these conditions, the cultural inclusiveness of universities is an important factor that offers tools for wider involvement of international students in higher education, their quick adaptation to the conditions of the educational and cultural environment of the university's country, and the creation of conditions for the development of their cultural identity in the conditions of the intercultural dialogue.

This research is valuable due to the presentation of a model of cultural inclusiveness of universities that are involved in the international market of higher education. Within this model, the main elements and stages of support for cultural inclusiveness are structured, and the sequence of measures that are necessary for the formation of intercultural communications competencies with students, growth of their creative potential, and expansion of the range of interests is described.

2 Concept of the Cultural Inclusiveness of Universities

The study of cultural inclusiveness of universities from the position of their international activities is based on the mix of several methodological approaches. It covers the spheres of international trade (export of educational services) [1], labour relations (migration of students and university employees) [2], international relations and institutionalism (formation of international organisations and institutions in the sphere of higher education) [3], social and cultural interaction (factor of soft power, intercultural dialogue, etc.) [4], pedagogical sciences (adaptation of foreign students to the life and studies in the conditions of another culture, creation of conditions for inclusion of the representatives of other cultures in higher education) [5]. The methods used allow interpreting concepts and vision of each approach and scientific sphere that belong to the general scientific methodological tools (analysis and synthesis, dialectics and critical analysis, observation and generalisation, graphical method and modelling).

The main works that were used during the preparation of this research form a set of theoretical and practical views. These views elaborate on different aspects of the studied problem, including such aspects as analysis of international higher education and mobility of students [6, 7]; theoretical aspects of international education management [8]; growth of competitiveness of higher education in Russian universities [9, 10] and prospects for its development [11]; specifics of cultural inclusiveness [12]; experience of intercultural communications in universities [13]; support for cultural inclusiveness in the system of higher education [14, 15], etc.

This research was aimed at developing and substantiating the model of cultural inclusiveness of universities from the position of development of international higher education. For this, the specifics of international higher education were analysed, conditions and factors of its development were determined, problems of intercultural communication in the conditions of internationalization of higher education were outlined, and the tools for ensuring cultural inclusiveness in universities from the position of their development were formed.

3 Proposed Model of Cultural Inclusiveness of Universities in the Sphere of the Internationalization of Education

The institute of international higher education belongs to well-organised and systemic factors of the civilizational development of the world. It ensures the transfer of knowledge, facilitates an increase in the level of intellectual capital, and is an important tool of cultural influence. Universities of the Russian Federation are deeply involved in the sphere of international higher education, which is manifested through the export of educational services and the formation of soft power regarding the adaptation of foreign students to the conditions of Russian culture and language environment.

The market of international higher education is very dynamic. At present, the main consumers (importers) of universities' services are mainly Asian countries, such as China, India, Hong Kong, Korea, Malaysia, etc. [6]. China and India also belong to the largest buyers of higher education services from Russian universities [10]. Such convergence forms good prospects for further increase in the internationalization of universities. An additional factor in its development is good traditions in inviting students from post-Soviet countries and African countries. Government support for the export of higher education services in Russia sets good preconditions for further growth of indicators and an increase in the level of internationalization of higher education [9].

The number of foreign students grows constantly, and geography of countries from which students arrive to study in Russian universities expands. Thus, the issue of intercultural communication and intercultural dialogue is very relevant. The creation of a multicultural environment in universities automatically forms a community that has large differences in many spheres, including the attitude towards joint studies, social activities, way of life, and different cultural traditions [13]. Therefore, it is important to understand the advantages and risks of such an environment and form mechanisms that are capable of not only ensuring favourable conditions for studies but also creating the appropriate mechanisms for efficient development and involvement of all people in the educational and social process in international universities.

From the position of perception, understanding, and cultural inclusiveness of universities in the conditions of an international educational market, there are certain inaccuracies in the key definitions. Thus, the notion of international higher education is often associated with international student exchange, and the notion of inclusion—with the creation of conditions for barrier-free education of people with special needs or disabilities [8]. As a matter of fact, these notions are much wider. Apart from the mobility of students, international higher education also includes the mobility of pedagogical workers and scholars, joint projects, coordination of educational plans, joint scientific publications, etc. Inclusion in its nature is the activities aimed at the maximum inclusion of all people in a certain process without focusing on differences and limitations [12].

Inclusion is one of the concepts of the future, which characterises the social mode of a new society with the corresponding sociocultural values. It involves the inclusion of all people in public processes—without differentiating people by features, including ethnicity or citizenship. This concept involves the use of the creative potential of all people for common development [15]. Thus, the implementation of inclusion in higher education implies not only acceptance and implementation of regulatory acts and the creation of the corresponding institutions. Inclusion needs acceptance in society at the level of human values to a larger extent.

Formation of a multicultural environment requires the creation of a system of intercultural communications, which is based on understanding, respect, and acceptance of other people and respect for them and their qualities. This process involves also the development of competence in productive co-existence with the representatives of other language, religious, or cultural groups with observation of the principles of respect, humanism, equality, and attention. Such a vision of a multicultural environment of higher education is the main factor in the cultural inclusiveness of universities. Ensuring a sufficient level of inclusiveness in this environment requires joint efforts from all participants of the educational process, from administration, teachers, and students to the representatives of public authorities in the sphere of culture and all representatives of the society [12].

In the conditions of the development of international higher education, the cultural inclusiveness of universities is one of the critical factors. Its task was to create conditions for joint and productive co-existence of the representatives of different cultural and ethnic groups, which often have contradictions of conflicts. According to this, when creating a system or model of a multicultural community, the primary task is the creation of a tolerant environment. However, the potential for cultural inclusiveness is much higher. Intercultural exchange allows forming new creative ideas and increasing the range of ideas and professional contacts. In the conditions of a globalised world and quick technological development, the skills of intercultural communication become more important. The general scheme of the model of cultural inclusiveness of universities is given in Fig. 1.

The model is based on the system of cultural inclusiveness. It contains provisions, norms, rules, instructions, and organisational aspects of intercultural communication and the involvement of the representatives of different cultures in the educational and social life of the educational institution. The educational process is given in the model in the form of "inputs" and "outputs." At the "input," there is the invitation of foreign students, which includes a range of information and communication actions that are aimed at informing and stimulating the foreign student to enter the university. Other elements, including the formation of an educational plan, creation of conditions for life and education, and migration, are preparatory stages, as a result of which foreign students start their education.

Given the character of international education, the result of inviting foreign students is the formation of a circle of education seekers who represent different countries and cultures. It is also possible to involve foreign lecturers in the educational process; together with students, they form a multicultural environment. From the position of cultural inclusiveness, within the multicultural environment, there emerge communication links, which require the appropriate coordination and agreement. Without targeted measures, chaotic communication links may provoke destructive processes, such as conflicts, division of the representatives of different cultures into separate exclusive groups, etc.

Therefore, it is necessary to stimulate contacts between the representatives of different ethnic groups and peoples (within the model of cultural inclusiveness of universities). These processes cover a range of actions aimed at the creation of interest in other cultures, realisation of their differences, and formation of a culture of acceptance of these differences, which eventually must ensure mutual enrichment for all representatives of the multicultural environment. Each of the above actions contains a concrete list of measures that can ensure the required result. Thus, to form an interest in other cultures it is important to perform presentations of cultures, their discussion and presentation. To realise the differences between cultures it is

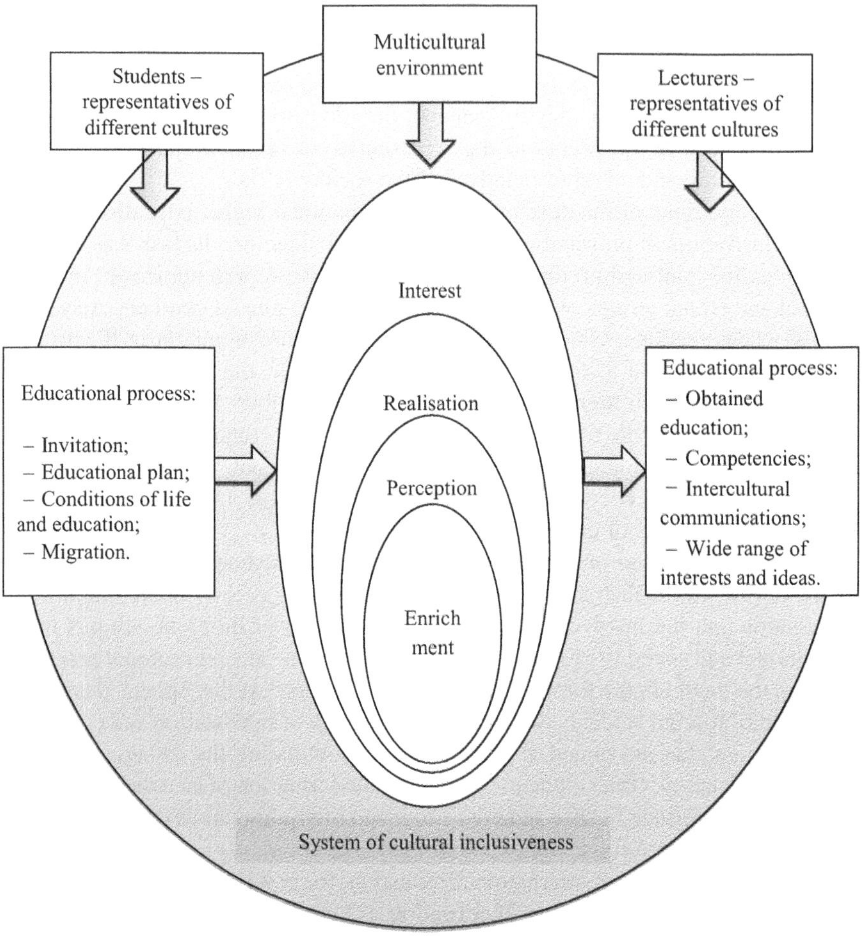

Fig. 1 The model of cultural inclusiveness of universities that work in the sphere of the internationalization of education. *Source* Developed by the authors

important to have discussion clubs, workshops, and other events aimed at a productive discussion. To ensure the perception of differences, apart from the above measures, it is important to form conditions for psychological support for students, intercultural dialogue, joint projects, etc.

All the above processes will allow ensuring the formation of a productive environment, which—at the "output" of the educational process—will produce graduates, who will receive the skills in intercultural communication, a wide range of professional contacts, as well as interests and ideas formed on the basis of cultural inclusiveness. Intercultural competencies, together with personal, professional, and social competencies, are among the key factors that form a good specialist in the given spheres of higher education [11].

4 Conditions for Cultural Inclusiveness in the Context of the Development of International Higher Education

Formation of the conditions for cultural inclusiveness in the context of the development of international higher education is an important task that is connected with many risks and limitations. They require additional research and discussion, since apart from the value manifestation, cultural inclusion has a large potential for the development of higher education, involvement of talents, and support for the influence of soft power. The most relevant problems within this process are determination of the role of the culture of the country that hosts students who are representatives of other cultures. The basis of this discussion lies in the striving to preserve cultural diversity and ensure intercultural exchange in the conditions of the dominating influence of the hosting country's culture.

With this in mind, an important task is to find balance between the influence of the dominating culture on other cultures, represented by the corresponding groups of students (Fig. 2).

The proposed components of the model of cultural inclusiveness of universities that are connected to interest in another culture, perception of differences, and achievement of the state of mutual enrichment, require additional discussion, the results of which will differ for each institution of higher education. In this context, it is important to take into account the number and representation of foreign students, the level of their involvement in the educational, scientific, and social life of the university, and the risks peculiar to the formed multicultural environment.

Fig. 2 The proposed components of the model of cultural inclusiveness of universities. *Source* Developed by the authors

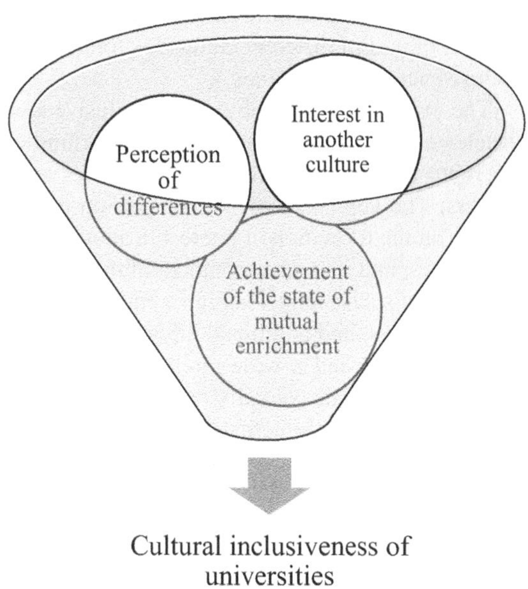

Apart from this, a separate direction of discussion is the problem of identification of the efficiency of cultural inclusiveness from the position of ensuring the development of higher education. In this context, it is necessary to pay attention to the issue of assessment of the potential of cultural inclusiveness, its manifestations, and character, and the possibility of measuring results that were achieved due to support for cultural inclusiveness of universities that work at the international market of education.

5 Conclusions

The main trends that define the state of international higher education are globalisation, institutionalisation, and dynamic development. The market of higher education in the Russian Federation is an important player at the global scale. The comparison of demand for higher education services and directions of international activity of Russian universities allow stating the presence of a potential for further development and growth of the internationalization of universities. At the same time, the trend of the growth of the number of international students and representation of countries in Russian universities forms certain risks that require detailed research and development of mechanisms for their neutralisation.

The multicultural environment, formed by international students, is characterised by large differences from the position of attitude to joint studies, participation in social life, and observation of traditions or rules of a certain lifestyle. According to this, there is a necessity to neutralise risks of the emergence of a possible conflict or formation of the features of the exclusiveness of the environment. A large potential for the development of universities and the use of the creative potential of the representatives of different cultures is found in the concept of ensuring the cultural inclusiveness of universities.

The proposed model of cultural inclusiveness of universities clearly structures the elements and processes that form opportunities for a comprehensive inclusion of the representatives of different nationalities and ethnic groups in the educational process. The key stages of such inclusion are measures that can be grouped into actions on the formation of interest in another culture, realisation and acceptance of differences, and mutual enrichment of the representatives of different cultures. Due to the use of such a model in higher education institutions, the quality of educational services grows, and additional effects in the form of competencies of intercultural communication and a wide circle of interests and ideas form. In aggregate, the above processes substantially raise the university's competitiveness and attractiveness in the market of international services of higher education.

References

1. Agostinelli, A. V., & Albert, L. R. (2025). University leaders' perspectives on international student support: An evaluation using the culturally engaging campus environments model. *Journal of Studies in International Education, 29*(1), 125–143. https://doi.org/10.1177/10283153241266966
2. Aldossari, A. S. (2025). Sense of belonging: International students' experience of intercultural and cultural inclusion in Saudi universities. *Language and Intercultural Communication, 25*(1), 61–75. https://doi.org/10.1080/14708477.2023.2257193
3. Çimşir, E., & Ünlü Kaynakçı, F. Z. (2024). Acculturative stress and depressive symptoms among international university students: A meta-analytic investigation. *International Journal of Intercultural Relations, 102*, 102041. https://doi.org/10.1016/j.ijintrel.2024.102041
4. Ellis, T., Jola, C., & Cameron, A. (2024). Cultural adaptation and transitions within international higher education: University students' experiences of studying abroad during the 2020 coronavirus pandemic. *PLoS One, 19*(10), e0308134. https://doi.org/10.1371/journal.pone.0308134
5. Lu, J., Wang Guénier, A. D., & Hird, D. (2025). An empirical study on intercultural contact and intercultural communication competence of Chinese international students in the UK: A case study of Lancaster University. *Journal of International Students, 15*(2), 147–167. https://doi.org/10.32674/xtpdzb48
6. Beghin, J., & Park, B. (2021). The exports of higher education services from OECD countries to Asian countries: A gravity approach. *The World Economy, 45*(4), 1050–1080. https://doi.org/10.1111/twec.13192
7. Brooks, R., & Waters, J. (2009). International higher education and the mobility of UK students. *Journal of Research in International Education, 8*(2), 191–209. https://doi.org/10.1177/1475240909105204
8. Egorova, I. A., & Makarova, E. L. (2020). International education management: A review of key concepts. *Bulletin of the Taganrog Institute of Management and Economics, 1*(31), 102–106. Accessed Mar 23, 2023, from https://cyberleninka.ru/article/n/upravlenie-mezhdunarodnym-obrazovaniem-obzor-osnovnyh-ponyatiy
9. Ebzeeva, Y. N., & Gishkaeva, L. N. (2023). Improving the competitiveness of education in Russia at the international level. *Logos et Praxis, 22*(1), 62–72. https://doi.org/10.15688/lp.jvolsu.2023.1.8
10. Kanunnikova, A. M., & Moga, I. S. (2022). Increasing the international competitiveness of higher education in Russia. *The Eurasian Scientific Journal, 14*(2), 20ECVN222. Accessed Mar 22, 2023, from https://esj.today/PDF/20ECVN222.pdf
11. Muraveva, A. A., Oleinikova, O. N., & Viktorova, A. O. (2019). Main vectors of development of higher education: International perspective. *Bulletin of Nizhny Novgorod University named after N.I. Lobachevsky. Series: Social Sciences, 4*(56), 311–320. Accessed Mar 23, 2023, from https://cyberleninka.ru/article/n/osnovnye-vektory-razvitiya-vysshego-obrazovaniya-mezhdunarodnaya-perspektiva
12. Diachkova, M. A., Tomiuk, O. N., Shutaleva, A. V., & Dudchik, A. Y. (2019). Inclusive organizational culture as a culture of acceptance of diversity and understanding. *Perspektivy nauki i obrazovaniia [Perspectives on Science and Education], 5*(41), 373–385. Accessed Mar 25, 2023, from https://cyberleninka.ru/article/n/inklyuzivnaya-organizatsionnaya-kultura-kak-kultura-prinyatiya-raznoobraziya-i-vzaimoponimaniya
13. Suchkov, M. A. (2020). Foreign experience of the ethnocultural factor in inclusive education. *Problems of Modern Teacher Education, 67-2*, 231–234. Accessed Mar 22, 2023, from https://cyberleninka.ru/article/n/zarubezhnyy-opyt-etnokulturnogo-faktora-v-inklyuzivnom-obrazovanii

14. Deltsova, I. A. (2020). Inclusiveness as a new educational focus. *Noosfernyye issledovaniya [Noospheric Studies]*, *2*, 55–58. https://doi.org/10.46726/NOOS.2020.2.55-58
15. Gokh, A. F., Shestakova, N. N., & Iurkov, D. V. (2019). Inclusive culture: From educational to sociocultural context. *Bulletin of the V. P. Astafiev Krasnoyarsk state Pedagogical University*, *2*(48), 189–198. https://doi.org/10.25146/1995-0861-2019-48-2-136

Gender Neutrality of the University as the Basis for its Inclusiveness

Aktam U. Burkhanov, Mavluda S. Akhmedova, Ainura A. Adieva, and Elena I. Litvinenko

Abstract This paper studied the issue of the role and influence of gender neutrality of universities from the position of an increase in their inclusiveness and support for equal rights for higher education applicants, regardless of sex, age, social origin, etc. The research methodology is multidisciplinary, it is based on the principles of social constructivism and a formal legal approach to the identification of human rights and conditions of support for their equality. The methodological framework of this work lies in gender sociology and is based on the methods of generalisation, analysis, synthesis, comparison, observation, and the graphical method. This work determines the essence of gender and gender neutrality and characterises their place in the system of support for system of ensuring the equality of human rights. Focus is made on global institutions of support for equality of rights for all genders, analysis of the access of men and women to higher education is performed, and the main risks for universities regarding the violation of the rules of gender policy in the process of universities' educational activities are determined. The novelty of this work lies in the conceptual substantiation of the role of gender neutrality for compliance with the terms of gender equality and inclusiveness in modern universities. Conclusions regarding the absence of critical problems from the position of discrimination of women in education in Russia and countries of Central Asia and Russia and regarding the substantiation of the necessity of taking measures for raising the role of women in science and education are made.

A. U. Burkhanov (✉)
Alfraganus University, Tashkent, Uzbekistan

M. S. Akhmedova
Tashkent State University of Economics, Tashkent, Uzbekistan
e-mail: m.akhmedova@tsue.uz

A. A. Adieva
International University of Kyrgyzstan, Bishkek, Kyrgyzstan

E. I. Litvinenko
Platov South-Russian State Polytechnic University (NPI), Novocherkassk, Russia
e-mail: litvinenko_eiu@npi-tu.ru

B. S. Sergi et al. (eds.), *University 4.0 and Educational Technology for Sustainable Development*, Education in the Asia-Pacific Region: Issues, Concerns and Prospects 75, https://doi.org/10.1007/978-981-96-7376-6_9

Keywords Gender · Gender equality · Gender neutrality · University · Inclusion · Inclusiveness

JEL Codes I21 · I23 · I24 · I28 · I32 · J16 · J24 · O15

1 Introduction

The modern world is changing very quickly. Traditional approaches to cultural and social self-identification gradually lose their topicality. The formation of new traditions takes place based on liberal values that are based on the striving towards freedom and equal opportunities. An important role in this process belongs to higher education, as a sociocultural and intellectual institution that accumulates and disseminates new values and meanings, forms the future elite, and determines the trends of social development.

The inclusiveness of universities is treated as an important component in the formation of new traditions. According to it, access to higher education and the use of its results in future professional life cannot be limited to any category of citizens, especially citizens that are representatives of vulnerable groups of the population or have special demands for education—for them, corresponding conditions must be created to avoid any discrimination.

A general recognition of inclusiveness as one of the values of modern universities in Russia and countries of Central Asia determines conditions for the creation of the corresponding cultural and educational environment, one of the elements of which is ensuring gender equality and neutrality [1, 2]. The main directions for the gender policy of universities, according to modern challenges, are ensuring the right for individual expression, avoiding discrimination by the gender principles, expanding conditions of access to higher education for vulnerable groups of society, upbringing of student youth in the spirit of the modern treatment of gender policy and its values, and wide cooperation in the issues of gender policy and inclusiveness with international institutions [3]. Thus, gender neutrality is one of the basic elements for ensuring the inclusiveness of universities, within which all persons, regardless of sex and social ethnic, or cultural environment, will have equal rights and cooperate for the achievement of the goals of the university community and the whole society [4, 5].

2 Gender Neutrality of Universities from the Position of Their Inclusiveness

This research is based on the combination of a range of approaches and theories within one multidisciplinary vision, which covers the aspects of sociology, the legal sphere, pedagogics, institutional theory, etc. Within the determination and interpretation of gender rights, the formal legal approach is the main one, for it allows for the correct determination of the contents of these rights, as well as the principles of social constructivism, which involve distinguishing men and women as separate heterogeneous groups with available gender differences. Most of the provisions of this work are given within the theory of gender sociology and, accordingly, the treatment of gender in different social groups and institutions. The main methods utilised in this work are general scientific methods of observation, comparison, generalisation, analysis, synthesis, and the graphical method, which allows for visual comparison of the specific features of gaps in higher education in Russia and countries of Central Asia.

The study of gender neutrality of universities from the position of their inclusiveness belongs to a wide range of problems that consider it from different aspects, including a study of the general understanding and treatment of gender equality [6–9] and its reflection for the UN Sustainable Development Goals [5]; study of the aspects of implementing gender policies in the world [10–12] and in global processes [13]; implementation of the policy of gender equality in education at the international level [14]; specifics of gender neutrality [15] and its treatment in the legal sphere [16]; study of the gender gap in education and the labour market [17], etc.

The main goal of this research is to determine the role of gender neutrality in ensuring the inclusiveness of universities, which involves the resolution of a range of tasks aimed at the specification of the categorical apparatus, study of the key aspects of gender policy and the role of gender neutrality in it, as well as determination of the state of the gender situation in the education of selected countries.

3 Inclusiveness of Universities Through Their Gender Neutrality in Russia and the Countries of Central Asia

In the context of this research, it is important to identify the key definitions of this work. The term "gender" is rather new in scientific literature. One of the first scholars to use this word in a meaning that is close to the modern one was R. Stoller, who studies the notion of sex and gender from the position of masculinity and femininity [8]. However, a clear definition of this category was given by R. Unger, who suggested differentiating the notions of sex and gender according to the following principle: sex is to be used when characterising special biological features of a human, while gender is to be used during when speaking of its social, cultural, or psychological parameters, primarily social roles associated with men or women by society [9].

Gender equality, which involves the avoidance of discrimination of women in all spheres is one of the priorities of UNESCO. It is closely connected with educational initiatives, within which not only equal access to higher education for boys and girls is ensured, but also equal opportunities for obtaining education and using its results during their whole life are provided [14]. New approaches to the treatment of women's rights in international legal acts are connected with the desire to strengthen and adopt the legal status of women in society. In these conditions,

Russia and countries of Central Asia face gender asymmetry, according to which, despite the declaration of equality of rights, there is domination of a certain gender in many spheres. A reaction to such a situation should be the incorporation of international legal standards, which would not only declare gender equality but would also form the corresponding culture [7].

The basis for the treatment of inclusiveness of universities through the lens of gender neutrality is biological and social differences, which identify different sexes and genders. Such an approach is a part of the general understanding of the theory of human resources, which is characterised by a certain structure based on objective and subjective differences between different representatives of society. As a result, the age and gender structure of employment and population is formed, by the level of education and residence, quality of life, etc. [13].

Relationships in the socio-gender sphere change and a new order of gender interaction is determined in the world [15]. The concept of avoidance of discrimination against women is more frequently interpreted not only in the context of equal rights but also in the context of the desire to avoid differentiation of people by sex or gender. The development of theories that characterise attitudes towards the equality of men and women, as well as close integration of the community of men and women, is clearly defined by the UN global gender agenda, which is reflected in the Sustainable Development Goals, in particular SDG 5 [18]. The corresponding calls are found in SDG 4, which deals with ensuring quality education and life-long learning.

Based on the data from the UNESCO portal, which covers the problems of inequality in education [18], it is possible to track gender differences in Russia and countries of Central Asia as to the involvement of the youth in university education (Fig. 1). The data are representative, for they take into account the entire studied region. However, the presented information belongs to the period of 2012–2021.

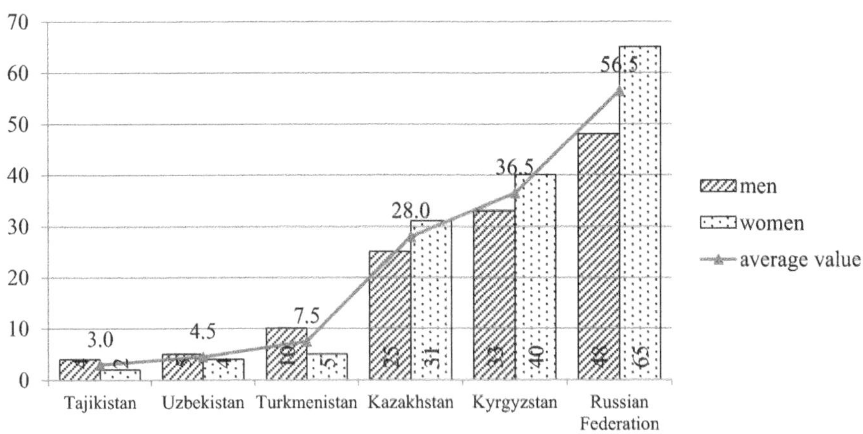

Fig. 1 Share of citizens of Russia and countries of Central Asia aged 18–22 who study in higher educational establishments, %. *Source* Prepared by the authors based on [5]

It is possible to see that the region does not have significant problems with inequality of rights of women in access to higher education. In regions where the share of women in higher education is very low, there is a generally low level of involvement of youth in the study at universities (3%–7.5%). This is peculiar to Tajikistan, Uzbekistan, and Turkmenistan. However, in Kazakhstan, Kyrgyzstan, and the Russian Federation, the level of the youth's involvement in higher education is much higher (28%–56.5%). These countries are also peculiar in the domination of the share of females who obtain higher education at ages 18–22. The difference is from 21% in Kyrgyzstan to 36% in the Russian Federation. Thus, it is possible to conclude that there is no discrimination as to the access of females to higher education in the region. In countries where females are less involved in university education, there is a general problem of access to higher education. In countries that do not have such a problem, there is a domination of females who obtain higher education.

In expert circles, the problem of females' access to education gradually loses its urgency. The right of women to education is still discriminated against only in several countries with strong patriarchal traditions. However, most countries face the problem of a gender gap in professions [17]. Studies show that women are poorly represented in high-tech spheres of activities, such as IT and engineering. Their representation in the social and cultural spheres is much higher. From the position of the perspective of human development, such imbalance may lead to women losing their influence on the key processes of social progress shortly. An important role in the reduction of this gap belongs to higher education establishments.

On the whole, universities perform many tasks that seriously influence social processes. Very often, these educational establishments are treated as institutions that prepare young people for professional activities, providing them with an opportunity to master relevant competencies and knowledge (Fig. 2).

Without entering a discussion regarding the wider role of universities, it should be noted that by forming professional competencies, educational establishments face the risks of violating of gender equality policy. The main aspects of such risks are as follows:

- Public stereotypes regarding purely male or female professions;
- Possible difference in attitude towards representatives of different genders in the educational process;
- Complexity in ensuring equal gender rights for obtaining certain professions;
- Absence of the formed social and information culture regarding the interaction with gender minorities.

Apart from this, universities, as scientific institutions, form—within the gender policy—preconditions for a deeper study of gender problems. These aspects are implemented within the three main directions:

- Increase in the share of women in scientific research and science;
- Increase in the number of women in managing positions in education and science;
- Implementation of gender provisions in educational programmes and topics of scientific research [10].

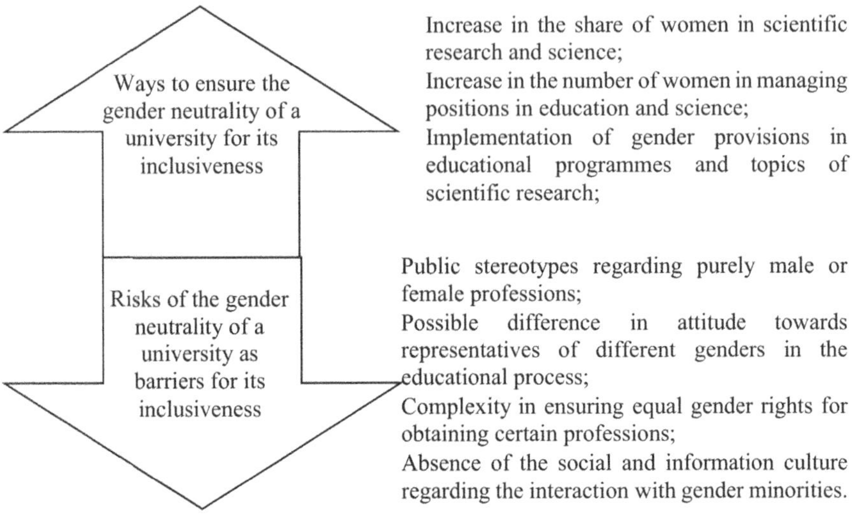

Fig. 2 Risks and ways to ensure the gender neutrality of a university for its inclusiveness. *Source* Prepared by the authors

Gender neutrality is an element of gender policy. It focuses not so much on ensuring the equality of rights as on the formation of conditions for avoiding the need to divide groups of people into genders according to their social roles. Thus, gender neutrality is an element of gender policy in ensuring equality and avoiding discrimination in rights. It implies the refusal of providing preference in language and/or determination of social roles to any gender or sex, to avoid discrimination or provision of unjustified advantage to a gender [6]. The issue of gender neutrality is posed most clearly in legal and lexical discourses.

4 Problems of Development of the Educational Policy and Measures of Gender Neutrality on the Way of Universities' Inclusiveness Achievement

The issue of gender neutrality of the university from the position of its inclusiveness lies in a wide field of discussions, which concern different aspects of the functioning of higher educational establishments, support for equality, formation of an inclusive environment, and culture. According to this, key provisions, which require a detailed discussion, are connected with the very treatment of gender neutrality in the society of the studied region's countries. In this context, it is important to clearly differentiate gender neutrality and gender equality—for the purpose of the use of correct means in the implementation of its policy and tools in universities.

Another relevant problem is the development of the methodology for measuring the level of gender neutrality in different spheres of universities' activities. Resolution of this problem will allow obtaining the tools for the objective assessment of the successfulness of the measures of gender policy and determination of the facts of violation of the conditions of gender neutrality at different stages of the educational process—during the development of work programmes, educational courses, or control of knowledge, the use of knowledge in practical activities, etc.

The obtained results, which point to the absence of critical facts of violation of the principles of gender equality, together with the actualisation of the problem of the formation of the culture of gender neutrality in universities, may be a basis for further scientific research on an increase in the inclusiveness of higher educational establishments. According to this, it is necessary to deal with the problems of development of the policy and measures of gender neutrality, their implementation in the educational environment and culture of the educational institution, training of academic staff in the new conditions of social and gender policy, and development of measures on the dissemination of information on these issues in the external environment of the university. The final result of the measures taken must facilitate the avoidance of problems of gender discrimination or imposing gender social roles within an educational establishment of higher education, as well as raise the level of positive perception of the university from the position of its ability to ensure accessibility and inclusiveness of education without dividing students and employees into groups by any criterion, including gender, age, ethnic origin, or property status.

5 Conclusions

In the modern conditions of the transformation of value and sociocultural provisions, the problem of gender equality is very important, given the announced global goals and priorities in ensuring equal access and eliminating discrimination by gender, including in the higher education environment. According to the definition of gender as a set of social roles that identify a person's behaviour or person's perception in society, gender neutrality has the potential to reach an important task of eliminating a stereotypical connection of the biological sex of a person to the person's gender identification. The use of the principles of gender neutrality in universities is a significant tool that can ensure equality of the representatives of both sexes and, accordingly, raise the level of their inclusiveness.

The modern state of gender policy in the world is rather well-regulated from the position of international and domestic norms and rules. According to this, there are practically no barriers to involving women in public processes, including obtaining higher education. However, against the background of an increase in innovative development, gender features are more often identified from the position of a gap in employment in high-tech sectors. This forms large risks for women from the position of participation in the most promising, from the position of mankind's development, processes. That is why gender neutrality—as one of the important factors of universities' inclusiveness—must be actively implemented by dealing with stereotypes on the designated social gender roles, as well as creating conditions to avoid reservations and stereotypical conclusions on gender differences.

References

1. Egitim, S. (2024). Promoting gender-inclusive leadership perceptions among Japanese university students: A nine-week critical inquiry-based action learning program. *International Journal of Educational Research Open, 7*, 100362. https://doi.org/10.1016/j.ijedro.2024.100362
2. Hinton-Smith, T. (2025). Positioning gender inequality awareness in higher education teaching internationally: Colonial legacies, cultural explanations and responsibility of universities. *Gender and Education, 37*(1), 1–15. https://doi.org/10.1080/09540253.2024.2404048
3. Khairani, A. Z., & Zhang, H. (2025). The influence of social cognitive factors on physical activity of Chinese university students: The moderating role of gender. *Humanities and Social Sciences Letters, 13*(1), 1–24. https://doi.org/10.18488/73.v13i1.3961
4. Michl, T., Wilson, J., Stookey, A., Chiou, K., Raque, T. L., & Kracen, A. (2025). There's more out there: Examining the study abroad experiences of American transgender and gender expansive university students. *Higher Education*. https://doi.org/10.1007/s10734-024-01381-0
5. UNESCO. (2023). *What you need to know about how UNESCO advances education and gender equality*. Accessed Mar 27, 2023, from https://www.unesco.org/en/gender-equality/education/need-know
6. Denisova, A. A., et al. (2002). *Dictionary of gender terms*. Informatsiia XXI vek. Accessed Mar 26, 2023, from http://www.owl.ru/gender/
7. Golovinov, A. V., & Golovinova, Y. V. (2022). International dimension of gender equality: Social and legal research. *Society and Security Insights, 5*(3), 37–46. https://doi.org/10.14258/ssi(2022)3-03
8. Stoller, R. J. (1964). A contribution to the study of gender identity. *International Journal of Psychoanalysis, 45*, 220–226.
9. Unger, R. K. (1979). Toward a redefinition of sex and gender. *American Psychologist, 34*(11), 1085–1094. https://doi.org/10.1037/0003-066X.34.11.1085
10. Palmén, R., Arroyo, L., Müller, J., Reidl, S., Caprile, M., & Unger, M. (2020). Integrating the gender dimension in teaching, research content & knowledge and technology transfer: Validating the EFFORTI evaluation framework through three case studies in Europe. *Evaluation and Program Planning, 79*, 101751. https://doi.org/10.1016/j.evalprogplan.2019.101751
11. Shadurskii, V., Zinchenko, Y., Bezbozhnaya, O., & Shapovalova, I. (2020). *Gender dialogue: Brief results of a gender audit at the Faculty of International Relations of the Belarusian state university*. UNIPAC. Accessed Mar 26, 2023, from https://elib.bsu.by/handle/123456789/241084
12. Shvetsova, A. V. (2020). Gender approach in inclusive education. In M. A. Galaguzova (Ed.), *Conceptual apparatus of pedagogy and education: Collection of scientific works* (pp. 143–149). Ekaterinburg.
13. Sillaste, G. G. (2022). The global gender landscape and country gender order in the context of global transformation of social systems. *Vek globalizatsii [The age of Globalization], 3*(43), 112–131. Accessed Mar 28, 2023, from https://cyberleninka.ru/article/n/mirovoy-gendernyy-landshaft-i-stranovoy-gendernyy-poryadok-v-usloviyah-globalnoy-transformatsii-sotsialnyh-sistem
14. UNESCO. (2021). Higher education attendance. *Indicators*. Accessed Mar 26, 2023, from https://tinyurl.com/2z5sr9kx
15. Brushkova, L. A. (2022). Gender neutrality as the norm of the new gender order: The attitude of Russians. *Bulletin of the Russian State University for the Humanities*. Series "Philosophy. Sociology. Art history", *1-3*, 383–392. https://doi.org/10.28995/2073-6401-2022-1-383-392
16. Fedotova, I. M. (2018). Gender neutrality in legal discourse (based on the UK budget law). *Bulletin of Moscow state linguistic university. Education and pedagogical sciences, 6*(814), 158–170. Accessed Mar 26, 2023, from https://cyberleninka.ru/article/n/gendernaya-neytralnost-v-pravovom-diskurse-na-materiale-zakona-o-byudzhete-velikobritanii
17. Weforum. (2020). *Global gender gap report 2020* [Online source]. Accessed Mar 26, 2023, from https://www.weforum.org/reports/gender-gap-2020-report-100-years-pay-equality
18. UN. (2023). *Sustainable development goals: 2030*. Accessed Mar 27, 2023, from https://www.un.org/sustainabledevelopment/ru/sustainable-development-goals/

Anti-Ageism and Openness to Lifelong Learning, Work-Life Balance in an Inclusive University

Gulnora K. Abdurakhmanova, Aidarbek T. Gyiazov, and Yury A. Tikhomirov

Abstract The processes of discrimination in society negatively influence the social climate and economic development of the country. Manifestations of discrimination in the modern world have many forms. To deal with them, governments develop and implement policies aimed at preventing their further development. Overcoming age discrimination, which is called anti-ageism, has been adopted in many countries, including the studied countries of Central Asia (Kazakhstan and Kyrgyzstan).

In this paper, we demonstrated the features of the formation of anti-ageism and its component—openness to lifelong learning in inclusive universities—and proved the positive influence of these processes on work-life balance. We revealed positive aspects of the normative and legal regulation of these processes in Kazakhstan and Kyrgyzstan, which gradually influence the achievement of positive results in the labour and education markets and the sphere of social support.

The goal of this paper was to systematise the specifics of the formation of anti-ageism and openness to lifelong learning as the precondition for the achievement of work-life balance in inclusive universities in the selected countries of Central Asia. To achieve this goal, we tried to determine the characteristics of the anti-ageism policy in certain countries of Central Asia and identify the specifics of lifelong learning in inclusive universities of Central Asia as a factor of achievement of work-life balance.

Keywords Anti-ageism · Lifelong learning · Inclusive universities · Work-life balance · Age groups · Discrimination · Employment

---

G. K. Abdurakhmanova
Tashkent State University of Economics, Tashkent, Uzbekistan
e-mail: g.abdurakhmanova@tsue.uz

A. T. Gyiazov (✉)
Batken State University, Batken, Kyrgyzstan

Y. A. Tikhomirov
RUDN University, Moscow, Russia
e-mail: Tikhomirov-yua@rudn.ru

B. S. Sergi et al. (eds.), *University 4.0 and Educational Technology for Sustainable Development*, Education in the Asia-Pacific Region: Issues, Concerns and Prospects 75, https://doi.org/10.1007/978-981-96-7376-6_10

JEL Codes E24 · J01 · J08 · J14 · J24 · J26

1 Introduction

Ageism is a phenomenon connected with manifestations of discrimination or depreciation of life experience and personal qualities of older people. It emerged due to the conflict of generations and dissemination of public discourse that the ageing of the population leads to an increase in the burden on the national budget. This, in turn, will negatively influence the well-being of the young generation due to the possible growth of taxes [1]. Also, ageism is considered as discrimination and the young generation's non-acceptance of the older generation in the context of the main spheres (culture, fashion, priorities, etc.). The main difference between ageism from many other types of discrimination is that many people face it when reaching a certain age.

Discussions on the socioeconomic and moral aspects of ageism as an inadmissible form of discrimination predetermined the formation of various practices and policies connected with the overcoming of its manifestations. This direction is called anti-ageism. This direction should be studied due to the necessity of revealing new possibilities for reducing the manifestations of ageism as a social problem [2, 3], which forms disunity in society by the age indicator, which influences the well-being of citizens and socioeconomic development of the country. Integration and reduction of discrimination against older people can be achieved due to their openness to lifelong learning in the aspect of the main or adjacent professions [4]. The formation of anti-ageism is related to the focus on lifelong learning of older generations in the conditions of inclusive universities and the need to support work-life balance [5]. Most older people cannot attend traditional universities because of socioeconomic and other reasons [6]. The younger generation, with the proper socioeconomic conditions, acquires traditional education, which conforms to the employers' needs [7]. The older generation should have the same knowledge, which is connected with the necessity for lifelong learning in inclusive universities and receiving new skills and information [8]. The focus on inclusiveness in learning allows reaching personal and professional goals, with the achievement of results in an increase in the level of knowledge and skills within the profession and corresponding spheres.

2 Anti-Ageism and Openness of Universities for Inclusive Lifelong Learning

The study of the formation of anti-ageism and openness to lifelong learning was performed based on the developed methodological framework, which includes the order and methods of assessment. The stages of the analysis include the determination of the character of the manifestations of anti-ageism in the studied countries of

Central Asia; identification of the level of openness to lifelong learning as an element of anti-ageism in the selected countries; evaluation of the influence of anti-ageism and openness to lifelong learning in the inclusive university on the state of work-life balance.

Also, within this methodological framework, we distinguished a range of methods used to implement the stages of the analysis. The statistical analysis is used to determine official data on the manifestations of anti-ageism and openness to lifelong learning in inclusive universities in the selected countries of Central Asia. The feedback method was used to analyse the reactions of these countries' citizens to the influence of anti-ageism and openness to lifelong learning in the inclusive university on the work-life balance. The comparative method was used to compare the level of formation of the considered phenomena in the studied countries.

The issues of anti-ageism and openness to lifelong learning in the inclusive university and the possibilities for their influence on the state of work-life balance were studied at the theoretical & empirical and analytical levels [1, 2, 9]. presented theoretical aspects and characteristics of the notions of ageism and anti-ageism [10]. presented the ranking results of the Global Life-Work Balance Index for 2022, with the determination of the level of this indicator in Kazakhstan. Energyprom.kz [11] contains analytical information on the labour market, demographic indicators, and problems of Kazakhstan, with recent statistics on the results of implementing the policy of overcoming age discrimination over 2018–2021. Despite the presence of a certain number of studies in this sphere, it is still necessary to dwell on the considered problems.

3 Lifelong Learning in Inclusive Universities in Kyrgyzstan

Let us analyse the character of the phenomena of anti-ageism and openness to lifelong learning in inclusive universities in the selected countries of Central Asia at the modern stage.

First, let us consider these processes in Kazakhstan (Table 1).

Based on the data from Table 1, it was determined that the implementation of anti-ageism measures in the labour sphere in Kazakhstan led to a slight increase in positive results in the reduction of age discrimination. Despite the absence of quick growth of positive results, the country works on implementing measures on an increase in the level of knowledge and skills of citizens of younger and older age groups and their attraction in large companies.

In Kazakhstan, at the normative and legal level, there has been a course towards lifelong learning in the inclusive university. This phenomenon is widely used within additional and informal education. This realisation of lifelong learning in an inclusive university in Kazakhstan conforms to UNESCO's approach to this issue. UNESCO considers this process as lifelong learning (within age categories), which can be implemented within informal and formal types through the use of flexible mechanisms [17]. The focus on the use of flexible mechanisms is predetermined by the necessity for considering the specifics of the life activities of people who

Table 1 Specifics of the manifestation of anti-ageism in the labour sphere and openness to lifelong learning in the inclusive university in Kazakhstan

	Indicator	Character and state of indicators
1	Normative and legal regulation of anti-ageism in the educational sphere	The law of Kazakhstan "on education" does not have a clear indication of the inadmissibility of age discrimination. According to sub-article 2 of article 37 of this law, additional education of adults is connected with the necessity of realising their needs in the educational sphere during their entire life, which is connected with the public transformations. Digitalisation and new requirements for professions imply constant improvement of knowledge and skills. Normative and legal regulation in the educational sphere demonstrates openness to lifelong learning within additional and informal education.
2	Normative and legal regulation of anti-ageism in the labour law	It has been regulated from the moment of adoption of the first edition of the labour code of Kazakhstan. The current norms of this law involve the prohibition of age discrimination.
3	Normative and legal regulation of anti-ageism in the sphere of labour law and education	Adoption of the normative and legal document that regulates the order of acknowledgement of informal learning and advanced training in state-funded and private institutions. An increase in the quality of labour resources of two age categories (younger age group (16–24) and older age group (45–64) is implemented by the initiative of companies that send personnel to advanced training and retraining in special organisations (including online).
4	Normative and legal regulation in the sphere of support of work-life balance	Introduction of four-day working week (starting 1 July 2023)
5	Dynamics of % of workforce to total population of the country	Values: 2000–51%, 2005–52.3%, 2010–53.5%, 2015–51.4%, 2020–49%, 2021–48.9%, 2022–49.1%. Dynamics of the change in the indicator: Insignificant, with a trend for reduction, connected with the phenomenon of "ageing" of the nation.
6	Measures for the improvement of corporate social responsibility of companies of various categories and sectors	Many large companies started a comprehensive adaptation of employees without work experience in the age group of 16–24 and older age group (45–64). Adaptation is implemented with the use of special programmes (mainly in the sphere of an increase in knowledge and skills with ICT).
7	Results of implementing the measures of anti-ageism in the labour sphere at the level of companies of various categories and sectors	Indicator: Level of employment depending on the age indicator. 2018: Younger age group (16–24)—11.2%; medium age group (25–44)—57.4%; older age group (45–64)—30.8%, age group 65+ – 0.6%. 2019: Younger age group—12%; medium age group—55.9%; medium age group—31.4%, age group 65+ – 0.7%. 2020: Younger age group—12%; medium age group—55.7%; medium age group—31.4%, age group 65+ – 0.9%. First quarter of 2021: Younger age group—11.9%; medium age group—55.3%; medium age group—32.1%, age group 65+ – 0.7%. Second quarter of 2021: Younger age group—11.9%; medium age group—54.9%; medium age group—32.4%, age group 65+ – 0.8%.

Source Created by the authors based on [3, 11–16]

implement individually determined programmes of lifelong learning, connected with the need to support work-life balance.

It is necessary to assess the influence of lifelong learning in inclusive universities on the factor of achievement of work-life balance in Kazakhstan. Analysis of the ranking Global Life-Work Balance Index for 2022 demonstrated that the country was ranked 52nd (in 2022) by the level of this indicator [10]. Though this is not a high value, the country was able to achieve significant results (including thanks to the implementation of anti-ageism measures and lifelong learning in inclusive universities).

All universities of Kazakhstan declare and implement the policy of anti-ageism. Some of them use such tools of inclusiveness as digitalisation to form the environment for the satisfaction of citizens' needs for lifelong learning. These organisations include the following:

- Turan University (Almaty), with a system of inclusive learning for citizens with special needs and elderly people, who can study according to individual schedules [18]. The measures facilitating the inclusive education of such people include the possibility of forming schedules as per students' requests, full online access to all educational and methodological materials from the university's library, functioning of the online support centre, the possibility of choosing remote and combined learning;
- Al-Farabi Kazakh National University (Almaty). The university has a remote form of education for bachelor's and master's programmes, as well as national and international educational programmes (courses) for different specialities [19]. Remote education is available for elderly citizens, who can study and work at the same time.

Assessment of the results from Table 2 showed that Kyrgyzstan has an active national policy in the sphere of anti-ageism and openness to lifelong learning in inclusive universities. There are no analytical and ranking data on the influence of these measures on the achievement of work-life balance, but it is possible to state that these measures positively impact the support of the qualification level of labour resources of the older age groups in Kyrgyzstan.

Among Kyrgyz digital inclusive universities that satisfy the need for education among different age groups of the population, we should distinguish Issyk-Kul State University named after K. Tynystanov. Since the academic year 2011–2012, the extramural form of studies was passed to the Centre of remote education and advanced training [24].

4 Connection Between Anti-Ageism and Openness to Lifelong Learning in Inclusive Universities and Achievement of Work-Life Balance

We revealed a positive connection between anti-ageism and openness to lifelong learning in inclusive universities and achievement of work-life balance in the selected countries of Central Asia.

Table 2 Specifics of the manifestation of anti-ageism in the labour sphere and openness to lifelong learning in inclusive universities in Kyrgyzstan

	Indicators	Characteristics of the indicators
1	Legal regulation of anti-ageism in the educational sphere	According to article 1 of the law of Kyrgyzstan "on education", there is a component of additional education for persons of the adult age group. This type of education ensures the possibility to meet citizens' demands in the spheres of lifelong learning and improvement of professional qualification due to the quick growth of requirements to specialities.
2	Legal regulation of anti-ageism in the labour sphere	According to sub-article 2 of article 2 of the labour code of Kyrgyzstan, it is prohibited to provide indirect and direct advantages and limitations of citizens' labour rights based on age indicators. According to sub-article 3 of article 9 of this law, it is prohibited to provide unequal wages based on discrimination approaches (including in the context of the age indicator)
3	General legal regulation of anti-ageism and framework regulation of access to professional training and education for elderly people	Law no. 133 dated 26.07.2011 sets the framework for the national policy of supporting and attracting elderly people to political and economic activities. Article 21 determines government guarantees for access to professional training and education for elderly people.
4	Measures on an increase in corporate social responsibility of companies of various categories and sectors	Programmes on advanced training and retraining of personnel (including representatives of older age groups) at the level of companies.
5	Results of the implementation of anti-ageism measures in the labour sphere at the level of companies of various categories and sectors	Indicator: Distribution of the level of employment by the age indicator. Age group 15–19: In 2017–3.32%, in 2022–3.01%, reduction of employment by 0.31%; Age group 20–29: In 2017–28.35%; in 2022–24.08%, reduction of employment by 0.27%; Age group 30–39: In 2017–27.32%, in 2022–29.64%, growth of employment by 2.32%; Age group 40–49: In 2017–21.36%, in 2022–20.89%, reduction of employment by 0.47%; Age group 50–59: In 2017–15.56%, in 2022–15.78%, growth of employment by 0.22%; Age group 60–69: In 2017–3.75%; Age group 70+: In 2017–0.35%, in 2022–0.63%, growth of employment by 0.28%

Source Created by the authors based on [20–23]

Given the demographic situation in Kazakhstan—domination of children and elderly people—the country has a course towards adaptation of all possible groups of able-bodied population to active labour activities, which are necessary to support the well-being and development of the national economy. This course is implemented to overcome of manifestations of age discrimination. Large companies and the government are involved in this, which leads to the creation of an attractive climate for human development. The selected policy and the efficiency of its implementation demonstrate the perspectives of its provisions and the possibility of further changes regarding more active involvement of various age categories of citizens in tasks aimed at an increase in the population's readiness for changes in technologies and competencies.

An important direction in this sphere is systemic monitoring of the level of labour resource quality in the context of the focus on age. The results of such analysis, which could be conducted by the government and private organisations, can stimulate the timely determination of problems aspects in this sphere. A special role in this direction can be assigned to the creation of consultation centres and provision of support to small and medium companies in the realisation of social responsibility programmes, aimed at an increase in the level of training and retraining of personnel. Interaction at the level of public-private partnership could be organised. This would envisage the creation of territorial centres of support for business and public and municipal establishments in the sphere of courses of training and retraining for the most relevant specialities, including in the sphere of ICT (Fig. 1).

Analysis showed that Kyrgyzstan has normative and legal standards that ensure government guarantees to citizens in the realisation of their labour rights and the right to education, regardless of the age indicator. Despite the significant results in work with the age category of 40–49, there is room for further improvement due to the policy of anti-ageism. Similarly to Kazakhstan, it is possible to engage in public-private partnerships, which would stimulate support for companies of various categories and sectors of the economy in the context of anti-ageism.

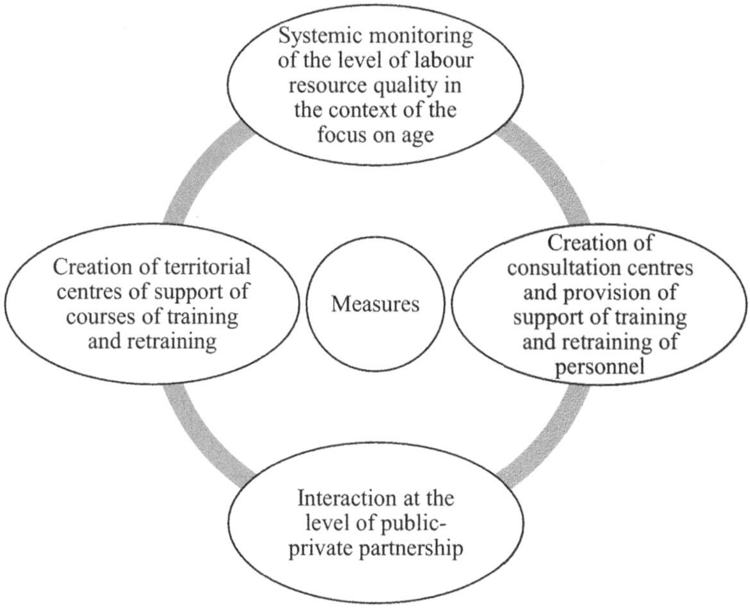

Fig. 1 Measures for further improvement of openness to lifelong learning in Kazakhstan by Inclusive Universities' policy of anti-ageism. *Source* Created by the authors

5 Conclusion

It is possible to state the gradual growth of the manifestations of anti-ageism and openness to lifelong learning in inclusive universities of the selected countries of Central Asia (Kazakhstan and Kyrgyzstan). Growth trends in this sphere demonstrate an increase in openness of these countries and the adaptation of their policies to global policy. Insufficiently quick implementation of these processes is connected with the country's ability to participate in the improvement of the quality of labour resources and the well-being of citizens. Certain positive results are shown by the effectiveness of national policies in this direction.

Further improvement of the policy of anti-ageism and openness to lifelong learning in inclusive universities depends on the implementation of public-private partnerships and the development of public projects in this sphere. Given the revealed positive influence of these manifestations on the level of work-life balance, it is possible to develop further pilot projects to create conditions for its achievement.

References

1. Fragoso, A., & Fonseca, J. (2022). Combating ageism through adult education and learning. *Social Sciences, 11*(3), 110. https://doi.org/10.3390/socsci11030110
2. Sánchez-Izquierdo, M., García-Sánchez, R., & Fernández-Ballesteros, R. (2022). Anti-ageism social actions: Lights and shadows. *Land, 11*, 195. https://doi.org/10.3390/land11020195
3. The World Bank. (2023b). *Population, total—Kazakhstan*. Accessed Dec 18, 2023, from https://data.worldbank.org/indicator/SP.POP.TOTL?locations=KZ&view=chart
4. Tachie-Donkor, G., & Ezema, I. J. (2023). Effect of information literacy skills on university students' information seeking behaviour and lifelong learning. *Heliyon, 9*(8), e18427. https://doi.org/10.1016/j.heliyon.2023.e18427
5. Tumuheki, P. B., Zeelen, J., & Openjuru, G. L. (2024). Towards a transformative lifelong learning agenda for non-traditional students at university. *Journal of Adult and Continuing Education, 30*(1), 22–38. https://doi.org/10.1177/14779714231196044
6. Walker, M., & Martinez-Vargas, C. (2025). How dignity matters for life and lifelong learning: Insights from participatory story-telling by university students in South Africa. *International Journal of Lifelong Education, 44*(1), 57–72. https://doi.org/10.1080/02601370.2024.2405543
7. Wang, H. (2025). Lifelong learning in rural communities in China and its impact on adult subjective Well-being. *International Journal of Educational Research, 129*, 102496. https://doi.org/10.1016/j.ijer.2024.102496
8. Williams, L. (2024). Indigenous intergenerational resilience and lifelong learning: Critical leverage points for deep sustainability transformation in turbulent times. *Sustainability, 16*(11), 4494. https://doi.org/10.3390/su16114494
9. Thu Nguyen, K. (2023). *Early reframing of aging and intergenerational relations for improved technology design and usage*. IntechOpen. https://doi.org/10.5772/intechopen.1002316.
10. Remote. (2023). *Global Life-work balance index*. Accessed Dec 18, 2023, from https://remote.com/resources/research/global-life-work-balance-index
11. Energyprom.kz. (2021). *Nation of the elderly and children: Less and less able-bodied population remains in Kazakhstan*. Accessed Dec 18, 2023, from https://www.energyprom.kz/ru/a/monitoring/naciya-starikov-i-detej-v-kazahstane-ostaetsya-vse-menshe-naseleniya-trudosposobnogo-vozrasta

12. Habr. (2023). *Kazakhstan to implement a four-day working week from July*. Accessed Dec 18, 2023, from https://habr.com/ru/news/731112/
13. Online.zakon.kz. (2023a). *Code of the Republic of Kazakhstan "Labour code of the Republic of Kazakhstan"* (23 November, 2015 No. 414-V, has been amended by the 2017). Accessed Dec 18, 2023, from https://online.zakon.kz/Document/?doc_id=38910832&pos=3646;-61#pos=3646;-61
14. Online.zakon.kz. (2023b). *Law of the Republic of Kazakhstan "On education"* (dated July 27, 2007 No. 319-III). Accessed Dec 18, 2023, from https://online.zakon.kz/Document/?doc_id=30118747&pos=3;-107#pos=3;-107
15. The World Bank. (2023a). *Labor force, total—Kazakhstan*. Accessed Dec 18, 2023, from https://data.worldbank.org/indicator/SL.TLF.TOTL.IN?end=2022&locations=KZ&start=1990&view=chart
16. Zakon.kz. (2023). *Rules of recognition of informal education results updated in Kazakhstan*. Accessed Dec 18, 2023, from https://www.zakon.kz/pravo/6412349-v-kazakhstane-obnovili-pravila-priznaniya-rezultatov-neformalnogo-obucheniya.html
17. UNESCO. (2023). *Analytical note. Life-long learning in Central Asia*. Accessed Dec 18, 2023, from https://unesdoc.unesco.org/ark:/48223/pf0000370956_rus
18. Turan.edu.kz. (2023). *Inclusive education*. Accessed Dec 25, 2023, from https://turan.edu.kz/en/ob-universitete/inklyuzivnoe-obrazovanie/
19. Kaznu.kz. (2023). *Remote education*. Accessed Dec 18, 2023, from https://www.kaznu.kz/ru/17843/page/
20. Cbd.minjust.gov.kg. (2023a). *The law of the Kyrgyz Republic "On education"* (dated April 30, 2003 No. 92). Accessed Dec 18, 2023, from https://cbd.minjust.gov.kg/1216
21. Cbd.minjust.gov.kg. (2023b). *The law of the Kyrgyz Republic "On elderly people in the Kyrgyz Republic"* (dated July 26, 2011 No. 133). Accessed Dec 18, 2023, from https://cbd.minjust.gov.kg/203373?cl=ru-ru
22. Cbd.minjust.gov.kg. (2023c). *The labour code of the Kyrgyz Republic* (dated August 4, 2004 No. 106). Accessed Dec 18, 2023, from https://cbd.minjust.gov.kg/1505
23. Stat.kg. (2023). *Number of employed population by gender and age*. Accessed Dec 18, 2023, from https://www.stat.kg/ru/statistics/zanyatost/
24. Main.iksu.kg. (2023). *Centre of remote education and advanced training*. Accessed Dec 18, 2023, from http://main.iksu.kg/?page_id=116

The Inclusion of People with Disabilities in Higher Education

Abrorbek Kozimjonov, Olga V. Dybina, Galina A. Dorofeeva, and Talgat T. Talantbekov

Abstract The authors conduct a conceptual study on the issue of inclusion of people with disabilities in higher education, determine the key directions for inclusion, find out the problems of its implementation in universities and specifics of the implementation of the process given the objective and subjective factors, and develop a range of suggestions regarding the improvement of the state of inclusion of people with disabilities in the system of higher education. The research methodology is rooted in the social sciences, integrating theoretical and methodological frameworks from psychology, philosophy, pedagogy, economics, and management. It employs general scientific methods such as observation, generalization, analysis, synthesis, and modeling. The modeling method, in particular, enables the reflection of the interconnected systems within the studied processes and phenomena. The research describes the essence and origin of the inclusion of people with disabilities in the higher education environment. The main directions and problems of such inclusion are determined. This phenomenon is considered from the position of empathy towards people with disabilities and the position of rational actions, which have the potential for raising the level of intellectualization of society and better perception of current human values. The authors offered actions and measures to improve the state of inclusion in the system of higher school. Comprehensive and systemic research on the problem of inclusion of people with disabilities in higher

A. Kozimjonov
Tashkent State University of Economics, Tashkent, Uzbekistan
e-mail: a.kozimjonov@tsue.uz

O. V. Dybina (✉)
Togliatti State University, Togliatti, Russia

G. A. Dorofeeva
Sebryakovskiy Branch of Volgograd State Technical University, Mikhailovka, Russia

T. T. Talantbekov
Batken State University, Batken, Kyrgyzstan

State University of Management, Moscow, Russia

B. S. Sergi et al. (eds.), *University 4.0 and Educational Technology for Sustainable Development*, Education in the Asia-Pacific Region: Issues, Concerns and Prospects 75, https://doi.org/10.1007/978-981-96-7376-6_11

education allows revealing specific features, problems, and reserves for its resolution.

Keywords Inclusion · University · Higher education · People with disabilities · People with disabilities

JEL Codes D63 · I21 · I25 · I28 · I32 · J16 · J71 · O15

1 Introduction

The inclusion of people with disabilities in the system of higher education is a complex phenomenon. It is implemented within many directions of the development of human civilization. Thus, is very relevant. Apart from correspondence to the global demands and support for human rights, the creation of conditions for barrier-free receipt of higher education by people with disabilities ensures the achievement of additional effects, which go beyond the limits of direct influence. These effects include a positive impact of such inclusion on the intellectual development of society, support for the wide involvement of talents in higher education and professional activities, and upbringing new generation in an environment of empathy, sympathy, compassion, tolerance, and cooperation. According to this, the inclusion of people with disabilities in higher education is more frequently treated as an act of free will towards people with disabilities and a tool for involving new talents in the intellectual and innovative sphere or as a rational action for a separate group of people with disabilities.

The available scientific studies demonstrate that a higher level of education for people with disabilities creates better conditions for full integration into the public space, successful employment, and financial support for their needs. The consequences of the inclusion of people with disabilities in higher education potentially form another effect, making it possible to reduce the social tension around the topic of financial support for people with disabilities and their achievement of financial independence.

Thus, the inclusion of people with disabilities in higher education is one of the most important tools for their integration into public processes. This is implemented to the greatest extent through an increase in the capabilities of people with disabilities in the sphere of professional development and self-realization, expansion of the circle of their social contacts and roles, development of communication skills, and critical thinking, and better preparedness for social challenges and obstacles.

Despite the long and targeted efforts in the sphere of support for equal rights and access of people with disabilities to higher education, there are many criteria according to which the achievement of full inclusion is not yet ensured. The most problematic aspects are limited access to the physical environment of universities and their educational materials, weakness of the system of support for the individual needs of students with special needs, poor social integration and prejudice in the student community, and insufficient level of the development of the regulatory basis and financial support.

2 Materials and Methods

The inclusion of people with disabilities in higher education belongs to the current tasks of a range of social and fundamental sciences [1, 2]. The methodological basis for such research is formed by the following sciences:

- Sociological sciences, which consider social behavior, interaction, and development;
- Psychology, which studies the problem of empathy and emotional interaction of people with disabilities with the educational environment;
- Pedagogics, which efforts are aimed at the search for effective tools for the full involvement of people with disabilities in the standardized educational process or performance of actions in adapting educational programs and technologies;
- Other scientific spheres, including philosophy, economics, management, etc. [3, 4].

The methodological apparatus of this research is comprised of the methods of synthesis, analysis, generalization, and modeling, which elaborate on the deep aspects of inclusion in higher education, with an emphasis on the determination and structuring of its directions and problems, as well as the development of tools and actions for raising the level of inclusiveness of higher education.

Given the multidimensionality and complexity of the considered problem, the papers considered in this research belong to different directions of research, including the following:

- The elaboration on the problems and experience of disability in higher education [5, 6];
- Assessment of the state of implementing the programs of inclusion in higher education establishments [7, 8];
- Identification of the regulatory barriers to inclusion in higher education [9];
- Development of specific measures to implement inclusion in the system of higher education [1, 10];
- Offers on the strategic development of inclusiveness in higher education [11];
- Assessment of the level of inclusiveness and its successfulness in universities [12].

As per the proposed methodological basis, the main objective of this research was to perform a comprehensive conceptual assessment of the state, implementation, problems, and prospects for further increase in inclusion in higher education, which identifying the promising directions of the development of inclusion and solving the current problems of its popularization, expansion, and implementation in educational establishments.

3 Key Provisions of Ensuring Inclusive Education for People with Disabilities

The foundation of inclusion in higher education started from the process of including people with disabilities in the full educational environment [13, 14]. These rights, designated in the UN Convention on the Rights of Disabled Persons [15], were gradually acknowledged around the world, including Russia and countries of Central Asia. The goal of the Convention was a legislative adoption of the processes of stimulation, support, and protection of all rights and liberties of people with disabilities in society. Article 24 of the Convention determined the key provisions of securing rights for education of this category of people in the form of responsibility to ensure inclusive education at all levels (Fig. 1).

The above provisions are based on the refusal of barriers and isolation of people with disabilities within the educational environment, the use of special devices to consider the individual needs of such people, and ensuring their support and socialization. These actions are strengthened due to better mastering of social and life skills by people with disabilities, who are integrated into the educational space. Additional factors of successful inclusion in higher education include training of teachers with the necessary knowledge for working with people with disabilities and the creation of conditions for their life-long learning.

A more detailed satisfaction of needs of people with disabilities in universities is implemented from the position of consideration of the specifics of their needs, particularly the following ones:

- Ensuring architectural accessibility for students with reduced mobility;
- Preparation of special educational and methodological materials for people with impaired vision and hearing;
- Creation of a special educational environment that would allow correcting disruptions in the development and social adaptation of people with disabilities;
- Development of special individual programs for the rehabilitation of people with disabilities [9].

Declaration of the above provisions and their inclusion into the regulatory and institutional environment allowed for a substantial improvement of the state of inclusion of people with disabilities in the system of higher education. However, too many aspects and specific needs remain unrealized. Thus, an important problem that has to be solved within the studied problem is the low level of socialization of people with disabilities in the conditions of higher school. The main reason for such a state is the absence of previous experience of active social interaction at lower levels of education. Accordingly, the success of inclusion in the higher education environment largely depends on the ability of primary and secondary schools to prepare people with disabilities for the next stage of studying.

Another equally urgent problem is the multidimensional character of inclusive education, which includes organized and well-coordinated work with different disorders and needs of people with disabilities. In this context, medical, compensating,

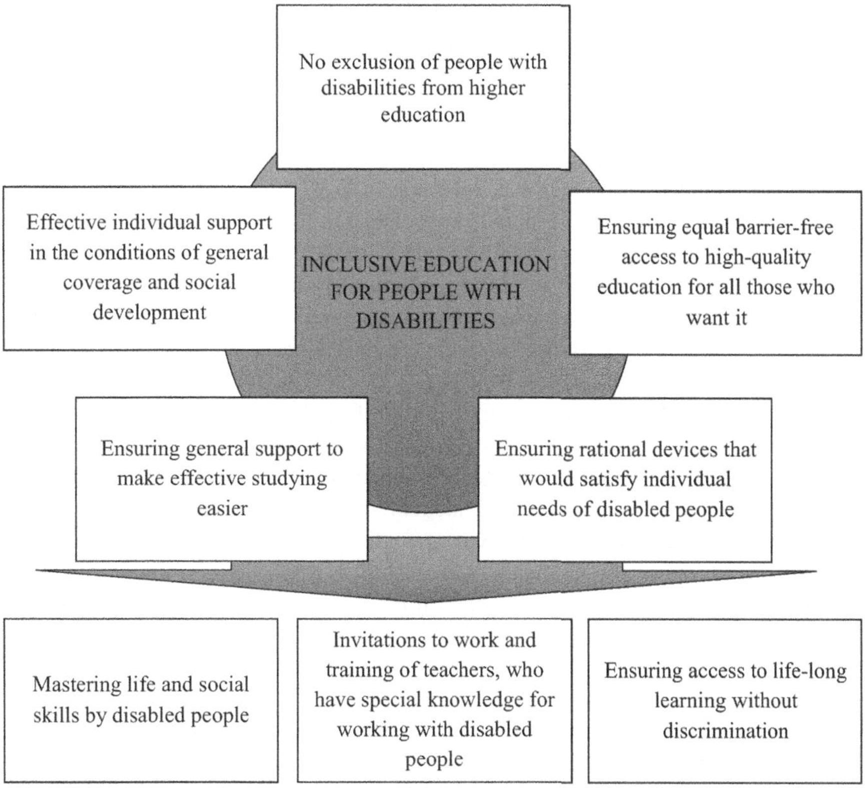

Fig. 1 Key provisions of ensuring inclusive education for people with disabilities. *Source* Compiled by the authors based on [15]

and rehabilitating measures must take place in coordination with the plans of studying and socialization, have systemic character, and last for a long period, from early childhood and, if necessary, during the whole life. However, inclusive education cannot be infinitely adapted to all needs of people with disabilities. Thus, even in the most successful implementation, it will have limitations [8]. Still, the general striving of inclusive education must be formed around the stage-by-stage, consequent educational program, which is developed for a specific person with special needs and which accompanies it during the whole period of studies, from pre-school establishments to receipt of scientific degree or employment [6].

4 The System of Assessment of the Successfulness of Organization of Inclusive Education

While in the context of the organization of inclusive education for people with disabilities, diverse initiatives are implemented in higher schools, which, to a different degree, are realized at the national, regional, or local levels, the system of assessment of the successfulness of such events is still at the primary stage. Among the existing studies [7, 10], the main focus is made on the comparison of the successfulness of students with disabilities compared to other categories of students, evaluation of efforts made by different groups of students for completing the curriculum, and a survey among the participants of the educational process as to their evaluation of inclusive education in the university.

Among the performed studies, the results of students with disabilities did not differ much from the results of the students of the general group. However, efforts made by students with disabilities for the achievement of such results were much higher. Very often, students of this category reach success due to full or partial refusal to participate in social or extracurricular activities [10]. Therefore, social integration, as one of the elements of inclusion, is often not coordinated with the successful completion of the curriculum through larger emotional, intellectual, and physical spending of the labor of students with disabilities, required for the execution of tasks that are similar to the typical program.

Another aspect of the assessment of the inclusiveness of higher education focuses on the attitude of the participants of the educational process towards its implementation. According to this, teachers treat inclusion mostly positively but focus on the necessity for the corresponding material and technical support for the educational process. Along with this, they acknowledge that an important factor of successfulness of inclusive education is the level of the teacher's empathy, their internal self-actualization, and the ability for self-realization through communication, support, and help [7].

Thus, the current stage of support for inclusion in higher education and at other levels is still far from the desired parameters. Many aspects of inclusion have an unsystemic, one-time, and contradictory character. The implementation of the comprehensive and continuous inclusion requires the elimination of a range of barriers (Fig. 2).

The proposed recommendations include elimination of the following of barriers:

- Physical limitations, connected with the absence of a special infrastructure and fragmentary inclusiveness of the educational space;
- Academic limitations connected with the lack of pedagogical workers, who can effectively work with people with disabilities;
- Cognitive limitations formed based on stereotypes, concerns, and negative attitudes towards people with disabilities, which are very often connected with a low level of culture or absence of the corresponding knowledge [9];
- Gaps in career-oriented work, which complicate the process of inclusion and adaptation of people with disabilities with the irrational choice of a profession [11].

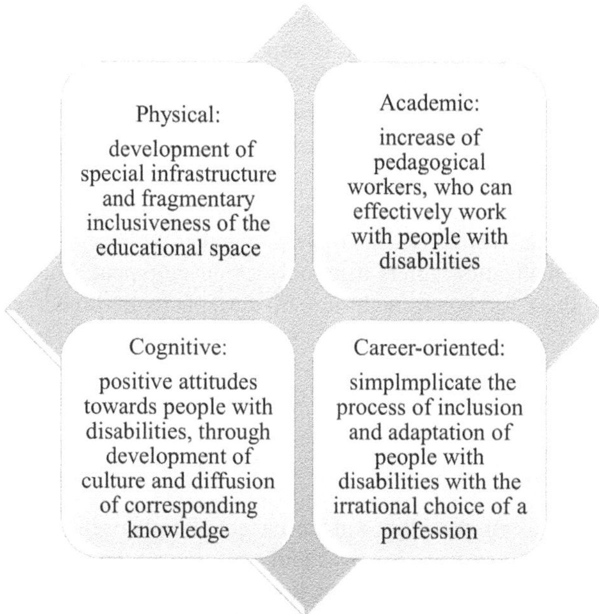

Fig. 2 Measures for supporting the inclusion in higher education. *Source* Compiled by the authors

5 Problems with the Formation of Inclusion in Higher Education

The conclusions drawn and results obtained leave many issues for the discussion. The key aspects requiring discussion and further elaboration include the problem of measuring successfulness and the level of involvement of students with disabilities in the social and academic life of the university. The main indicators of such assessment are an average score of successfulness and percent of successful graduation from the educational establishment. To ensure the evaluation of the successfulness of studies of the considered category of people from the position of qualitative indicators, the methods of self-assessment have been used recently. They allow considering the personal attitude of the respondents towards their involvement in the educational process.

Another problem with the formation of inclusion in higher education is the involvement of people with disabilities in the management of this process [16]. From this point of view, it is believed that only those who understand problems from within can understand them in full and, accordingly, solve them in the best possible way. That is why an important direction for further elaboration on the problem is studying the leadership of people with disabilities in the formation and development of inclusive higher education. It is supposed that this approach will allow revealing completely new aspects of the problem and offer original tools to solve them.

Given the large differentiation of the manifestations of disability and its character, full inclusion in higher education is impossible. Thus, it is important to determine areas in which its implementations have limits and offer compensation mechanisms that would allow involving people with disabilities in the widest range of opportunities in the receipt of higher education, self-realization, scientific development, and successful employment.

The above aspects of scientific discussion are the stage of development and active consideration. Technological progress, together with the digital economy and the spread of globalization, offers new tools for development. Thus, the search for reserves and ideas for better inclusion in the sphere of higher education will remain open.

6 Conclusions

Inclusion of people with disabilities in the system of higher school is a current task, which is derived from global initiatives and conforms to the values of the modern development of human civilization. In scientific and expert circles, inclusion is often also considered as a rational action, which allows improving the intellectual potential of society, positively influences the socio-emotional interaction in society, and forms conditions for self-realization of people with disabilities in society.

Inclusion is based on global norms and institutions that determine the key provisions for ensuring inclusive education for people with disabilities. A critical consideration of these provisions shows the large problems, which have to be solved to ensure real inclusion in higher education. The main problems of this character include ensuring physical accessibility for low mobility groups of students, formation of the corresponding personnel potential in the environment of pedagogical workers, formation of a comprehensive inclusive educational and social environment in the university, and development of educational programs, which would be adapted to different needs of the studied category of citizens. Specific factors that go beyond higher education and include the necessity of involving people with disabilities in the educational process at all levels of education, as well as a large differentiation of needs, not of which can be taken into account within the educational establishment, are also important.

An equally significant factor is the consideration of the trends in the society's development. Here the most important directions of the development of the methodology of assessment of successfulness and satisfaction of people with disabilities, who are included in the educational process, development of inclusive leadership, and constant monitoring of scientific and research environment for the search for new tools, ideas, and solutions.

References

1. Naeemy, M. I., & Yoneda, H. (2024). Lecturers' perspectives on disability and inclusion in afghan higher education institutions: A study toward achieving the sustainable development goals. *Journal of Lifestyle and SDG'S Review, 4*(4), e02504. https://doi.org/10.47172/2965-730X.SDGsReview.v4.n04.pe02504
2. Osuna-Juárez, A., & González-Castellano, N. (2024). Understanding professors' and students with disabilities' perceptions of inclusive higher education: A systematic literature review. *European Journal of Special Needs Education, 39*(5), 774–788. https://doi.org/10.1080/08856257.2023.2294238
3. Pierrès, O., Darvishy, A., & Christen, M. (2024). Exploring the role of generative AI in higher education: Semi-structured interviews with students with disabilities. *Education and Information Technologies., 30*, 8923–8952. https://doi.org/10.1007/s10639-024-13134-8
4. Römhild, A., & Hollederer, A. (2024). Effects of disability-related services, accommodations, and integration on academic success of students with disabilities in higher education. A scoping review. *European Journal of Special Needs Education, 39*(1), 143–166. https://doi.org/10.1080/08856257.2023.2195074
5. Allan, J., & Omarova, T. (2022). Disability and inclusion in Kazakhstan. *Disability & Society, 37*(7), 1067–1084. https://doi.org/10.1080/09687599.2020.1867073
6. Nevzorov, B. P., Zaguzina, N. N., & Bokov, A. V. (2017). Continuity as the basis for the quality of inclusive education. *Bulletin of Kemerovo State University Series: Humanities and Social Sciences, 3*(3), 26–34. Accessed Mar 30, 2023, from https://cyberleninka.ru/article/n/nepreryvnost-kak-osnova-kachestva-inklyuzivnogo-obrazovaniya
7. Abdulpatakhov, M. A., Omarov, M. O., & Abdulova, P. M. (2021). Higher education in the process of inclusion of people with disabilities: Role and development prospects. *Regional Problems of Economic Transformation, 10*(132), 82–92. Accessed Mar 30, 2023, from https://cyberleninka.ru/article/n/vysshee-obrazovanie-v-protsesse-inklyuzii-lyudey-s-ovz-rol-i-perspektivy-razvitiya
8. Kostromina, T. A., & Fediukina, I. I. (2019). On the issue of inclusive education. *Analytical Review. Humanities Research Collection, 3*(18), 25–31. Accessed Mar 30, 2023, from https://cyberleninka.ru/article/n/k-voprosu-ob-inklyuzivnom-obrazovanii-analiticheskiy-obzor
9. Balashov, A. E., Krasnova, E. A., & Khristoforova, L. V. (2020). Legal barriers in the system of university inclusive education. *The Education and Science Journal, 1*(22), 59–83. https://doi.org/10.17853/1994-5639-2020-1-59-83
10. Evans, C., & Zhu, X. (2022). A checklist for making disability inclusion a reality in higher education. THE*: Times Higher Education*. Accessed Mar 29, 2023, from https://www.timeshighereducation.com/campus/checklist-making-disability-inclusion-reality-higher-education
11. Grabchuk, K. M., Oganesian, L. L., & Oganesian, T. L. (2018). Strategic directions for the development of inclusive education in universities of the Krasnodar territory. *Bulletin of the Academy of Knowledge, 5*(28), 96–103. Accessed Mar 30, 2023, from https://cyberleninka.ru/article/n/strategicheskie-napravleniya-razvitiya-inklyuzivnogo-obrazovaniya-v-vuzah-krasnodarskogo-kraya
12. Harpur, P., & Szücs, B. (2022). Where are the leaders with a disability in higher education? *THE: Times Higher Education*. Accessed Mar 29, 2023, from https://www.timeshighereducation.com/campus/where-are-leaders-disability-higher-education
13. Smith, P. S., & Myers, B. (2024). Instructor experiences providing accommodations and modifications for students with intellectual disability in inclusive higher education. *International Journal of Inclusive Education.*, 1–14. https://doi.org/10.1080/13603116.2024.2305653
14. Smith, P. S., & Myers, B. (2025). Narrating access and agency: Students with intellectual disability share their experiences with higher education. *Remedial and Special Education, 46*(1), 31–41. https://doi.org/10.1177/07419325231226109
15. UN. (2006). *UN Convention on the rights of disabled persons*. Accessed Mar 28, 2023, from https://www.un.org/ru/documents/decl_conv/conventions/disability.shtml
16. Sachs, D., & Schreuer, N. (2011). Inclusion of students with disabilities in higher education: Performance and participation in student's experiences. *Disability Studies Quarterly, 31*(2), 31. https://doi.org/10.18061/dsq.v31i2.1593

Accessibility of Higher Education for People with Different Income Levels as a Manifestation of the Inclusiveness of the University

Aktam U. Burkhanov, Dilobar M. Mavlyanova, Victoria N. Ostrovskaya, and Svetlana E. Karpushova

Abstract This study dwells on the state of the accessibility of higher education in countries of Central Asia and Russia depending on the level of income and social status of citizens. The focus is on the problem of the necessity of forming inclusive higher education, which would be accessible without limitations by financial and material capabilities. We used ideological provisions and concepts on the equality of all people and the necessity of observing their rights in the sphere of higher education. These provisions are based on a comprehensive combination of theoretical and practical aspects, which are peculiar to the sphere of philosophy, sociology, pedagogics, economics, state regulation, etc. This study contains characterisation of equality and non-discrimination of people by different features, including gender and nationality, as well as physical capabilities and language. The main emphasis is placed on the necessity of ensuring equal rights for obtaining higher education with a low level of income. The role of inclusiveness according to this feature is described, which contains the potential for multiplication of the positive impact of people with higher education in this environment. Analysis of obtaining higher education by groups of people with different financial capabilities in countries of Central Asia and Russia was performed. The results obtained demonstrated the direct connection between the level of education and income, which points to the necessity of supporting less well-off groups of the population in their desire to obtain higher education. However, this vision needs additional research.

A. U. Burkhanov (✉)
Alfraganus University, Tashkent, Uzbekistan

D. M. Mavlyanova
Tashkent State University of Economics, Tashkent, Uzbekistan

V. N. Ostrovskaya
Center for Marketing Initiatives LLC, Stavropol, Russia

S. E. Karpushova
Sebryakovskiy Branch of Volgograd State Technical University, Mikhailovka, Russia

B. S. Sergi et al. (eds.), *University 4.0 and Educational Technology for Sustainable Development*, Education in the Asia-Pacific Region: Issues, Concerns and Prospects 75, https://doi.org/10.1007/978-981-96-7376-6_12

Keywords Accessibility · Higher education · University · Income · Inclusion · Inclusiveness

JEL Codes H52 · I22 · I23 · I24 · I26 · I28 · J24 · O15

1 Introduction

The inclusiveness of universities is a relevant concept that reflects the current paradigm of social interaction and conforms to the modern liberal system of values. According to this, the notion of equality and ensuring the rights of humans and involvement of all representatives of society, regardless of their specific features, in education in the system of higher school are the key categories of the current scientific discourse on the issues of equality and justice.

In the conditions of the development of liberal values, the notion of equality gradually acquires a larger role and dissemination. Its essence is realised through dialectical foundations of social life and is manifested in the form of equal access, equal opportunities, equal conditions, etc. Continuing the movement for the equality of people, it should be realised that many of its aspects have been already implemented in the regulatory and institutional fields. In most countries of the world, women and men have equal constitutional and economic rights, people with disabilities gradually receive access to a larger number of inclusive spaces, and religious beliefs and ethnic origin less and less are reasons for conflicts and discrimination.

However, among the cases of discrimination from the position of access to education for sustainable development [1], apart from the factor of disability, gender, ethnic or language identification, and belonging to the category of refugees or migrants, the level of material support is also taken into account [2]. Low-income families are considered a factor of violation of equality and accessibility of higher education. That is why the policy of global institutes, the public sector, and most countries of the world is aimed at ensuring the conditions for access to higher education for a wider circle of citizens, including talented people from families with low levels of income.

2 Interdisciplinaryty of the Problem of Inclusiveness of Higher Education from the Position of People with Different Levels of Income

The methodology of this research is formed by institutional provisions, provided and generalised in the Declaration of Human Rights, as well as numerous provisions and acts aimed at the formation of an equal and fair society. Hence global institutionalism, which is intertwined with social sciences, psychology, pedagogics, philosophy, public administration, and economics forms an interdisciplinary basis for the study of the problem of inclusiveness of higher education from the position of people with different levels of income. Within this methodology, the main concepts are the concept of sustainable development, social justice, and the "third mission" of the university. The first concept is manifested through the designated list of goals, one of which directly concerns the support for equal rights for education by all categories of people [3, 4]. The second concept is based on worldviews on the problem of human rights and ensuring their equality [5, 6]. The third concept is aimed at reconsideration of the mission of universities—not only as a source of knowledge but also as a tool for the achievement of social success and a centre of public development and evolution [7, 8]. The research methodology is based on the system approach, which elaborates on the problem posed in its interconnection with public processes and phenomena and is based on the methods of observation, comparison, generalization, graphical method, grouping, and analysis of time rows.

The problem of the research of the inclusiveness of higher education from the position of ensuring equal rights and opportunities for people with different levels of income, together with a wider study of inclusiveness and the general problem of equality in society, forms a separate direction of scientific discourse. The key issued that are studied within the problem posed include the achievement of the mission of higher education by observing the principles of sustainable development [9]; formation of an intellectual society based on an inclusive economy and education [10, 11]; influence of social origin of students on their successfulness and integration in the educational environment and social processes [12]; retrospective analysis of adhering to the equal conditions in the sphere of higher education in the conditions of socialist and capitalistic state systems [13, 14]; ensuring justice in higher education [15]; direct support for inclusive education in higher educational establishments [16]; reconsideration of the role of universities in social development [17].

The goal of this article was to find the main factors that influence the inclusiveness of universities from the position of ensuring equal conditions for access to higher education for people with different levels of income. The tasks to be solved were as follows: conceptual substantiation of the notion of inclusiveness of universities, research on the factor of material inequality os one of the factors of discrimination, and analysis of the state of accessibility of higher education in Russia and countries of Central Asia given the level of income.

3 The "Third Mission" as the Principle of Inclusiveness in Universities its Effects for People with Different Income Levels until 2030

A current trend in higher education is the realisation of the values and mission of universities. According to this, together with the categories of corporate social responsibility and sustainable development, the term "third mission" is considered more often. It involves the redirection of a part of universities' resources to the resolution of socially important problems, as well as the formation and popularisation of trends aimed at the protection of ecology, prevention of climate change, and formation of equal and inclusive society [18]. In this context, higher education is considered as a factor or space that can create the most favourable conditions for people with different features for the complete realisation of their rights, liberties, and talents. Within such space, students receive an opportunity to get rid of aggravating mental barriers and thus free the entire society of stereotypes, narrow views, and non-productive theories [9].

The notion of inclusion, from the position of ensuring equal access to higher education, reflects current public demands and is implemented at all levels, reflecting social injustice and striving towards the development of talents. At that, a wide range of measures is used, which are aimed at the reduction and liquidation of obstacles on the path of involving all social groups in the educational process in universities [16].

Such a vision conforms to the sustainable development concept and its goals. Thus, according to SDG 4, to ensure complete development of potential by all children they must have equal conditions for obtaining quality education at all levels. Sub-goals, which specify the achievement of this effect, include four effects (Fig. 1), which are implemented due to compliance with the principle of inclusiveness in universities and involve the achievement of the following effects until 2030:

1. Equal access to higher education for men and women;
2. Liquidation of gender disproportion in education and support for equal access to higher education for all vulnerable social groups;
3. Increase in scholarship support for developing countries, to create conditions for obtaining higher education. Scholarship support involves also the stimulation of

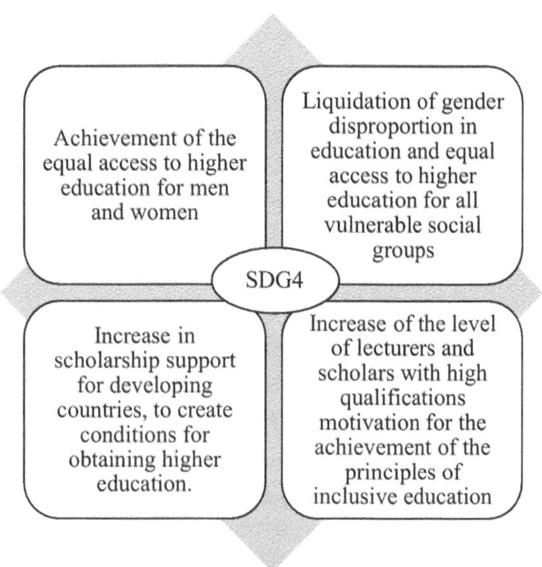

Fig. 1 The positive effects that can be achieved in higher education until 2030 due to compliance with the principle of inclusiveness in universities. *Source* Created by the authors

this category of people for active scientific & technical and inventing activities, which allows expecting a multiplication effect, scaled at the level of the countries of the origin of such scholars;
4. Increase in the level of support in the sphere of higher education by lecturers and scholars with high qualifications and motivation for the achievement of the principles of inclusive education [1].

The notion of inclusiveness is based on the concept of equality and non-discrimination. According to it, any person, regardless of gender, ethnicity, religion, or financial conditions, must have equal rights and equal access to a full social life. This includes the chance for obtaining an education at all levels of study, a right for successful employment, barrier-free social environment, etc. The main criteria for creating such an environment include elimination of discrimination and the expansion of the ideas of inclusion. The term "discrimination" has a very negative meaning nowadays. In any sphere, it is treated as a limitation, a violation of rights, etc. In the conditions of domination of liberal values, discrimination by any criterion is a trigger that initiates the process of ensuring equality and justice. This problem is particularly urgent in sectors that have strategic value for the development of society: education, science, culture, etc.

Higher education is a complex strategic sphere that defines the prospects for society's development. It develops according to different models, based on the concepts of economic development of a certain country, combining, in different proportions, private and government-funded, higher educational establishments. In any case, higher education is an expensive service which requires long training by personnel that tool long and expensive training as well. Characterising higher education in this way, we often face the factor of cost. Depending on the models of the educational environment, payment for education in universities is performed by students, their parents, the government, or other interested parties. In this context, the notion of motivation and ability to obtain higher education by citizens with different levels of income and social origin becomes very relevant.

From the position of social origin, a different level of citizens' incomes creates inequality of opportunities in education. This inequality forms two main effects:

1. Dependence of accessibility, motivation, and success in the sphere of higher education given the status, social origin, and cultural and educational environment surrounding the person. These factors influence the level of preparedness and adaptability of the person from an intelligent environment to university studies;
2. Higher level of financial affordability of education for people with good social background (origin, belonging to an intelligent environment, availability of accommodations, etc.).

Besides, an important factor that facilitates better accessibility of higher education is parents' experience. In this context, children of parents who have university or college diplomas try to reach or exceed their success and thus are better motivated to obtain higher education. In many cases, the experience and connections of the parents influence the choice of profession and future employment, which raises the level of predictability of further development, contrary to uncertainty which is peculiar to people from families without higher education.

Thus, apart from the level of income, an important factor that influences inequality in the university environment is the level of education of parents, the type of their jobs, and financial capabilities, which, in their totality, form the socioeconomic position of the higher education applicant [12].

Accessibility of higher education for people with different levels of income in Russia and countries of Central Asia is rather differentiated. People with a higher level of income have better access to education (Fig. 2). However, this connection is not very high. In the studied countries, the coefficient of paired correlation (r) between the level of income and the share of people with this level of income with university diploma is within the tange of r = 0.732–0.995. The lowest value of this indicator is observed in Kyrgyzstan, where the share of the poorest people with higher education is higher than the similar indicator for groups of people with medium or low income. The connection in Russia is not very high (r = 0.850). The correlation indicator that is close to one is observed in Tajikistan (0.947) and Kazakhstan (0.995).

The incline of the graphs in Fig. 1 is a sign of a gap between categories of people with different levels of income from the position of their access to higher education. The lowest polarization is seen in Russia, where the share of people who obtained higher education is within the range of 26%–43%. The gap is 1.7 times. The gap

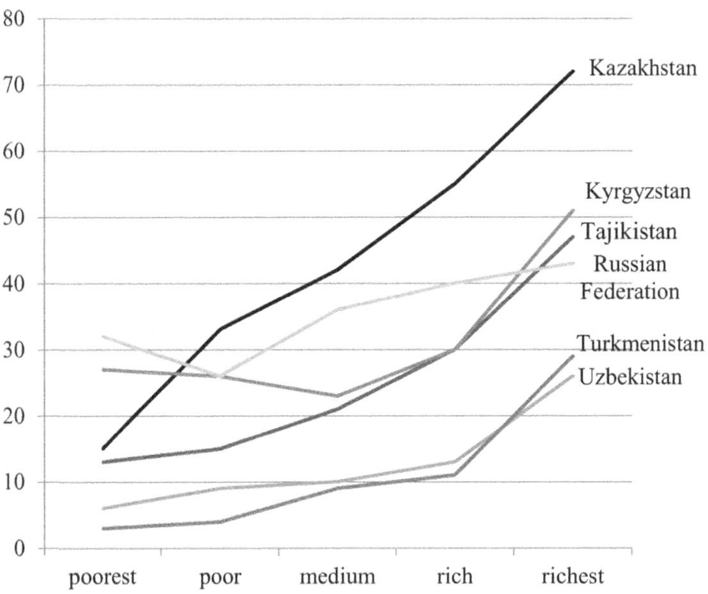

Fig. 2 Share of people aged 30–34 by the level of income, who have at least 4 years of higher education. *Source* Created by the author based on [2, 19]

between the poorest and the richest who obtained higher education is also small in Kyrgyzstan (two times). The largest polarization is peculiar to Turkmenistan, where the share of the richest with a university diploma is almost ten times higher than the share of the poorest. Other countries of the region have an average indicator of difference of 3.6–4.8 times between the maximum and minimum values.

In the context of the analysis of the level of accessibility of higher education in countries of the region, three groups could be distinguished. The first group includes Kazakhstan, which has the highest level of education in the categories from the poorest to the richest, but which also has the largest gap. Another group is Kyrgyzstan, Tajikistan, and the Russian Federation, which have medium indicators by the share of people with education in the sector of the richest. Tajikistan has a large gap between the accessibility of higher education for the poorest and the richest and direct dependence on the level of income and share of people with a university diploma. Russia and Kyrgyzstan have the lowest values of such connection and the smallest gap. The third group is comprised of Turkmenistan and Uzbekistan, with the lowest level of higher education in all groups differentiated by the levels of income. In these countries, the share of people with higher education is below 30%.

From the position of the research results, the reasons for unequal access to higher education for the citizens of the Russian Federation should be considered. These include the reorientation of a large number of applicants for professional education [14], as well as the desire of Russian citizens to receive free higher education [11], which levels the material differentiation in case of availability of the sufficient number of places at universities.

4 Prospects of Increase of the Inclusiveness of the Universities for People with Different Income Levels in Russia and Countries of Central Asia

To reveal deep reasons for the differences in access of different categories of people by the level of income to higher education in Russia and countries of Central Asia it is necessary to conduct further research. The main attention should be paid to the development of a universal methodology, which would allow confirming certain assumptions based on the results of mathematical or sociological modelling. Revealing the cause-and-effect character of the studied phenomena and processes will allow for a more precise interpretation of the conclusions drawn, characterising the impact of the growth or reduction of the level of involvement of different categories of people in higher education on the general state of intellectual potential, and forecasting further development of the problem.

While discrimination based on ethnicity, religion, and gender belongs to objective factors, caused by reasons that do not depend on the higher education applicant, the factors of the financial position often depend on the applicant or their parents. Therefore, the problem of revealing such dependencies and their correct assessment from the position of inclusiveness of higher education will be additional arguments in the determination of the level of motivation of different groups of people and the study of the possibility of financing higher education from non-government resources.

Determination of the causes of the absence of direct dependence between the number of people who received a university diploma in different groups by the level of income, which is peculiar to Russia, allowed for the correct interpretation of the results obtained. According to this, it is necessary to research these causes in other countries of the sample, with a special emphasis on the possibility of creating conditions for receiving education by the group of people with low financial capabilities.

As with any other empirical study, this paper has its limitations, which can be eliminated in the process of further scientific work. Thus, focus should be made on the development of a wider range of indicators that would reflect not only the financial capabilities of different groups of people in obtaining higher education but would also take into account social origin, the level of motivation and interest, etc.

5 Conclusions

Ensuring equal rights and accessibility to higher education is an important direction for global and national policies in most countries of the world. The popularity and dissemination of this idea formed based on the worldview principles of equality of people, regardless of race, gender, religion, ethnic origin, or financial position. According to this, support for obtaining higher education by people with a low level of income allows involving a wide range of people with diverse talents in the sphere of better opportunities. The effect of obtaining higher education by such people is considered from the position of dissemination of values and motivation in the corresponding environment and stimulation of the multiplication effect on the popularisation of higher education and the corresponding way of life.

The performed analysis of the dependence of the number of people with higher education on the financial position in countries of Central Asia and Russia allowed drawing several conclusions. A range of countries (Kazakhstan and Tajikistan) are peculiar for the direct dependence of the number of people with higher education on their financial position. In Russia and Kyrgyzstan, such dependence is the lowest. In Turkmenistan and Uzbekistan, there is a problem of a low level of population education in all groups by the financial conditions. The reasons for the presence of a weak connection between the studied factors in Russia include the population's striving towards obtaining free education and the reorientation of a large number of applicants to professional education with the refusal from university studies. According to this, a deeper study of the causes of the presence of a strong or weak connection between the financial standing of the population and the level of education from the position of the inclusiveness of higher education is required.

References

1. UN. (2023). SDGS 4: Ensure inclusive and equitable quality education and promote life-long learning opportunities for all. Department of Economic and Social Affairs. *Sustainable Development*. Accessed Nov 2, 2023, from https://sdgs.un.org/goals/goal4#undefined
2. UNESCO. (2021a). Global education monitoring report, 2021, central and Eastern Europe, Caucasus and Central Asia: Inclusion and education: All means all. *UNESCO: UNESDOC Digital library*. Accessed Nov 2, 2023, from https://unesdoc.unesco.org/ark:/48223/pf0000375490
3. Hu, X., Wan, G., & Zuo, C. (2025). Education development and income inequality: Evidence from China. *Journal of Economic Inequality*. https://doi.org/10.1007/s10888-024-09656-3
4. Sadaf, F., Bano, S., & Rahat, R. (2024). First-generation female professors from low-income families in Pakistan: The influence of parents on access to and involvement in higher education. *Sociological Inquiry, 94*(4), 1025–1040. https://doi.org/10.1111/soin.12601
5. Bernal, G. L., Abadía, L. K., Álvarez-Arango, L. E., & De Witte, K. (2024). Financial aid uncertainty and low-income students' higher education preferences. *Higher Education, 87*(6), 1845–1863. https://doi.org/10.1007/s10734-023-01094-w
6. Senkevics, A. S., Barbosa, R. J., Carvalhaes, F., & Ribeiro, C. A. C. (2024). Decomposing heterogeneity in inequality of educational opportunities: Family income and academic performance in Brazilian higher education. *Sociological Science, 11*, 854–885. https://doi.org/10.15195/V11.A31
7. Bolhaar, J., Kuijpers, S., Webbink, D., & Zumbuehl, M. (2024). Does replacing grants by income-contingent loans harm enrolment? New evidence from a reform in Dutch higher education. *Economics of Education Review, 101*, 102546. https://doi.org/10.1016/j.econedurev.2024.102546
8. Zeidan, R., de Almeida, S. L., Inácio, B. Ó., & Lewis, N. (2024). Racial and income-based affirmative action in higher education admissions: Lessons from the Brazilian experience. *Journal of Economic Surveys, 38*(3), 956–972. https://doi.org/10.1111/joes.12564
9. Chankseliani, M., & McCowan, T. (2021). Higher education and the sustainable development goals. *Higher Education, 81*, 1–8. https://doi.org/10.1007/s10734-020-00652-w
10. Dubovik, M. V. (2019). Model of smart, sustainable, inclusive economic growth in the context of inequality in Russia. *International Journal of Civil Engineering and Technology (IJCIET), 10*(2), 1704–1711. Accessed Nov 3, 2023, from http://www.iaeme.com/ijciet/issues.asp?JType=IJCIET&VType=10&IType=2
11. Shaikhutdinov, N. R. (2021). Accessibility of higher education as the basis for sustainable development of the state at the present stage. Modern political science about the trajectories of development of the state, business and civil society. In *The world in the post-Covid era: From disunity to unity: Collection of articles of the II International Scientific and Practical Conference. (pp. 159–162)*. Belarusian State Economic University. Accessed Nov 3, 2023, from http://edoc.bseu.by:8080/bitstream/edoc/91871/1/Shaykhutdinov_N.R._s_159_162.pdf
12. Esina, G. K. (2020). Social representations of higher education of students from families with different socio-economic status. *Sotsial'naya psikhologiya i obshchestvo [Social Psychology and Society], 11*(3), 99–113. https://doi.org/10.17759/sps.2020110307
13. Ibragimova, Z. F., & Frants, M. V. (2021). Inequality of educational opportunities in the soviet and post-soviet periods: An empirical analysis. *Educational Studies Moscow, 2*, 43–62. https://doi.org/10.17323/1814-9545-2021-2-43-62
14. Malinovskii, S. S., & Shibanova, E. I. (2022). Reasons and factors for completing the expansion of higher education in Russia. *Sotsiologicheskii zhurnal [Sociological Journal], 3*, 8–37. https://doi.org/10.19181/socjour.2022.28.3.9149
15. Kargapolova, E. V., & Davydova, I. A. (2020). Equity in higher education: Research discourses. *Sotsiologicheskii almanakh [Sociological Almanac], 11*, 36–46. Accessed Nov 5, 2023, from https://cyberleninka.ru/article/n/spravedlivost-v-sisteme-vysshego-obrazovaniya-issledovatelskie-diskursy

16. Salmi, J. (2020). *Higher education and inclusion. UNESCO: UNESDOC Digital library.* Accessed Nov 2, 2023, from https://unesdoc.unesco.org/ark:/48223/pf0000373689
17. Smetanina, M. D. (2021). University as a cultural center. *Obshhestvo: filosofija, istorija, kul'tura* [Society: Philosophy, History, Culture], 1(81), 93–95. Accessed Nov 4, 2023, from https://cyberleninka.ru/article/n/universitet-kak-tsentr-kultury
18. Voevodina, E. V. (2022). Analysis of the "third mission of universities" in the aspect of educational risks of digital divide. *Digital Sociology, 5*(1), 54–63. https://doi.org/10.26425/2658-347X-2022-5-1-54-63
19. UNESCO. (2021b). *Tertiary completion rate.* Indicators. Accessed Nov 1, 2023, from https://www.education-inequalities.org/indicators

Influence of Science on the Development of the Moral Consciousness of Students

Gulayim A. Dzhumabaeva and Gulnur M. Amanova

Abstract The research examines the concept of consciousness from a materialistic and idealistic position and clarifies the essence of the concept of moral consciousness. The authors give a brief overview of the problem of forming moral consciousness of schoolchildren. The authors describe the influence of the humanities and natural sciences on the development of students' consciousness, the role of pedagogical science in developing consciousness, the features of the development of consciousness, the formation of a sequence of moral behavior, and the characteristics of morality. The authors also describe the problems of pedagogy and school practice in forming moral consciousness. To develop the collective consciousness, it is necessary to develop and change personal consciousness first. Education should not be limited to the school; the whole world should become a platform for educating students. Children must go beyond the school and visit museums, fields, workshops, gardens, libraries, etc. Children need to live, not sit idle. Educational lessons should be practice-oriented and lively. Real events taken from life should be given during lessons, around which all knowledge is gathered.

Keywords Science · Moral consciousness · Development features · Education · Knowledge · Problems · Students

JEL Codes I2 · I21 · I31

1 Introduction

The problems of the moral consciousness of students and the formation of pedagogical conditions that contribute to their development have always been relevant in pedagogical science and practice [1].

G. A. Dzhumabaeva · G. M. Amanova (✉)
Kyrgyz Academy of Education, Bishkek, Kyrgyzstan

B. S. Sergi et al. (eds.), *University 4.0 and Educational Technology for Sustainable Development*, Education in the Asia-Pacific Region: Issues, Concerns and Prospects 75, https://doi.org/10.1007/978-981-96-7376-6_13

The development of democracy was accompanied by increased freedom of personal manifestations, which, in turn, negatively affected the moral consciousness of adolescents [2].

The development of the student's moral consciousness depends on the level of the teacher's consciousness and pedagogical skills [3]. The morals and actions of schoolchildren are associated with the thinking, views, and development of teachers' consciousness [4, 5]. A timely solution to the problem of the development of moral consciousness today contributes to tomorrow's development of society [6].

The research aims to identify the influence of science on the development of the moral consciousness of students, identify the omissions and errors of previous educational work, and give recommendations for their elimination.

The theoretical foundations of this research are the achievements of contemporary Kyrgyz and foreign science and practice, the educational concepts of the Kyrgyz Republic and neighboring countries, the cultural heritage of the peoples, and "The concept of educating schoolchildren and youth of the Kyrgyz Republic for 2021–2030."

2 Influence of Science and Education on the Moral Development of Students

Consciousness is one of the problems that most scientists have thought about. In different historical periods, different statements were made about the nature of consciousness, depending on the possession of scientific knowledge. Gradually, it developed as a philosophical category that characterizes consciousness: materialistic, idealistic, objective, subjective, cognitive, emotional, rational, social, biological, scientific, vital, personal, public, atheistic, religious consciousness [7].

"Consciousness is the reflection of being, truth in the thoughts of a person" [7].

If a child is not brought up and separated from the outside world, his or her consciousness cannot develop. Under the influence of the outside world, the organs of the receptor analyzer (ears, eyes, nose, tongue, and skin) receive information and deliver it to the brain. The image of the reflected object (image, thought, sensation, etc.) remains in the brain. The reflection of truth in the brain is called consciousness.

Materialistic philosophy recognizes consciousness as derived from the material external world. People do not know each other's thoughts; everyone knows what is on their own mind. They communicate their thoughts to each other. This is a materialistic view.

"Consciousness and the spirit of man live separately," say idealist thinkers. "Awareness of current life problems, one's own situation and what others know is called consciousness" [8].

Moral consciousness is the student's ability to distinguish (feel) good from bad, good from evil, selfishness from altruism, spirituality from lack of spirituality, compassion from cruelty, love from hatred, etc.

The problem of morality is one of the oldest in philosophy and psychology. Even pre-revolutionary democrats in Russia expressed the need to prepare moral consciousness for the individual's moral development. Almost every school and conception of philosophy, psychology, and even physiology dealt with questions of the nature and content of moral consciousness. Morality is defined as the highest human characteristic, one of the aspects of human spirituality, and a multidimensional essential property of a person.

The level of moral education of students is considered as the level of moral consciousness. This is the breadth of outlook, the depth of moral and ethical knowledge and beliefs, knowledge of moral norms and rules, etc. The ethical-psychological-pedagogical sciences study the development of moral consciousness. At the ethical level, the relationship between people is studied (A. Huseynov) [9], at the pedagogical level, it determines the content and principles of the moral education of the younger generation (G. I. Shchukina) [10]. Research by S. L. Rubinstein, D. I. Feldstein, and V. M. Kharlamov were devoted to studying the motivation of moral behavior [11].

Moral consciousness includes all other characteristics of morality: beliefs, feelings, orientation, activity, installation properties, needs, motives, abilities, volitional properties, value orientations, character properties, and others. Moral actions are manifestations of moral consciousness.

Moral education is always associated with the fight against negligence in work, rudeness, malevolence, etc.

Science is a sector of human activity, the function of which is the development and theoretical systematization of objective knowledge about reality [12, p. 866]. Science is trying to find new knowledge. The conclusions of science can be a form of development of social consciousness. Science is associated with social consciousness and other forms (art, morality) and fundamentally differs from them. The purpose of science is the knowledge of the laws and relations between nature, society, and thinking. During historical development, science becomes a productive force. Consciousness, thought, spirit, and representation form the basis of the world and reality and determine its origin. This is the classic direction.

The upbringing of moral consciousness is a very complex multifaceted process. The success of the educational process depends on each teacher, the cohesion of the team, and the creative contribution of each participant of this team. It is impossible to talk about the development of moral consciousness without solidarity in the teaching staff. Children feel the relationship of teachers and disharmony in the relationship of teachers causes disgust in the child [13, p. 33].

The development of the spiritual life of a developed society is unthinkable without the constant influence of science on all forms of social consciousness and practical human activity. The direct influence of science on the moral development of children is carried out through education. Education is a tool for assimilating knowledge and helps a person form a scientific worldview and the ability and skills of labor and social activities.

The role in developing the moral consciousness of students of those sciences that have direct access to moral education (humanities, ethics, pedagogy, and psychology) is great. Science creates momentum by improving the human intellect, consciousness, thoughts, and feelings, as well as by creating a harmonious relationship with the environment. There is no high moral consciousness without knowledge.

In contemporary conditions of the development of science and technology, the role of natural and technical sciences is growing: without specifically considering the problems of moral education, they contribute to their solution by forming a scientific view of the world. Expanding a person's knowledge of the surrounding world, they deepen the scientific understanding of the world and the person. The natural and social sciences develop clarity and rigor of thinking, adherence to principles, and instill moral responsibility for life on earth in a schoolchild.

Developing under the influence of natural and social science, they constitute the ideological basis of morality, moral assessments of the phenomena of reality, and human behavior. Even literary works have an educational impact. The purpose of teaching humanities to students is to explain the history of human development, develop students' thinking, teach them to think, form the right worldview, raise awareness, teach humanism, etc. The humanities have a great influence in educating an honest citizen who loves the Motherland.

Science provides guidelines for true values and understanding of ethical rules. From scientific books, students receive information that motivates thinking, satisfies curiosity, teaches self-esteem, and develops self-knowledge skills. Scientific books can assist in choosing a profession. The information received at school age determines the way of life, helps the formation of moral and spiritual qualities, and increases consciousness (M. D. Smorodinskaya) [14].

Science is the most powerful tool in the development of the mind. Knowledge is a source of enrichment of the spiritual world. Scientific books are a collection of knowledge and wise thoughts. Scientific books are food for the mind, a powerful tool in shaping a teenager as a person, like a compass necessary for determining moral guidelines. Every word and sentence in the book contain a thought, which is a product of the mind. The book is a good tool for developing the moral consciousness of children. Reading scientific books, students improve themselves, learning and improving.

Studying the laws of the historical development of humankind, students strive to learn the moral norms and spiritual heritage created by humankind. This enriches the hearts and minds of children. The spiritual world of humankind, peoples, individuals, their aspirations, goals, dangers, and mistakes are multifaceted fields of knowledge. Reading impressive historical events, the child worries and thinks. It is desirable that the student understands the importance of establishing humanity in relations between people, not being indifferent to scientific knowledge about the development of society, as a person devoted to the prosperity of people and the planet, wanting solidarity, peace, and friendship between peoples, and avoiding injustice and ignorance.

Studying the biography of heroes, thoughts and actions of sages, searches and discoveries of scientific geniuses and politicians, children awaken a feeling of love for their people, fatherland, gratitude, and, maybe, even shame or pride in the past.

Fascism, racial discrimination, chauvinism, terrorism, genocide, apartheid, repression, war, and other destructive forces create fear in students.

A moral conviction is formed based on scientific concepts and judgments. Moral conviction is not just knowledge about moral norms—their exact understanding is what is experienced, felt, and proved, which is a guide to action.

Moral consciousness develops and changes along with the individual's spiritual development. If one does not cultivate moral feelings, they can degrade. Bad manners of feelings lead to cynicism, lack of a sense of shame, the manifestation of inadequate reactions, when, for example, a person shows contempt at the sight of grief, laughter at the sight of concern, etc.

A different orientation also characterizes the different polar levels of moral education of individual students and the entire classroom groups. In this, a significant role is played by the moral ideal that children want to imitate. Schoolchildren find a moral ideal when studying humanitarian subjects, in films and serials, or from life.

A special place in the moral education system is occupied by moral habits committed in any condition. Without moral habits, there is a visible gap between attitude and behavior, between word and deed.

The question of the criteria for moral upbringing is one of the most difficult. In the theory of moral education, there are different points of view on this matter. The measure of moral education is the social activity of the student, the attitude of the student to the requirements of teachers, parents, and educators, actions based on rational assessments, and the presence of formed habits, moral personality traits, and the moral orientation of the student. People with high moral consciousness do not have fear and aggressive feelings. They are not saddened by what happened in their lives, not humiliated by poverty, and are characterized by peace of mind and spiritual activity [10, pp. 93–112].

3 Problems of Development of the Moral Consciousness of Students in Science and Education and their Solving

In pedagogy, there are already many studies that reveal an understanding of moral norms. The desire to give as much knowledge as possible does not automatically make students kind, honest, and sincere (i.e., morally educated) (Fig. 1).

Humanity's incomplete knowledge of moral rules has led to an ecological crisis. The ecological crisis is a global problem of humankind. The state of nature begins to threaten the existence of humanity. People are already beginning to pay for violating the laws of the biosphere with their health. The only salvation is to hear the voice of nature, obey its laws, and abandon the consumer attitude towards the Earth

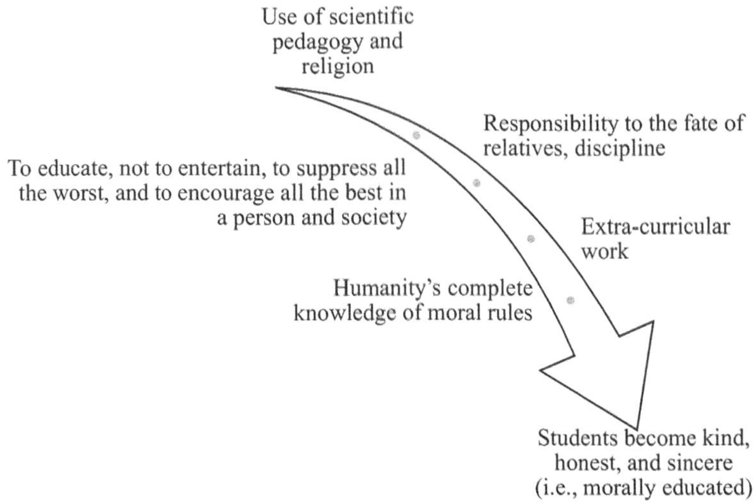

Fig. 1 Development of the moral consciousness of students in science and education. *Source* Created by the authors

and its riches. People have no right to violate the relationship in nature. The salvation of humankind lies in ecological culture [13, pp. 71–75].

If the moral consciousness of people is not awakened, it will continue to create chaos in nature. If humanity continues to use natural resources ruthlessly, these resources will run out sooner or later.

Science developed in the twentieth century. Family divorces in the USSR have been increasing every year since 1970 (in 1989 they were 35.5%). However, this problem has not been given due attention. The number of children born out of wedlock increased (more than 1500 illegitimate children were born daily in the CIS). In 1991, the All-Russian Institute of Public Opinion found the following:

- 31% of Russians stated a sharp decline in morals;
- 35% believed that the problems had been accumulating but not noticed in recent years;
- 45% noticed an increase in violence and cruelty;
- 40% indicated indifference to the fate of others;
- 32% noticed greed;
- 26% believed there was a drop in labor discipline;
- 26% noticed an increase in criminal violations and promiscuity of young people;
- 24% said about indifference to the fate of the Motherland [15, pp. 198–199].

Special fears and moral condemnation of the world community are caused by the experiments of scientists who are searching for the creation of new types of mass destruction weapons. Only those scientific studies that are not directed against a person and humanity are morally justified.

In vitro cultivation of human embryos, genetic engineering methods to improve the human race, experiments to modify people's behavior, and much more raises the question of the ethical permissibility of such research.

Scientists from different countries are guided by different criteria when creating a school curriculum. For example, the demands of society are considered when creating a school curriculum in the USA. We consider the requirements of the global educational space. Therefore, the program is too complicated; children get tired of mental work. Because of the overload that the new century brings forward, children suffer from stress, depression, nervousness, etc. Even before leaving the school bench, nervous children go to aggression and violence. There are already cases of children being killed by children. The position of students is similar.

Such types of people as human-robot, human-machine, and human without value appeared in the pedagogical literature. Such children are present in almost all schools. "These teenagers are ready to destroy without thinking or understanding what they are doing, if they are ordered to destroy, break, or knock down." "It is necessary that he be a "necessary person" for society, thick-skinned, ready for active participation in underground work." Such characters have become the main characters on television. They travel all over the world [13, p. 36].

In the middle of the twentieth century, science gave humanity a new means of communication – the Internet. One can talk about the benefits of the Internet and computer technology for a long time. Programmers provided invaluable assistance to teachers. However, the Internet has brought a new type of person. A student who is addicted to the Internet. Although these students participate in life, they are indifferent to it, feeling unique. There are computer games with high aggression, in which a terrible creature destroys cities, tramples people, and brings total destruction. They affect the psyche of children. "We are now digging a well with water that is not suitable for drinking, but capable of killing people" [9].

Where were the mistakes? "In the history of pedagogy lie the roots of cardinal and contemporary problems, as well as many keys to their solution" [16, p. 30].

The theory of nurturing education, developed by the German teacher I. F. Eerbert (1776–1841), formed graduates of losers, groundless, short-sighted in body and spirit, overworked, and incapable of energetic action. This theory was substantiated in the Soviet Union (and Kyrgyzstan) [17, pp. 50–51]. Children studied an average of 30 h a week, of which only one hour was devoted to an educational lesson.

Extra-curricular work existed formally. It was not paid. "Bringing up one generation after another on sedentary education, we launched the mechanism of the psychobiological degradation of peoples, we deprived the cornerstone on the basis of which the spiritual, moral, and rational life of people is built and maintained. The school began to form an instructively-programmed zombie intellect" [18, p. 60].

Soviet pedagogy tried to distance itself from scientific pedagogy, considering it bourgeois and unacceptable. Many of that educational system stemmed from ignoring scientific pedagogy and religion.

The school was not closely connected with life. The educational process was subordinated to the upbringing of obedient and non-initiative children. As a result, they prepared graduates who were removed from life, unprepared for anything, infantile, and lost in the maelstrom of life. Hence, the indifference, indifference to the fate of relatives and others, as well as the fall of discipline. Thus, it deprived the personality and the very ability to develop moral consciousness, which objectively led to the cutting of the cultural roots of the nation and nationalities, the intellectual potential of society; we all obeyed pedagogy and instructions descended from above. Another important reason why we have reached this state is related to the methodology on which our psychological and pedagogical science is based.

Many authors criticize the new social paradigm: "to entertain, not to educate, not to suppress all the worst, and not to encourage all the best in a person and society, but to indulge the basest human instincts and needs" [11, p. 14].

4 Conclusion

The development of collective consciousness will be the main impulse in science. To develop the collective consciousness, it is necessary to develop and change personal consciousness. The methods of educating others are not sufficient.

It is vital to involve parents and the public in the educational process. In many countries, parents are not only teachers' assistants. Parents even organize extracurricular activities with their children and participate in various projects, charity events, and community work days [19].

Education should not be limited to the school; the whole world should become a platform for the education of students. Children must go beyond the school, be it a museum, a field, a workshop, a garden, a library, or other places. Children need to live, not sit idle. Educational lessons should be practice-oriented and lively (i.e., real events taken from life should be given there). In educational lessons, students should live an exciting life.

References

1. Aldridge, J. M., & McLure, F. (2024). Investigating the influence of the school climate of church-based schools on students' moral identity development and hope for the future. *Learning Environments Research, 27*(3), 971–993. https://doi.org/10.1007/s10984-024-09515-y
2. Bart, V. K. E., Sharavdorj, E., Boldbaatar, E., Munkhnasan, D., & Rieger, M. (2024). What shall I do? Similarities and differences in moral judgements between Austrian and Mongolian students. *Journal of Cultural Cognitive Science, 8*(1), 13–30. https://doi.org/10.1007/s41809-024-00141-x
3. De Asildo, N. Z. S., Yasin, M., Mokhtar, M. M., & Arshad, M. M. (2024). Former students' perspectives on moral education learning. *Journal of Institutional Research South East Asia, 22*(1), 1.
4. Hanim, W., & Marjo, H. K. (2025). Students' moral apprehension: Implementation of character education in the university curriculum. *International Journal of Learning in Higher Education, 32*(1), 21–37. https://doi.org/10.18848/2327-7955/CGP/v32i01/21-37

5. Huang, A., & Zhang, R. (2024). An empirical analysis of the effectiveness of hands-on inquiry-based teaching on the moral development of college students. *Journal of Educational and Social Research, 14*(5), 68–82. https://doi.org/10.36941/jesr-2024-0122
6. Reig-Aleixandre, N., Obispo-Díaz, B., de Dios-Alija, T., & de la Calle-Maldonado, C. (2024). Moral values as mediators in the relationship between prosocial behaviour and social responsibility among university students. *International Journal of Educational Psychology, 13*(3), 167–183. https://doi.org/10.17583/ijep.11704
7. Bobrov, V. (2000). *Introduction to philosophy: Textbook*. INFRA-M.
8. Stone, R. B. (2000). How to contact the good fairy waiting to grant your unlimited wishes. In *Celestial 911: Call with your right brain for answers!* (pp. 14–19). (Original work published 1997).
9. Huseynov, A. (1988). *The golden rule of morality* (3rd ed.). Molodaya gvardiya.
10. Shchukina, G. I., Kazakina, M. G., Akhaln, T. K., Semyonov, V. A., & Vinogradov, Y. A. (1977). Labor education. In G. I. Shchukina (Ed.), *Pedagogy of the school: Textbook* (pp. 93–112). Prosveshcheniye.
11. Melnikova, N. V. (2009). *The development of the moral sphere of a preschooler's personality* (Synopsis of Dissertation of Candidate of Psychological Sciences). Kazan State University.
12. Prokhorov, A. M., Gilyarov, M. S., Zhukov, E. M., Inozemtsev, N. N., Knunyants, I. L., Fedoseev, P. N., … Khrapchenko, M. B. (Eds.). (1980). Science. In *Soviet encyclopedic dictionary* (1st ed.) (p. 866). Soviet Encyclopedia.
13. Zhumabaeva, G. A. (2009). *Pedagogical and psychological aspects of moral education*. Gulchynar.
14. Smorodinskaya, M. D. (1979). *Traveling on the ships of thought*. Knowledge.
15. Kukushin, V. S. (2002). The origin of morality. In *Theory and methods of educational work: Textbook* (pp. 197–199). Rostov-on-Don, Russia.
16. Dneprov, E. D. (2006). *Education and politics, the latest political history of Russian education* (Vol. 1).
17. Dobaev, K. D. (2023). *New school – New thinking*. Fast Print.
18. Bazarny, V. F. (2009). *Human child*. Psychophysiology of development and regression.
19. Amonashvili, S. A. (1989). Foundations of pedagogy of cooperation. In A. V. Petrovsky (Ed.), *New pedagogical thinking* (pp. 146–148). Pedagogy.

Macroeconomic Implications and Prospects for Increasing Inclusiveness of the National Science and Higher Education System

Shavkat I. Ilkhamov, Timur M. Israilov, Victoria S. Ustenko, Elena V. Makarova, and Natalya A. Levchenko

Abstract Inclusiveness is the indicator of the government and its institutions' participation in support of human development for all categories of people. Obtaining higher education and scientific activities are spheres that were hardly accessible to people with special needs in countries of Central Asia in the past. At the current stage of development, countries of this region are oriented towards international integration in all spheres, including the provision of inclusiveness of the national system of science and higher education. In this paper, we dwell on the relationships between an increase in the inclusiveness of the system of science and higher education and the macroeconomic implications and prospects for their changes in the selected Central Asian countries (Kazakhstan and Kyrgyzstan). In this paper, we substantiate the influence of the measures on the creation of the conditions of inclusiveness in the studied spheres on the factual and prospective state of the selected macroeconomic results, which include employment of the population, GDP, and GDP per capita. We revealed that a further increase in inclusiveness in these spheres will allow ensuring technological growth and will raise the level of foreign investments.

S. I. Ilkhamov
International School of Finance and Technology, Tashkent, Uzbekistan

T. M. Israilov
Osh State University, Osh, Kyrgyzstan
e-mail: israilov@oshsu.kg

V. S. Ustenko (✉)
Center for Women's Empowerment and Gender Equality, Amrita Vishwa Vidyapeetham, Amritapuri, India

E. V. Makarova
Sebryakovskiy Branch of Volgograd State Technical University, Mikhailovka, Russia

N. A. Levchenko
Platov South-Russian State Polytechnic University (NPI), Novocherkassk, Russia

B. S. Sergi et al. (eds.), *University 4.0 and Educational Technology for Sustainable Development*, Education in the Asia-Pacific Region: Issues, Concerns and Prospects 75, https://doi.org/10.1007/978-981-96-7376-6_14

Keywords Macroeconomic implications · Inclusiveness · National system of science and higher education · People with disabilities · Digitalisation · Central Asia · GDP

JEL Codes H52 · I23 · I24 · I26 · J14

1 Introduction

Inclusiveness, as a principle of ensuring the functioning of a certain sphere given the needs and capabilities of all groups of society, gains wide implementation in all spheres of life activities [1, 2]. Education and science facilitate the quality of the population's life and lead to the development and growth of the competitive advantages of labour resources, while inclusive innovations allow the attraction of foreign students, scholars, and teachers interested in receipt, exchange, transition, and creation of knowledge [3]. This stimulates the receipt of revenues in foreign currency from the provision of educational and scientific services (higher education, postgraduate studies, doctoral studies) [4, 5]. Therefore, it can be assumed that an increase in the inclusiveness in science and higher education directly impacts transformations of the country's macroeconomic indicators. Given the fact that countries of Central Asia have been demonstrating quick rates of population growth in recent decades (55.939 million people in 2000, 63.615 million people in 2010, 74.724 million people in 2020, and 78.165 million people in 2023) [6], countries of the region implement certain measures for creation of adequate conditions for high living standards, including through the use of accessibility of higher education and science for all categories of citizens. Accordingly, a study of the specific features of this process based on the example of Central Asian countries poses a large scientific interest at the current stage of the development of science.

The objective of this article was to determine a link between an increase in the inclusiveness of science and higher education and macroeconomic implications and prospects for their changes in the selected countries of Central Asia. According to this goal, the following tasks were formulated: assessment of the improvement of the inclusiveness of science and higher education in the studied countries; determination of the link at the level of growth of this indicator and macroeconomic implications, as well as prospects for their transformations.

2 The Core of Inclusiveness in the Sphere of Science and Higher Education

Some studies dwell on implementing inclusiveness in the spheres under consideration. These studies contain theoretical, methodological, and empirical materials on the formation of this process [7] presented a set of measures for the improvement of inclusiveness in the sphere of science and higher education в Kazakhstan [8] contains an analysis of the trends and state of inclusiveness in the system of Kazakhstan's education [9] formulated the directions and trends of the formation of digitalisation in the sphere of higher education in Kyrgyzstan and determined measures aimed at the improvement of the quality of this process, which took place in the country due to the introduction of limitations and social distancing during the COVID-19 pandemic [10] formulated characteristics of inclusive education in Kyrgyzstan and assessed its influence on the state of the public and political life in the country.

Even though there are studies in this direction, there remains the necessity of determining the effect of the growth of inclusiveness in the context of macroeconomic implications and prospects.

To achieve this goal, we determined a set of methods that enable us to prove the direct link between the considered variables. The method of deduction allowed for the systematisation of isolated facts on the studied phenomenon and their presentation as a formalised provision. The factor method allowed determining the connection between the growth of the implementation of the inclusiveness of science and higher education and macroeconomic implications and prospects. The method of comparison was utilised to compare this phenomenon and the dependence of variables.

3 Implementation of Inclusiveness in the Central Asian System of Science and Higher Education

Let us find the level of improvement of the inclusiveness of science and higher education in the selected countries of Central Asia.

First off, let us consider Kazakhstan.

As shown in Table 1, Kazakhstan has taken measures to enhance the inclusiveness in the studied spheres, with the participation of the government and subjects of science and education. Due to the improvement of legal regulation and the government's participation in the international framework agreements, as well as cooperation with the university environment, there were formed the conditions for the growth of access of people with special needs to higher education and science. This was due to the achievement of a certain number of students with special needs in the total number of university students in the country thanks to the performed measures and growth of the number of published scientific papers.

Let us perform an assessment of the state of inclusiveness in higher education and science in Kyrgyzstan.

Level of implementing inclusiveness in Kyrgyzstan's system of science and higher education [9, 14–18]:

1. Indicator of inclusiveness of science and higher education: Regulatory framework of inclusiveness in higher education and science.
   - Factual state: Law of Kyrgyzstan "On education" (2003). Regulation of the general principles of inclusive education.

Table 1 Level of implementation of inclusiveness in Kazakhstan's system of science and higher education

No.	Indicator of inclusiveness of science and higher education	Factual state	Change
1	Regulatory support for inclusiveness of education as the basis of equal accessibility in this sphere for citizens given their individual capabilities and needs for the organisation of the educational process	Article 21.7 of the Law of Kazakhstan "On education (2007). Three regulatory acts (adopted in 2002 and 2005), aimed at ensuring the inclusiveness of education, and social protection of people with special needs, including access to all spheres of life activities (science, higher education). Participation of Kazakhstan in three global framework agreements.	Educational reforms in Kazakhstan in 2011–2020 and 2016–2019. Focus—access to the sphere of science and higher education.
2	Level of involvement of university students and postgraduates who have special needs	No official data	3520 students (as of December 2020).
3	Tertiary Education Enrollment Gross %, Value	35% (2007)	36% (2008), 45% (2009), 51% (2010), 59% (2011), 60% (2012), 62% (2014), and 63% (2017). The growth of tertiary education enrollment directly influences the growth of the number of people with special needs in the national system of higher education. Over 2007–2017, this indicator grew by 28%.
4	Scientific And Technical Journal Articles, of Articles	190 articles (2000)	216 articles (2007), 231 articles (2008), 323 articles (2010), 934 articles (2014), 1170 articles (2015), 1600 articles (2016), 1960 articles (2017), 2370 articles (2018). The growth of the number of scientific and technical papers shows the growth of access to science and postgraduate education.

(continued)

Table 1 (continued)

No.	Indicator of inclusiveness of science and higher education	Factual state	Change
5	Digitalisation of education and science	The foundations of digitalisation in these spheres were set by the national programme "Digital Kazakhstan": increase in the level of the readiness of personnel for the economy, development and implementation of innovations with the use of digital technologies	Introduction of new courses and pilot projects in the sphere of an increase in digital literacy of students and academic staff. Annually, the programme "Digital Kazakhstan" provides training for 21,400 specialists in ICT.

Source Formed by the authors using [7, 8, 11–13]

- Change:
  - Framework agreement on the focus on support of the inclusive development of society (Salamanca Declaration, 2008);
  - Implementing changes in the Law of Kyrgyzstan "On education" in 2021;
  - The new Law of Kyrgyzstan "On education" from 2021 contains the following provisions on the inclusiveness in these spheres:
  - Equal access to education for people with special needs and people with difficult life situations;
  - Improvement of the mechanism of financing of inclusive education;
  - Regulation of bonuses for teachers and universities which use the methods and principles of inclusiveness in education.

2. Indicator of inclusiveness of science and higher education: Number of undergraduates and postgraduates with special needs
   - Factual state: Since the adoption of the Law of Kyrgyzstan "On Education" (2003), 100 quotas for people with special needs have been allocated
   - Change: Individual interactions with undergraduates and postgraduates with special needs at the level of universities.
3. Indicator of inclusiveness of science and higher education: Tertiary Education Enrollment Gross %, Value
   - Factual state: 46% in 2007
   - Change: 50% in 2008, 51% in 2009, 53% in 2012, 55% in 2015, and 62% in 2017. The increase in this indicator directly depends on the improvement of regulatory support and corporate decisions at the level of universities and research organisations.

4. Indicator of inclusiveness of science and higher education: Scientific and Technical Journal Articles, of Articles

    - Factual state: 28 articles in international scientific journals were published in 2000;
    - Change: 33 articles in 2004, 44 articles in 2005, 51 articles in 2006, 52 articles in 2008, 61 articles in 2013, 106 articles in 2016, 107 articles in 2017, and 137 articles in 2018. The growth of this indicator is connected with the development of science and technology at the national level and the growth of the quality of human resources, including among talents from the circle of scholars with disabilities.

5. Indicator of inclusiveness of science and higher education: Digitalisation of education and science

    - Factual state: Implementation of electronic versions of programmes for individual courses in the period before 2018
    - Change: 2018 saw the adoption of the Concept "Digital Kyrgyzstan 2019–2023", developed based on the strategy of sustainable development of the country. At the beginning of the COVID-19 pandemic, there was an acceleration of the digitalisation of educational materials, which facilitated an increase in access to higher and postgraduate education for people with disabilities.

Kyrgyzstan demonstrates positive results in implementing inclusiveness in the national system of higher education and science. Namely, there is improvement in the regulatory and institutional support in these spheres, as well as additional synergetic effect for the country in the context of the development of science and the level of labour resources' readiness for new challenges in the economy and digital sphere.

Let us assess the influence of the growth of inclusiveness of the system of science and higher education on the change in macroeconomic indicators (implications and prospects).

We assume that these indicators are in direct dependence on each other. It is necessary to determine macroeconomic variables, which are subject to changes due to the impact of the transformations of inclusiveness in the considered spheres. Indicators that depend the most on the change in the inclusiveness in these spheres are as follows:

- GDP (current $), which is the aggregate income of the subjects of economic activities of the country. It is achieved also using the economic activity of citizens who received inclusive education or scientific degrees with a focus on inclusiveness, as well as revenues from the provision of inclusive education;
- GDP per capita (current $), which shows the quality of life and the level of economic activity of citizens. Economic activity is connected also with knowledge and skills in a certain sphere; its level depends on higher education and the university's ranking. Quality of life, set in this indicator, is achieved due to purchasing power that is ensured in the country. The analysed indicator has a high value

in developed, medium value in quickly developing countries, low value in developing countries, and very low value in underdeveloped countries. Growth of the indicator in the context of the connection with an increase in the inclusiveness in higher education and science is connected with the growth of GDP, which is received from the provision of educational services and services in postgraduate education to the country's citizens and foreign citizens; growth of employment; growth of incomes from individual economic activity (which also includes persons with special needs);

- Level of population's employment, which is connected with the demand of the for labour resources and satisfaction of the population's needs for employment (which also includes persons with special needs who were able to find jobs due to a high level of education).

Based on the results in Table 2, we can state the growth of the considered macroeconomic indicators in the two countries, which is connected also with an increase in the inclusiveness in the sphere of higher education and science, as predictors of the growth of economic activity of citizens with special needs.

4 Influence of the Inclusiveness of the National System of Science and Higher Education on the Macroeconomic Indicators of Countries

We confirmed the hypothesis on the influence of the improvement of the inclusiveness of the national system of science and higher education on the state of certain macroeconomic indicators of countries (Fig. 1).

The assessed macroeconomic variables, which are directly connected with inclusiveness in these spheres, were growing continuously, while a certain decrease in 2020 was caused by the crisis in the world economy due to restrictions on the economic activity of the population and subjects of entrepreneurial activities due to the COVID-19 pandemic. Therefore, it is possible to say that such indicators as GDP, GDP per capita, and employment could be corrected due to the national and institutional policy in the sphere of growth of inclusiveness in higher education and science, providing additional benefits for the country through the development of science, higher school, and technology due to support for talents from among people with special needs.

Participation of the government and subjects of the national system of science and higher education in raising inclusiveness in these spheres in the selected countries of Central Asia (Kazakhstan and Kyrgyzstan) has further prospects. In our opinion, the creation of attractive conditions for receiving services in the sphere of higher and postgraduate education and services in the sphere of science and technology for people with special needs will allow raising income and taxes which go to the budgets of different levels. Support for the development of talents is the basis of the innovative development of the country, which can be the priority of

Table 2 Macroeconomic indicators in Kazakhstan and Kyrgyzstan in 2000–2022

Indicator	Value											
	2000	2002	2003	2005	2007	2009	2016	2017	2018	2020	2021	2022
Kazakhstan												
GDP (current $), Billion	18.29	24.6	30.8	57.1	104.8	115.3	137.3	166.8	179.3	171.1	197	220.6
GDP per capita (current $)	1229	1658	2068	3771	6771	7165	7714.8	9247.6	9812	9121.6	10,373.8	11,243.7
% of employment of labour resources (number of the employed workforce/number of labour resources) * 100	No data	90.7	91.2	91.9	92.7	93.4	95.1	95.1	95.1	95.1	95.1	95.1
Kyrgyz Republic												
GDP (current $), Billion	1.37	1.61	1.9	2.5	3.8	4.7	6.8	7.7	8.3	7.8	8.7	10.9
GDP per capita (current $)	279.6	321.7	380.5	476.6	721.8	871.2	1120.7	1242.8	1308.1	1182.5	1306.2	1606.7
% of employment of labour resources	No data	No data	No data	No data	92	92	92.8	93.1	93.8	94.2	94.7	95.1

Source Formed by the authors using [19–24]

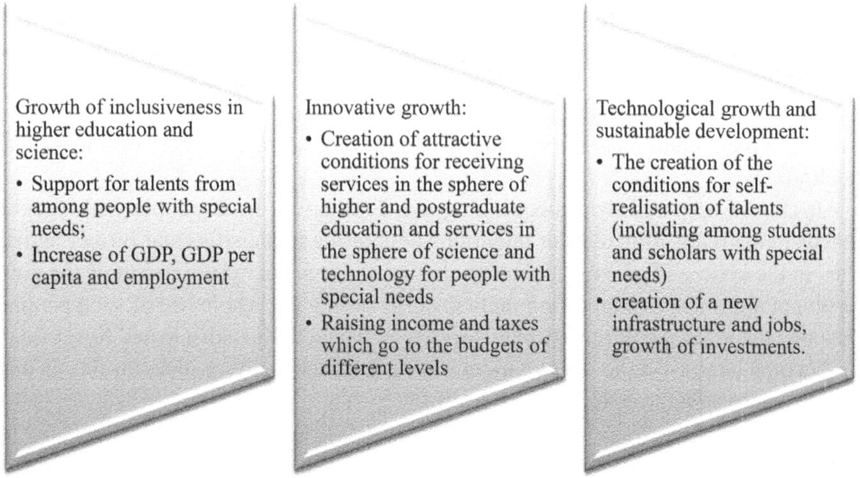

Fig. 1 Influence of the inclusiveness of the national system of science and higher education on the macroeconomic indicators of countries. *Source* Formed by the authors

transformation of the studied countries' economies. The creation of the conditions for self-realisation of talents (including among students and scholars with special needs) for Kazakhstan and Kyrgyzstan can be implemented within the formation of international centres for innovations, stimulating quick economic growth and technological progress. Potential innovative growth will allow ensuring the creation of a new infrastructure and jobs. In this case, it will be possible to consider hypotheses on the effect of inclusiveness in the studied spheres on such macroeconomic indicators as the growth of investments.

Besides, Kyrgyzstan ensures the improvement of inclusiveness in the studied spheres with a focus on the strategy of sustainable development. This approach to inclusiveness may be applied more comprehensively in the case of a focus on the achievement of specific goals of technological growth, which are also considered within sustainable development.

5 Conclusion

This research demonstrated the necessity and importance of raising the inclusiveness of science and higher education in the example of two countries in Central Asia. Support for people with special needs through the creation of attractive and optimal conditions for development in the direction of improvement of the level of education and access to sciences allows solving complex tasks of social support for citizens of this category with the help of the development of their economic activity.

The growth of macroeconomic indicators and the development of education and science in Kazakhstan and Kyrgyzstan are connected with the creation and improvement of the optimal system of human progress, including for people with special needs. A wide range of further prospects for the improvement of the national and institutional policy of raising inclusiveness in this sphere is implemented in the studied countries within the regulatory and programme support and at the level of the initiative of individual universities and organisations in the sphere of science and technology.

It can be assumed that encouraging good initiatives from people with special needs in the realisation of their intentions of entering universities and further activities in the sphere of science and technologies requires the improvement of the government approach to supporting inclusiveness during the whole life of such people. Raising the level of complexity in this sphere will enable Kazakhstan and Kyrgyzstan to become centres of the development of talents in the territory of Central Asia and ensure innovative growth in the long term.

References

1. Ugur, S., Dincer, G. D., & Bas, D. P. (2024). An evaluation of the managerial context for digital transformation in the context of open education in higher education. *Turkish Online Journal of Distance Education, 25*(3), 225–242. https://doi.org/10.17718/TOJDE.1417658
2. Varga, A., Horváth, G., & Trendl, F. (2024). Roma youth's perspective on an inclusive higher education community: A hungarian case study. *Education Sciences, 14*(7), 679. https://doi.org/10.3390/educsci14070679
3. Wang, X., Luo, R., & Wuji, S. (2023). Index and demonstration of international openness and influence of new research universities. *Journal of Modern Information, 44*(9), 137–153. https://doi.org/10.3969/j.issn.1008-0821.2023.09.012
4. Wang, X., & Wan, X. (2024). Towards inclusive classrooms: Shaping South African higher education for employability. *Asia-Pacific Education Researcher.* https://doi.org/10.1007/s40299-024-00902-w
5. Wick, M., Erickson, D., Hoffman, J., Johnson, L., & Angradi, T. (2024). Navigating university openness in research policy inconsistent with indigenous data sovereignty: A case analysis. *Ethics and Human Research, 46*(2), 2–15. https://doi.org/10.1002/eahr.500202
6. Worldometer. (2023). *Central Asia Population.* Retrieved from https://www.worldometers.info/world-population/central-asia-population/ (Accessed 10 November 2023).
7. Nadirova, G. (2022). *Approaches to inclusive education in Kazakhstan.* Retrieved from https://www.eurasian-research.org/publication/approaches-to-inclusive-education-in-kazakhstan/ (Accessed 10 November 2023).
8. Abdina, A. K., Uyzbaeva, A. A., & Zhanarstanova, M. B. (2023). Philosophy of independent living: The state of inclusive education in Kazakhstan (overview analysis of modern research). *Bulletin of the L.N. Gumilyov ENU. Historical sciences. Philosophy. Religion Series, 142*, 1. https://doi.org/10.32523/2616-7255-2023-142-1-135-157
9. Erasmusplus.kg. (2020). *Digitalisation in the system of higher professional education.* Retrieved from https://erasmusplus.kg/wp-content/uploads/2020/05/webinar_1_mes.pdf (Accessed 10 November 2023).
10. Suchkov, M. (2020). Inclusive education development in Kyrgyzstan in the context of socio-political modernization. *KANT, 3*(36), 393–396.
11. Egov.kz. (2018). *"Digital Kazakhstan" State Program.* Retrieved from https://egov.kz/cms/en/digital-kazakhstan (Accessed 10 November 2023).

12. Tcdata360.World Bank. (2023a). *Tertiary education enrollment gross %. Kazakhstan.* Retrieved from https://tcdata360.worldbank.org/indicators/hb89c4e88?country=KAZ&indicator=566&viz=line_chart&years=2007,2017 (Accessed 10 November 2023).
13. Tcdata360.World Bank. (2023c). *Scientific and Technical Journal Articles, of Articles. Kazakhstan.* Retrieved from https://tcdata360.worldbank.org/indicators/IP.JRN.ARTC.SC?country=KAZ&indicator=2015&viz=line_chart&years=2000,2018 (Accessed 10 November 2023).
14. Cbd.minjust.gov.kg (2023a). *The Law of the Kyrgyz Republic "On education"* (dated 30 April 2023, No. 92). Retrieved from http://cbd.minjust.gov.kg/act/view/ru-ru/1216 (Accessed 10 November 2023).
15. Cbd.minjust.gov.kg (2023b). *The Law of the Kyrgyz Republic "On education"* (dated 11 August 2023, No. 179). Retrieved from http://cbd.minjust.gov.kg/act/view/ru-ru/112665?cl=ru-ru (Accessed 10 November 2023).
16. Cbd.minjust.gov.kg. (2023c). *The concept of the development of inclusive education in the Kyrgyz Republic for 2019–2023* (adopted by the Decree of the Government of the Kyrgyz Republic dated 19 July 2019, No. 360). Retrieved from http://cbd.minjust.gov.kg/act/view/ru-ru/14591 (Accessed 10 November 2023).
17. Tcdata360.World Bank. (2023b). *Tertiary education enrollment gross %. Kyrgyz Republic.* Retrieved from https://tcdata360.worldbank.org/indicators/hb89c4e88?country=KGZ&indicator=566&viz=line_chart&years=2007,2017 (Accessed 10 November 2023).
18. Tcdata360.World Bank. (2023d). *Scientific and Technical Journal Articles, of Articles. Kyrgyz Republic.* Retrieved from https://tcdata360.worldbank.org/indicators/IP.JRN.ARTC.SC?country=KGZ&indicator=2015&viz=line_chart&years=2000,2018 (Accessed 10 November 2023).
19. Stat.gov.kz. (2023). Main indicators of the labour market in the Republic of Kazakhstan. Retrieved from https://stat.gov.kz/ru/industries/labor-and-income/stat-empt-unempl/. (Accessed 10 November 2023).
20. Stat.kg. (2023). *Unemployment level.* Retrieved from https://www.stat.kg/ru/statistics/zanyatost/ (Accessed 10 November 2023).
21. World Bank Group. (2023a). *GDP (current $). Kazakhstan.* Retrieved from https://data.worldbank.org/indicator/NY.GDP.MKTP.CD?locations=KZ (Accessed 10 November 2023).
22. World Bank Group. (2023b). *GDP (current $). Kyrgyz Republic.* Retrieved from https://data.worldbank.org/indicator/NY.GDP.MKTP.CD?locations=KG (Accessed 10 November 2023).
23. World Bank Group. (2023c). *GDP per capita (current $). Kazakhstan.* Retrieved from https://data.worldbank.org/indicator/NY.GDP.PCAP.CD?locations=KZ (Accessed 10 November 2023).
24. World Bank Group. (2023d). *GDP per capita (current $). Kyrgyz Republic.* Retrieved from https://data.worldbank.org/indicator/NY.GDP.PCAP.CD?locations=KG (Accessed 10 November 2023).

University 4.0 in Russia and Central Asia as Centers of International Higher Education

Preface to the Third Part of the Book

The third part of the book focuses on the internationalization of science and higher education in Russia and Central Asia through the development of Universities 4.0. The main idea is that the primary drivers for internationalizing the activities of Universities 4.0 are their global communications and international reputation.

The chapter "Life Values of Universities in Russia and the Latest Trends in Their Transformation in the Globalizing World" by Zoya V. Popkova, Anna Yu. Kozlova, Azamat M. Shadzhe, and Tatiana I. Barsukova reached an important scientific conclusion: the processes of globalizing higher education in Russia are accelerated by integrating life values into the development strategies of Russian University 4.0, such as autonomy, humanistic independence of faculty and researchers, and the formation of a professional elite.

The key conclusion of the chapter "Problems and Prospects of Intercultural Dialogue in a Multinational Classical University in a Globalizing World: The Experience of Modern Russia" by Angelina I. Litvinova, Mukaddas A. Dzhorobaeva, Bakhtiyar E. Yuldashev, and Zuriet A. Zhade is that there is a need to further develop intercultural communications in Russia's multinational classical University 4.0 by adapting students to a multicultural educational environment and creating conditions for effective intercultural dialogue.

The chapter "Problems of Higher Education in Modern Russia in the Context of the Large-Scale Process of Internationalization of Universities" by Barno S. Akbarova, Zara A. Mamisheva, Inna N. Gaidareva, and Aleksandr V. Bespyatyh contributes to the discussion by grouping the challenges of higher education in Russia as they emerge during the extensive internationalization of universities.

In the chapter "Image Communications of Universities: Current Technologies" by Yulia N. Zemskaya, Evgeniya A. Kuznetsova, and Anna V. Osmolovskaia, the authors argue that educators, as communicators, influence the image of University 4.0 and its international activities, further advancing the topic of the third part.

It is also worth noting the recommendations for the global promotion of University 4.0 using influencer imagery, developed in the chapter "Features of

Influence Marketing Technologies in the Field of the Promotion of Higher Education Institution" by Nina V. Trubnikova and Viktoria V. Shlenskaya.

The chapter "Study of the Impact of Reputation of Higher Education Institution on its Ranking" by Natalya A. Aleksandrova and Julia A. Aleksandrova expands on this topic by supplementing the criteria and methods for evaluating the reputation of University 4.0 and constructing its global reputational profile with digital components.

The value of the chapter "Features of Promoting the Image of Higher Educational Institutions in the Media Space" by Irina Yu. Glinskaya and Olga N. Popova lies in the examination of how the image of University 4.0 in social networks impacts the international promotion of their educational services.

Supporting the topic of this part, the chapter "The Product Approach in Marketing Control: The Practice of Implementation in Educational Institutions of Higher Education" by Svetlana A. Kucheryavenko presents a scientific concept for applying a product approach within marketing control over the international activities of University 4.0.

Thus, each of the eight chapters significantly contributes to the development of the topic of the third part, reinforcing the evidence for enhancing the inclusiveness of University 4.0 to boost their international engagement. Collectively, these chapters provide a comprehensive scientific understanding of University 4.0 in Russia and Central Asia as centers of international higher education.

Life Values of Universities in Russia and the Latest Trends in Their Transformation in the Globalizing World

Zoya V. Popkova, Anna Yu. Kozlova, Azamat M. Shadzhe, and Tatiana I. Barsukova

Abstract In this paper, we considered the current problems of the development of the concept of life values of Russian universities in the conditions of higher education's globalization. We determined the conceptual value of this approach, outlined its institutional basis, characterized it, and analyzed and assessed the declarations of life values of Russian leading universities. We used the institutionalized provisions of the "Living Value" project, offered by the Great Charter of Universities and the designated principles of autonomy of universities, humanism, and combination and independence of the teaching and research work of the employees of higher education institutions. We also analysed official sources and statutory documents of the best Russian universities to reveal the positioning of their life values. We generalized and analyzed the conceptual provisions of the "Living Value" project and the level of its popularity among Russian universities, emphasized the relevance of globalization processes in higher education and the need for Russian universities to take them into account in the context of their development strategy and identified the level of universities' interest in the positioning of life values and influence of this level on the value orientations of students and the future professional elite of the society. In many universities, especially RUDN University, the positioning of their values conforms to the vision of the Great Charter of Universities and covers the ideas of autonomy of higher education institutions, humanism, and development of moral values, quality of education, and quality professional training and creation of conditions for self-realization of students.

Z. V. Popkova (✉)
Volgograd State University, Volgograd, Russia

A. Y. Kozlova
Togliatti State University, Togliatti, Russia

A. M. Shadzhe
The Adyghe State University, Maykop, Russia

T. I. Barsukova
North-Caucasus Federal University, Stavropol, Russia

B. S. Sergi et al. (eds.), *University 4.0 and Educational Technology for Sustainable Development*, Education in the Asia-Pacific Region: Issues, Concerns and Prospects 75, https://doi.org/10.1007/978-981-96-7376-6_15

Keywords Globalization · Life values · Higher education · University · Transformation of values · Great Charter of Universities

JEL Codes A14 · I23 · I28 · J24 · O15 · O30 · P16 · P52

1 Introduction

In the conditions of a globalized world, which is developing very dynamically, strategies planning of universities' activities acquires a special sense. Global institutions that regulate the activities of higher education institutions allow coordinating the directions and conditions of the activities, while technological development increases uncertainty. In this context, the change in society's attitude to higher education assigns the problem with new features, which further increases the uncertainty and hinders the development of a clear strategic plan that could adequately consider all possible options for the situation's development.

In these conditions, the main element of the strategic planning of universities' activities is the determination of their life values. Values form an ideological basis for university employees and students and serve as key criteria during managerial decision-making. They also allow universities to declare themselves at the national and global levels. They influence the corporate culture of higher education institutions, determine priorities in the educational, scientific, creative, and other types of activities and are benchmarks for future elites and society.

Russian universities function in the conditions of a globalized market of higher education. They actively cooperate with educational establishments of different countries and take into account the global principles of international educational organizations during planning and implementing their activities. As a result, Russian universities acquire larger autonomy and are actively involved in international projects connected with environmental protection, social responsibility and justice, inclusion, etc. Such transformation requires proper planning and organization of the educational process, which involves the formation of a comprehensive conceptual apparatus of university management, which includes mission, principles, goals, and values. In this list, values are assigned a role of formation of the cultural and ethical attitude of university students, graduates, and employees towards the surrounding world and themselves. A clear identification of life values allows strengthening the influence of socio-psychological methods in university management and aiming their development in the desired direction.

2 The Concept of Life Values and Its Implementation in Russian Universities

The problem of development, discussion, adoption, and observation of life values of universities belongs to important scientific tasks. It has a clear methodological basis, which origins from the Great Charter of Universities and its project "Living Values" with such designated principles of activities of higher education institutions as autonomy of universities, combination of teaching and research work, freedom of teaching and research activities, and preservation and expansion of the traditions of humanism [1–5]. These principles are a starting point for the development of life values by higher education establishments.

When studying the life values of Russian universities, we revealed values declared by them that could be found in the statutory documents or on the official websites of the universities. Thus, we determined the level of the universities' involvement in the process of "Living Values" and identified specific features of each university.

The main methods used in this research include systemic analysis, which allowed considering the values of universities in the context of the system of their management; observation and generalization, which allowed determining the available values and interpreting them; grouping, which enables us to find common features and unifying the research objects according to them; and other methods and approaches.

The issues of life values of Russian universities in the context of the project of the Great Charter of Universities do not belong to the top-priority research topics. Most of the studies in this direction are focused on the issues of development and social interaction of students from the position of value orientation [6–8]. Life values were directly studied in the works that also contain the basic aspect of the formation and positioning of life values of universities [9, 10], their implementation from the position of the structure of personality [11], cultural development of the university [12], and strategic interaction of student-centric learning and life values of the university [13].

This research was aimed at studying the state of implementation of the concept of life values in Russian universities, which takes place under the influence of globalization and involves their transformation. The main tasks solved in this context are disclosing the theoretical foundation of the transformation of Russian universities' life values under the influence of globalization, determining the role of the concept of life values for society, and identifying the list of the declared values of Russian universities with the best rankings.

3 Life Values of the Russian Universities Included in Times Higher Education (THE) Ranking

Most of the acting universities in Russia were founded before 1970. Universal values that were set in their statutory documents remained unchanged—quality higher education, scientific development, and activities in the society's interests. However, society and the world changed a lot over this time. Most of the processes connected with society's development acquired a global character, with a rethinking of problems and directions of development and society's interests. Given this, the traditional values of universities require the procedure of audit, clarification, and correction (Fig. 1).

Apart from the change in the circumstances of the functioning of universities, the relevance of the problem of reconsideration of their life values is connected with the growing differences between stakeholders of universities, the government, and university administrations. Each party has a different attitude towards universities and their internal and external processes, such as financing, formation of educational plans, technological development, internationalization, inviting foreign students, increase in competition in the market of higher education, etc. The main problem is still the change in attitude to higher education in society. It is not treated as a guarantee of achievement of well-being or a successful career anymore. Therefore, universities have to offer an actual alternative for society, which would conform to social interests and values.

The Great Charter of Universities became probably the most important attempt to ensure the rethinking of the life values of universities. As of November 2023, 975 higher education establishments from 94 countries joined it. These include 17 Russian universities, in particular Lomonosov Moscow State University, Saint Petersburg State University, and RUDN University [9].

To study the specific features of the development and implementation of life values in Russia, we considered the values of Russian higher education institutions

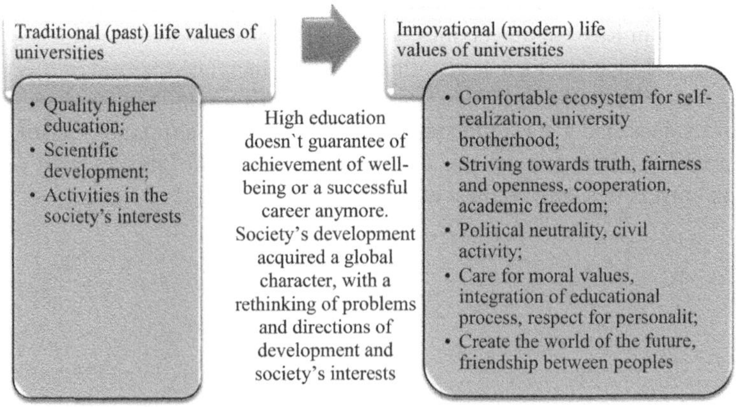

Fig. 1 Life values of universities in Uzbekistan and Kazakhstan, which strive for digital inclusion in the globalizing world. *Source* Made by the authors

that are among the best universities in the world according to Times Higher Education [14]. The research methodology involved the search for values mentioned on the universities' websites, statutory documents, the strategy of development or other documents that can be found in open access. Analysis showed that all universities declared certain value ideas, but, in most of them, they are not presented as values and are included in the mission, principles, goals, and tasks.

The list of the universities of the sample and a description of their declared values are shown in Table 1.

The results of the analysis showed that many universities have not openly declared their values on their official website, in their strategy, or statutory documents. Only RUDN University presented its values in the context of the "Living Values" concept.

Insufficient attention to the development and declaration of values by many universities is a sign of a lack of understanding of modern processes, according to which values are the main criteria of higher education institutions' development. Insufficient coverage of information on the values forms circumstances in which students and lecturers, when making decisions, are based on their beliefs, which could be very different. The absence of the university's influence on them during decision-making leads to an imbalance in the system of education and different lecturers offering different value orientations, which leads to the treatment of values as something unimportant and ambiguous. That's why development and, in particular, declaration of values by universities is an important task of their strategic development not only from the position of their vision but also from the position of public development.

As for the values that are presented in Table 1, it should be noted that almost all of them combine professional striving and social priorities. Large attention is paid

Table 1 Values of the selected universities from the top 15 universities of Russia according to Times Higher Education

University	Values
1. Peter the Great St. Petersburg Polytechnic University	Technological leadership, smart power, multidisciplinarity, social responsibility, comfortable ecosystem for self-realization, relevance, patriotism, university brotherhood
2. National Research University "Higher School of Economics"	Striving towards truth, fairness and openness, cooperation, academic freedom, political neutrality, professionalism, responsibility, civil activity
3. Tomsk State University	Care for moral values, receipt and dissemination of quality knowledge, training of intellectual elite that surpasses its time, integration of educational process, innovative approach
4. ITMO University	Respect for personality, openness, good faith, academic freedom, love
5. National University of Science and Technology MISiS	Create the world of the future, improve the world, cooperate, see globally, act bravely
6. RUDN University	Friendship between peoples, professionalism, culture of research, academic freedom, social responsibility

Source Compiled by the authors based on [15–20]

to social development and technological development and responsibility, technological vision of the future, and the issues of academic freedom of universities, multidisciplinarity, and moral and humanistic values. The declared life values of Russian universities were formed in the context of the Great Charter of Universities. However, the focus on the combination of lecturing and research work, as well as their freedom in the formation of educational courses and selection of topics and methods of the research, should be strengthened.

The mechanism of declaring, implementing, and using life values in universities is primarily aimed at the maximum involvement of academic staff. The next step is convincing students and constant emphasis on the values for the purpose of dissemination of the ideas of autonomy, humanism, and comprehensive development of personality in professional activities. By influencing the youth through such actions, universities receive an opportunity to form value orientations of the future professional elite in different spheres of activities.

RUDN's example shows that such an approach to the implementation of the "Living values" concept at the level of the university allows influencing the strategic development of the university. The creation of a complex system of teaching and education in higher education establishments, which rely on a high level of professionalism and motivation for professional activities, is the main task of the university from the position of its influence on public processes in the country and the world.

4 Specifics of Russian Universities and their Life Values in Conditions of Globalization of Higher Education

The research of any processes connected with the globalization of higher education poses large interest, for it is implemented in the conditions of a multinational, multicultural, and heterogeneous society. According to this, ignoring the current differences regarding the conditions of development and implementation of life values in universities requires active discussion and scientific substantiation of the expedience and possibility of realizing certain components of the process in a given country or region.

In this context, the specifics of Russian universities are connected with the multinational structure of the population of the Russian Federation and the high level of internationalization of many universities. An increase in multiculturality and multinationality, on the one hand, requires the development of universal life principles for Russian universities, on the other, requires checking each implemented measure for compliance with the cultural values of a certain representative.

The issue of the perception of global and European values in Russia is also debatable. Many scholars express their doubts regarding the effectiveness of the proposed educational system in the conditions of Russia. They support the necessity to preserve traditional values, which allows obtaining goods results in the past. However, the dynamic development of the conditions of the economy's functioning sets clear demands on the necessity for constant reconsideration of conceptual and value provisions of universities' development.

Thus, the problem of implementing the system of life values in Russian universities from the position of globalization does not have a common vision and understanding. That is why it requires a critical assessment of different views and their study and substantiation as to expedience and form of realization of life values in Russian universities.

5 Conclusions

In the globalized world, universities cannot remain aside from public processes and events. In these conditions, institutional execution of value orientations of the universities' functioning within the "Living Values" project of the Great Charter of Universities is an important tool for strategic management. The wide involvement of higher education institutions in this project in the world demonstrates the interest and importance of the project's declared provisions on the autonomy of universities' activities, planning of humanistic values in the training of specialists, and the combination and freedom of teaching and scientific activities of academic staff.

Among Russian universities, the acceptance of life values, which were designated by this project, requires further improvement. However, among independently determined and formed values, most universities demonstrate similar visions. Therefore, in the context of the transformation of values of Russian universities under the influence of the processes of globalization, there is agreement. However, many universities do not develop or openly declare their values. This leads to ambiguous attitudes of students to the system of values on the whole. The absence of correctly set values at the stage of learning might have a negative effect on the professional activities of a person, who, during making decisions, will not have a strong criterial basis. Even in case of incomplete acceptance of life values within the Great Charter of Universities project, Russian universities must form and declare their values, based on professionalism, morality, and humanism. These values will surely influence the formation of the future elite of the country.

References

1. Pugatch, T., & Thompson, P. (2024). Excellence for all? University Honors programs and human capital formation. *Educational Evaluation and Policy Analysis*. https://doi.org/10.3102/01623737241232971
2. Raghavan, S., Subramaniam, N. K., & Awang, A. I. (2025). Unboxing micro-credentials for ODL universities: Competency development for human capital. *Turkish Online Journal of Distance Education, 26*(1), 1–15. https://doi.org/10.17718/tojde.1408308
3. Ríos, L. P. B. (2024). The anthropocentric bias of humanism in university philosophy curricula: Anti-humanism, decolonialism, anti-speciesism, transhumanism, and posthumanism. *Universitas Philosophica, 41*(83), 151–205. https://doi.org/10.11144/Javeriana.uph41-83.sahf
4. Vilca, L. E. V., & Yucra, R. M. C. (2025). Integration of human rights in the university curriculum: An analysis of its implementation and results. *Multidisciplinary Reviews, 8*(6), e2025150. https://doi.org/10.31893/multirev.2025150

5. Xia, J., Zhang, M. M., Zhu, J. C., & Fan, D. (2024). Reconciling multiple institutional logics for ambidexterity: Human resource management reforms in Chinese public universities. *Higher Education, 87*(3), 611–636. https://doi.org/10.1007/s10734-023-01027-7
6. Ionina, O. S. (2021). Studying the life values of students (using the example of a pedagogical university). *Bulletin of Shadrinsk State Pedagogical University, 1*(49), 105–109. Retrieved from https://cyberleninka.ru/article/n/izuchenie-zhiznennyh-tsennostey-studentov-na-primere-pedagogicheskogo-vuza (Accessed 2 December 2023).
7. Kriazheva, E. V., & Vinogradskaia, M. Yu. (2020). Analysis of the dynamics of life values and value orientations of university students. *Problems of Modern Teacher Education, 68–3*, 332–334. Retrieved from https://cyberleninka.ru/article/n/analiz-dinamiki-zhiznennyh-tsennostey-i-tsennostnyh-orientatsiy-studentov-vuza (Accessed 2 December 2023).
8. Radevskaia, N. S., Radevskii, A. V., & Gorelysheva, N. V. (2019). Life values as a factor in the socialization of students in a modern university. *Chelovek i obrazovanie* [Man and Education], *1*(58), 68–71. Retrieved from https://cyberleninka.ru/article/n/zhiznennye-tsennosti-kak-faktor-sotsializatsii-studentov-v-sovremennom-vuze (Accessed 2 December 2023).
9. MCU. (2023). *Signatory Universities. Magna Charta Universitatum.* Retrieved from https://www.magna-charta.org/magna-charta-universitatum/signatory-universities (Accessed 1 December 2023).
10. RUDN. (2023a). Magna Charta Universitatum. *Peoples' Friendship University of Russia Website.* Retrieved from http://www.rudn.ru/about/traditions/magna-charta (Accessed 1 December 2023).
11. Mezhman, I. F., & Mezhman, S. S. (2021). Formation of a system of life values in accordance with the personality structure. *Avtonomiia lichnosti* [Personal Autonomy], *2*(25), 30–35. Retrieved from https://cyberleninka.ru/article/n/formirovanie-sistemy-zhiznennyh-tsennostey-v-sootvetstvii-so-strukturoy-lichnosti (Accessed 2 December 2023).
12. Smetanina, M. D. (2021). University as a cultural center. *Society: Philosophy, History, Culture, 1*(81), 93–95. Retrieved from https://cyberleninka.ru/article/n/universitet-kak-tsentr-kultury (Accessed 2 December 2023).
13. Kulikovskaia, G. A., & Kulikovskaia, N. A. (2020). Student-centered learning and the "life values" of the university: Strategic interaction. *Bulletin of the Peoples' Friendship University of Russia. Series: State and Municipal Administration, 7*(4), 387–399. Retrieved from https://cyberleninka.ru/article/n/studento-orientirovannoe-obuchenie-i-zhiznennye-tsennosti-universiteta-strategicheskoe-vzaimodeystvie (Accessed 2 December 2023).
14. THE. (2023). World University Rankings 2024. *Times Higher Education.* Retrieved from https://www.timeshighereducation.com/world-university-rankings/2024/world-ranking#!/length/25/locations/RUS/sort_by/rank/sort_order/asc/cols/stats (Accessed 3 December 2023).
15. HSE. (2023). Declaration of values. *Higher School of Economics Website.* Retrieved from https://www.hse.ru/info/statement. (Accessed 2 December 2023).
16. ITMO. (2023). Mission and values of the university. *Institute of Precision Mechanics and Optics Website.* Retrieved from https://itmo.ru/ru/page/209/missiya_i_cennosti_universiteta.htm (Accessed 2 December 2023).
17. MISiS. (2023). Updated brand of MISIS University. *National University of Science and Technology MISiS Website.* Retrieved from https://misis.ru/university/rebranding/#values (Accessed 2 December 2023).
18. RUDN. (2023b). "Project living value". *Peoples' Friendship University of Russia Website.* Retrieved from http://www.rudn.ru/about/traditions/magna-charta/living-values (Accessed 4 December 2023).
19. SPBSTU. (2023). Mission and values. *Peter the Great St. Petersburg Polytechnic University Website.* Retrieved from https://www.spbstu.ru/university/about-the-university/mission-and-values/ (Accessed 2 December 2023).
20. TSU. (2023). TSU mission. *Tomsk State University Website.* Retrieved from https://www.tsu.ru/university/mission.php (Accessed 2 December 2023).

Problems and Prospects of Intercultural Dialogue in a Multinational Classical University in a Globalizing World: The Experience of Modern Russia

Angelina I. Seliankina, Mukaddas A. Dzhorobaeva, Bakhtiyar E. Yuldashev, and Zuriet A. Zhade

Abstract We considered the problems, prospects, and tools for intercultural dialogue in the universities of Russia from the position of their generalization and identification of the successful experience for its extrapolation and implementation in other higher education establishments. The methodology of this research is comprised of complex tools, which lie in the dimension of a range of sciences of the sphere of sociology, political sciences, cultural sciences, pedagogics, ethnopsychology, etc. Focus was made on the concept of polyphony and mutual self-cognition, and the potential of intercultural dialogue from the position of its influence on the development and effectiveness of socio-cultural systems. The methodology of this research consists of empirical and theoretical methods of analysis of scientific and expert works and information from open sources. We studied theoretical and institutional aspects of international migration, including in the sphere of higher education. We determined the components intercultural dialogue from the position of achieving mutual understanding and interest in intercultural exchange. We identified the specific features and problems of intercultural communications in Russian universities and generalized and characterized the experience of successful measures in adapting students to the multicultural educational environment and creating of conditions for effective intercultural dialogue in multinational classical universities of Russia. The value and originality of this work lie in the generalization of the practical experience of the organization of space for intercultural dialogue in multi-

A. I. Seliankina
RUDN University, Moscow, Russia

M. A. Dzhorobaeva
Osh State University, Osh, Kyrgyzstan

B. E. Yuldashev
Tashkent State Transport University, Tashkent, Uzbekistan

Z. A. Zhade (✉)
The Adyghe State University, Maykop, Russia

B. S. Sergi et al. (eds.), *University 4.0 and Educational Technology for Sustainable Development*, Education in the Asia-Pacific Region: Issues, Concerns and Prospects 75, https://doi.org/10.1007/978-981-96-7376-6_16

national classical universities of Russia from the position of its extrapolation and dissemination in other educational establishments.

Keywords Intercultural communication · Intercultural dialogue · Adaptation of students · Multinational university · Globalization · Russia

JEL Codes F22 · F59 · I21 · I23 · J15 · J24 · O15 · O32 · P37 · Z10 · Z12

1 Introduction

At present, globalization is one of the decisive factors in the development of the higher education environment. Its influence is combined with the factors of the development of international trade, labor migration, digitalization, formation of the system of open knowledge, and unification of the higher education systems. In these conditions, the role of borders between countries is erased, and the mobility of students grows. Thin, in turn, forms a range of advantages and problems that determine the character of higher education.

These trends influence the market of educational services, especially in the segment of higher education. The factor of internationalization of universities is a factor of their competitiveness, which stimulates educational establishments to create conditions for attracting foreign students. It should be taken into account that actions aimed at internationalization influence not only the educational process (specialties, educational plans, learning tools) but also the everyday life of students (creation of conditions for comfortable life and studies, support for inclusiveness) and intercultural communication (formation of the culture of international dialogue, social interaction, etc.).

Given this, universities develop the direction of internationalization of studies differently, focusing on individual components of the educational process, everyday life, or intercultural communication. Multinational classical universities stand out among educational establishments. Their main purpose is to provide a wide range of services of higher education to students of different nationalities. The uniqueness of such educational establishments lies in the opportunities that they offer to the world and society, emphasizing the advantages of diversity and the problems that emerge as a result of this diversity.

Thus, the experience of intercultural dialogue in multinational classical universities of Russia poses a large scientific interest and forms a field for the active discussion on the use of the potential of such dialogue and the resolution of problems that are peculiar to interethnic communities. Generalization and rational interpretation of this experience is very valuable from the position of its use in other educational establishment in Russia and other countries.

2 Intercultural Dialogue in the Multinational Classical Universities

The problem of intercultural dialogue belongs to the important directions of research. Its methodological basis lies in the sphere of cultural sciences and philosophy, which originates from the dialectical methodology of sciences on communication. Attention should be given to the concept of "polyphony" and "mutual self-cognition" of cultural values and diversities [1, 2]. The key principles that allow considering interethnic dialogue as a form of communication in the conditions of cultural differentiation and diversity are equality of rights and the search for balance between absorption and identity [3–5].

The methodology is based on the use of empirical and theoretical methods of study and analysis of scientific works in the sphere of psychology, pedagogics, ethnopsychology, social sciences, etc.

Scientific and expert sources, which were used in this research, cover a wide range of connected problems, which shed light on the relevant issues of the study of theoretical provisions of intercultural education [6]; development of value-based and conceptual provisions of intercultural dialogue in a multicultural society [7]; study of the role of pedagogues and educational establishments in the formation of the experience of intercultural communication [8–10]; generalization of foreign and Russian experience of adaptation of foreign students [11, 12]; study of the programs of adapting foreign students to the conditions of studies in Russian universities [13, 14]; study of the directions of ensuring successful intercultural interaction in a multicultural space [15], etc.

The goal of this research was to identify and substantiate the problems and prospects for intercultural dialogue, which exists in multinational classical universities of the Russian Federation, as well as to interpret the parameters and mechanism of this dialogue in the conditions of globalization and connected processes. This involves the tasks of determining the essence and theoretical features of intercultural dialogue and studying its components and conditions of implementation in a multinational classical university from the position of the revealed problems and further prospects.

3 Russian Experience of Intercultural Dialogue in the Multinational Classical Universities

The problem of intercultural communication is one of the most relevant ones from the position of characterizing the socio-cultural development. Mass migration of the population led to the necessity of studying the problem of intercultural communication of people with different ethnic, language, or religious belonging. This problem is most relevant in modern Europe, which experienced a powerful wave of migration at the beginning of the twenty-first century. Given the integration processes in

Europe, and inflow of refugees, and the active movement of labor migrants, the EU government adopted the principles of intercultural dialogue as the basis for an open and fair exchange of thoughts between the representatives of different public and sociocultural groups, which takes place based on mutual agreement, respect, and perception [7].

Similar processes also covered higher educational establishments located mostly in developed countries. Globalization processes allow the unification of educational processes and the creation of conditions for free academic mobility of students of all levels of studies [16]. Russian universities also joined these initiatives, which enabled them to combine international potential that was accumulated in the USSR era with modern trends, initiatives, and processes of the multicultural dialogue of the multinational Russian Federation (Fig. 1).

The objectives of higher education in the conditions of a multinational community are developing skills, knowledge, and perception with students that are necessary for complete self-realization in the conditions of cultural diversity [6]. The main demands that are required for effective development in these conditions are understanding and respect for another culture, a skill of peaceful and productive cohabitation, and agreement with representatives of different nationalities, religions, or races.

In the system of views of education in multinational universities, there is a thought of the potential of intercultural dialogue, which forms better opportunities and conditions for the creation of new values and content. However, even though intercultural communications do not have a confirmed direct influence on economic or socio-cultural development, the idea of a dialogue between different cultures

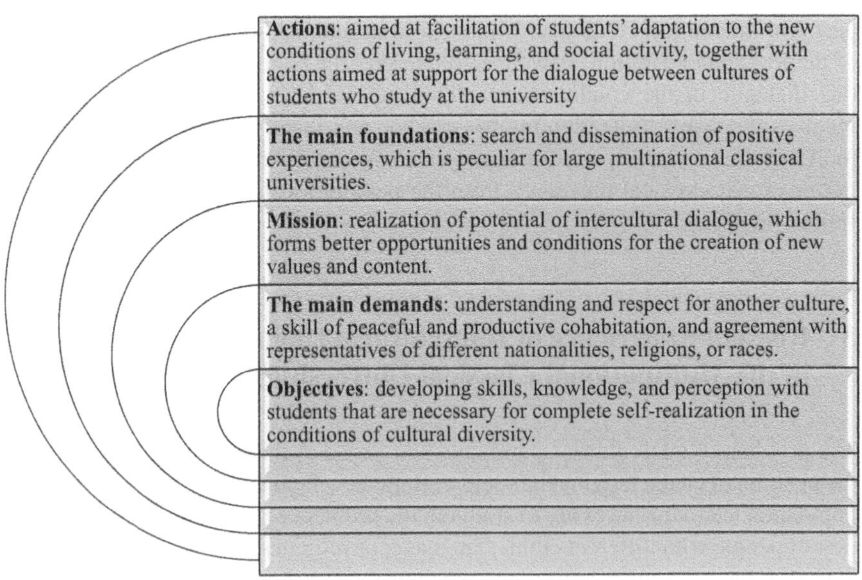

Fig. 1 Intercultural dialogue in a multinational classical university in Russia during globalization. *Source* Done by the authors

remains topical [17] and is widely used in the system of determination of the values and instruments of the development of Russian establishments of higher education.

The problem of intercultural dialogue in Russian universities is rather relevant. This is connected with the government initiatives on the export of educational services and the subsequent growth of the number of foreign students, an increase in intercultural communications, and growth of the number of intercultural conflicts due to differences in language environment, lack of understanding of the conditions of studies or characteristics of life. Such problems negatively influence the quality of international education in Russian universities and lead to the reduction of their potential in the sphere of educational services export [16].

In these conditions, the main foundations for developing an intercultural dialogue lie in the search and dissemination of positive experiences, which is peculiar for large multinational classical universities. One such university is Patrice Lumumba Peoples' Friendship University of Russia (RUDN University). This university is among the leaders in the number of foreign students and positions in international rankings. Around 10,000 foreign students study in RUDN, representing 155 countries of the world. For RUDN University, intercultural interaction is not something new and unclear—it is a norm that predetermines the necessity of intercultural dialogue and the formation of intercultural competency [12].

It should be noted that the basis for intercultural dialogue is often considered through the lens of foreign students adapting to the learning conditions in Russian universities [8]. In this context, the focus is on the following aspect: for foreign students, studies in Russian universities are a large challenge, which emerges due to the specifics of the educational process and the presence of cultural diversity. According to this, the development of efficient actions, aimed at facilitation of students' adaptation to the new conditions of living, learning, and social activity, together with actions aimed at support for the dialogue between cultures of students who study at the university, form the basis for an effective dialogue between the representatives of different ethnic, language, and other groups.

Examples of measures and tools for the adaptation of foreign students to learning in Russian universities in the context of the creation of conditions for effective intercultural dialogue in multinational classical universities of Russia are presented in Table 1.

Significant efforts in the formation of intercultural dialogue are mainly aimed at ensuring the adaptation of foreign students to the conditions and demands of the educational process and everyday life. This is facilitated by the measures of personal character (remote tutorship and pre-university adaptation, free psychological support, etc.), group support (language courses, development of corporate culture, organizational support for students in the educational space, etc.), program on learning Russian culture and history as an educational space (excursion program, development of interactive skills in learning historical sites of Russia, etc.), as well as measures for preserving own cultural identity and intercultural exchange (cultural program, creation of conditions for preservation of identity and culture, formation of the skills of intercultural communication through the corresponding educational program, development of corporate culture based on uniting people of different cultures, races, and nationalities).

Table 1 Measures and tools of intercultural dialogue in the selected multinational classical universities of Russia

University	Specifics of organization of intercultural dialogue	Measures and tools of intercultural dialogue
Russian State Vocational Pedagogical University	Focus on the adaptation of foreign students in the educational environment of the university, based on intercultural communications, respect, openness, and cooperation	– Additional educational courses (mainly language); – Individual psychological service; – Excursion program; – University Saturdays (meetings with famous people); – Cultural program
Tomsk National University	Focus on adaptation of foreign students to conditions of learning and living, formation of an intercultural environment	– Organizational support for students in the university's space; – Language training for foreign students; – Initial adaptation of foreign students to the urban environment; – Creation of conditions for the preservation of identity and cultural adaptation
RUDN University	Formation of convenient space for learning and self-realization of foreign students, creation of conditions for intercultural communication and dialogue	– Development of the system of adaptation (bilingual) training "Understanding each other" and "Intercultural dialogue and emotional intellect"; – Development of interactive programs on learning about historical sites of Russia; – Creation of the office of free psychological support; – Opportunities for adaptation to the environment and labor activities through volunteer activities; – Formation of skills in intercultural communication through an educational program of the master's program "Philosophy: dialogue of cultures"; – Development of corporate culture based on uniting people from different cultures.

Source Done by the authors based on [8, 14, 18–20]

4 The Problem of the Development of Intercultural Dialogue in Universities in the Conditions of an Increase in the Trends of Globalization

In the conditions of an increase in the trends of globalization of the educational environment and the world, the problem of the development of intercultural dialogue in universities is very relevant. Given the interpersonal and communicational character of the problem, its complete resolution belongs to tasks that do not have an unambiguous solution. That is why each case of the emergence of such intercultural dialogue, its development and popularization require a separate study for the determination of general, group, and individual socio-psychological features, trends, and visions.

In a more general aspect, the problem of intercultural dialogue in universities is viewed from the position of intercultural communication, which involves the development of the corresponding communicational channels and the creation of conditions of communication between the representatives of different cultural groups of students and teachers. However, the notion of dialogue—as a tool of not only communication but intercultural exchange—places certain demands on scientific discussions. These demands are aimed at substantiating its essence and potential from the position of intercultural enrichment and strengthening the tools for complete self-realization of the educational process's participants, which is achieved due to the exchange of thoughts, visions, and joint actions.

Measuring of effectiveness of intercultural dialogue belongs to the sphere of non-formalized processes. Thus, the detailed study of this problem has a large potential for strengthening the methodological basis of the issues considered. Focus must be made on the aspects of the dialogue as a communicational process. Here it is important to form a methodology for measuring the level of its favorability, effectiveness, etc. The next stage in the preparation of the appropriate methodology would be the tools of measuring the influence of communications and joint actions within the intercultural dialogue on the state of the studied system.

The prospects for intercultural dialogue in universities, which are implemented in the conditions of globalization and take place under the influence of digital technologies, also require discussion on the study of the experience of using the method of remote education, joint scientific developments within an open science, the introduction of the notion of digital academic mobility, etc.

5 Conclusions

At present, intercultural dialogue is one of the factors of the educational process in higher school. The growth of the influence of globalization, together with the policy of the Russian government, increased opportunities for the internationalization of universities. The result of this was the growth of the number of foreign students in educational establishments. This led to an increase in the number of conflicts and situations of non-acceptance of the representatives of other cultures and nationalities, with the growth of general tension in the educational process. The problems of intercultural communications actualized the demand for active actions of university administrations, which should be aimed at solving problem situations and creating conditions for an effective dialogue between the representatives of different cultures within a joint educational environment.

The experience of multinational classical Russian universities in the studied context has a large potential and is characterized by successful cases in the integration and adaptation of students to studying in a multicultural space. The main measures and actions that could be used in other educational institutions include adaptation of foreign students to the cultural environment of an educational establishment, creation of conditions for their language and social support, and formation of the space to preserve their cultural identity, adaptation to the work environment through the system of volunteering, and specific actions, aimed at the search for dialogue between the representatives of different cultures. Such measures should result in the formation of competencies of intercultural communication, reduction of conflict level between the representatives of different cultures, nationalities, and races, and expansion of opportunities and own intellectual level due to the intercultural exchange and participation in joint programs.

References

1. Qin, X. (2025). Evaluating intercultural learning materials from the RICH-Ed project through a non-essentialist perspective: Chinese University students' and instructors' perceptions. *Asia-Pacific Journal of Teacher Education, 53*(1), 17–49. https://doi.org/10.1080/1359866X.2024.2411598
2. Saidi, B. (2025). Intercultural education in the Global South: Decolonising canonical intercultural models in Moroccan University MA program courses. *Language and Intercultural Communication, 25*(1), 170–184. https://doi.org/10.1080/14708477.2024.2341772
3. Sarwari, A. Q., Javed, M. N., Mohd Adnan, H., & Abdul Wahab, M. N. (2024). Assessment of the impacts of artificial intelligence (AI) on intercultural communication among postgraduate students in a multicultural university environment. *Scientific Reports, 14*(1), 13849. https://doi.org/10.1038/s41598-024-63276-5
4. Tang, F., & Calafato, R. (2024). Exploring plurilingualism, ethnicity, and intercultural effectiveness: Insights from university students. *European Journal of Education, 59*(4), e12738. https://doi.org/10.1111/ejed.12738
5. Wang, X., & Shen, Y. (2024). Is internationalization abroad a prerequisite for university students' intercultural competence development? *Asian Education and Development Studies, 13*(3), 253–266. https://doi.org/10.1108/AEDS-02-2024-0040
6. Banks, J. A. (2002). *An introduction to multicultural education* (5th ed.). University of Washington. Retrieved from https://ucarecdn.com/d82827b2-9e39-40fe-a717-a1aca81ce4a9/ (Accessed 14 November 2023).
7. Council of Europe. (2008). *White paper on intercultural dialogue "Living together as equals in dignity"*. Retrieved from https://www.coe.int/t/dg4/intercultural/source/white%20paper_final_revised_en.pdf (Accessed 14 November 2023).

8. Kazarenkov, V. I., Rangelova, E. M., & Kazarenkova, T. B. (2014). The role of the teacher in shaping students' experience of intercultural interaction. *Bulletin of the Peoples' Friendship University of Russia. Series: Psychology and Pedagogy*, *2*, 5–12. Retrieved from https://cyberleninka.ru/article/n/rol-pedagoga-v-formirovanii-u-studentov-opyta-mezhkulturnogo-vzaimodeystviya (Accessed 14 November 2023).
9. Kazarenkov, V. I., Mikhailova, E. L., & Busel-Kuchinskaia, E. N. (2021). Formation of ethnic tolerance of future social sector specialists in a multicultural university environment. *Bulletin of Polotsk State University. Series E. Educational Sciences*, *7*, 14–21. Retrieved from https://rep.vsu.by/bitstream/123456789/31053/1/%d0%9a%d0%b0%d0%b7%d0%b0%d1%80%d0%b5%d0%bd%d0%ba%d0%be%d0%b2_2021-7.pdf (Accessed 16 November 2023).
10. Troinikova, E. V. (2019). Convergence of multicultural and multilingual education in the practice of foreign language training of students. *Iazyk i kultura* [Language and Culture], *45*, 299–311. https://doi.org/10.17223/19996195/45/21.
11. Aldakimova, O. V. (2020). Scientific and methodological conditions for adaptation of foreign experience in internationalization of higher education. *Samara Scientific Bulletin*, *9*(3), 245–249. Retrieved from https://cyberleninka.ru/article/n/nauchno-metodicheskie-usloviya-adaptatsii-zarubezhnogo-opyta-internatsionalizatsii-vysshego-obrazovaniya (Accessed 15 November 2023).
12. Novikova, I. A., & Maslova, O. V. (2020). T. G. Stefanenko's contribution to the development of ethnopsychology at RUDN University: Inspiring leadership. *Bulletin of the Peoples' Friendship University of Russia. Series: Psychology and Pedagogy, 17*(1), 159–167. https://doi.org/10.22363/2313-1683-2020-17-1-159-167
13. Evtyugina, A. A. (2019). University adaptation program for foreign students. *Contemporary Higher Education: Innovative Aspects, 11*(3), 46–52. https://doi.org/10.7442/2071-9620-2019-11-3-46-52
14. Kostiukova, T. A., Liventsova, E. I., Larionova, A. V., Fakhretdinova, A. P., & Asiutina, O. N. (2020). Development of mechanisms for adaptation of educational migrants: Problems and solutions (based on the example of Tomsk universities). *Bulletin of Tomsk State University, 456*, 188–196. https://doi.org/10.17223/15617793/456/22
15. Koriakina, A. A. (2022). Intercultural interaction in a multiethnic university space. *Obrazovanie i pravo* [Education and Law], *3*, 262–265. https://doi.org/10.24412/2076-1503-2022-3-262-265
16. Korotkova, E. G. (2021). Developing the competence of professional communication between university staff and foreign students in a multicultural educational environment. *Bulletin of the South Ural State University. Series: Education. Pedagogical Sciences*, *13*(3), 80–89. Retrieved from https://cyberleninka.ru/article/n/razvitie-kompetentsii-professionalnogo-obscheniya-sotrudnikov-universitetov-s-inostrannymi-studentami-v-polikulturnoy (Accessed 16 November 2023).
17. Sukhorukikh, A. V. (2020). Educational tradition as a factor in the formation of intercultural dialogue. *Intellekt. Innovatsii. Investitsii* [Intelligence. Innovation. Investments], *1*, 74–81. Retrieved from https://cyberleninka.ru/article/n/obrazovatelnaya-traditsiya-kak-faktor-stanovleniya-mezhkulturnogo-dialoga (Accessed 14 November 2023).
18. RUDN. (2018). The Institute of Foreign Languages of RUDN hosted an Open Day for Master's Degrees. *"It's a beautiful, beautiful, beautiful world!"*. RUDN. Retrieved from https://www.rudn.ru/media/news/about_rudn/v-institute-inostrannyh-yazykov%2D%2Drudn-sostoyalsya-den-otkrytyh-dverey-magistratury-its-a-beautiful-beautiful-beautiful-world (Accessed 15 November 2023).
19. RUDN. (2023a). *Philosophy: Dialogue of cultures. Educational program*. RUDN. Retrieved from https://www.rudn.ru/education/educational-programs/32fa872e-22fa-11ea-80e8-00155dfbaa09 (Accessed 16 November 2023).
20. RUDN. (2023b). *Corporate culture*. RUDN. Retrieved from https://www.rudn.ru/about/corporate-culture (Accessed 16 November 2023).

Problems of Higher Education in Modern Russia in the Context of the Large-Scale Process of Internationalization of Universities

Barno S. Akbarova, Zara A. Mamisheva, Inna N. Gaidareva, and Aleksandr V. Bespyatyh

Abstract This paper dwells on the conceptual foundations and problems of the development of higher education in Russia in the conditions of popularization of the idea of internationalization of universities. We provided a general definition of the essence and role of internationalization, characterized its influence on socioeconomic and cultural processes, and revealed and grouped the main problems of higher education. The research methodology is based on the process approach, retrospective analysis, philosophical reconstruction, observation, generalization, and grouping. Internationalization is an objective process that accompanies the development of universities and higher education around the world. It ensures a positive influence on higher education, which involves an increase in the level of intellectual capital in society, the formation of students' competencies in global communications and openness for cooperation, and an increase in the level of methodological and scientific potential of universities. We characterized the influence of the government on the full-scale internationalization in Russian universities and emphasized the possibility of interpreting it from the position of export and import of educational services. We generalized and characterized the main problems of higher education in Russia in the conditions of large-scale internationalization of universities. We also proposed the grouping of the problems of higher education in Russia, which are manifested in the context of large-scale internationalization of universities. These problems cover a wide range of public processes, including the commercial-

B. S. Akbarova
Tashkent State University of Economics, Tashkent, Uzbekistan
e-mail: b.akbarova@tsue.uz

Z. A. Mamisheva · I. N. Gaidareva
The Adyghe State University, Maykop, Russia

A. V. Bespyatyh (✉)
Vyatka State University, Kirov, Russia
e-mail: bespyatyh@vyatsu.ru

B. S. Sergi et al. (eds.), *University 4.0 and Educational Technology for Sustainable Development*, Education in the Asia-Pacific Region: Issues, Concerns and Prospects 75, https://doi.org/10.1007/978-981-96-7376-6_17

ization of education and autonomy of universities, improvement of scientific and methodological levels of universities, expansion of horizontal connections, and increase in the level of general competence.

Keywords Higher education · Internationalization of higher education · Problems of higher education · University · Globalization · Export of higher education

JEL Codes F22 · I23 · I28 · J24 · N35 · O15 · P33 · Q56

1 Introduction

Dynamic public development assigns the system of higher education a special role, which is manifested through the influence on socioeconomic and cultural processes. Apart from this, the development of the system of higher education is one of the factors in the formation of the country's competitiveness, the main tool for raising its intellectual capital and the level of innovativeness and well-being. That is why any processes connected with the development and modernization of university education must be the focus of scientific research.

One of the processes that actively influence the state and perspectives of higher education is its internationalization. It is manifested through the active integration of university processes, academic staff, and students into the global educational environment, which has its principles and norms. According to them, higher education establishments are considered subjects of the educational services market that have autonomy in making managerial decisions and implementation of the policy of formation of educational programs, international activities, etc. However, government financing is an important source of financing for Russian universities and a significant tool for influencing their activities.

Large-scale internationalization of Russian universities is, to a great extent, a reaction to government initiatives, rather than an independent decision on determining the goals of the policy of international activities of universities. Together with other processes, this forms limitations and problems in the development of Russian higher education establishments in the conditions of a globalized educational space. These problems have different origins and manifestations. On the one hand, they restrain the active progress of the internationalization of higher education in Russia, and, on the other hand, form unfavorable conditions for its further development due to lack of coordination of many processes, conditions, and principles. Identification and resolution of these problems have a large potential for the transition to a higher level of development of university education in the conditions of globalization and the digital economy.

2 The Main Aspects of Higher Education Internationalization

The study of the problems of higher education in the context of internationalization has its origin in the range of basic provisions in the sphere of pedagogics [1], social sciences [2, 3], international relations [4], educational establishments management [5], and public administration [6]. These provisions determine the goal of internationalization, which may vary depending on context and offer methodological tools for studying internationalization from the position of historical process, public administration, influence on socio-cultural contacts between different countries, etc.

The use of the process approach allows tracking the states of implementing the internationalization of higher education in Russian universities, from independent contacts and isolated cases of academic mobility to full-scale integration of different aspects of cooperation between universities of different countries, international organizations, and governments. According to this, the methodology of this work is comprised of retrospective analysis, philosophical reconstruction of the socio-historical transition between different stages of internationalization of higher education, and general scientific methods of observation, generations, grouping, etc.

The considered provisions and used materials cover a wide range of aspects connected with the research on the process of internationalization of higher education, including the formation of its conceptual provisions [7–10]; determination of the specifics of internationalization from the position of universities [11]; identification of its problems [12] and factors of influence on the competitiveness of universities [5]; generalization of the impact of internationalization on the modern public development [13, 14], and substantiation of its further prospects [15, 16]. Each direction of the research, depending on the context, forms a list of problems and proposed measures to solve them.

The purpose of this work was to reveal, group, and systematize the current problems faced by the establishments of higher education in Russia in the context of the large-scale process of their internationalization. Achievement of this goal involves several scientific tasks, which include the necessity of determining the essence and conceptual specifics of internationalization and studying the stages of its gradual implementation in Russia with the simultaneous determination of important problems, which hinder internationalization or emerge as a result of its implementation in the country's educational space.

3 Problems of Higher Education in Russia in the Conditions of Large-scale Internationalization of Universities

The conceptual foundations of internationalization, as a separate process in the environment of higher education, were initially formed based on determining its positive influence on the quality of education and the effectiveness of universities' activities [7]. They were largely opposed to globalization and treated as processes aimed at preserving the national & cultural language uniqueness of partners in the process of interaction [8, 9]. In this context, internationalization was treated as an effective integration of global, international, and intercultural interaction in the process of provision of educational service [15].

On the whole, the internationalization of higher education is an inseparable element of economic and social development, which ensures the survival and development of mankind in conditions of constant complication of living conditions. Given the characteristics of higher education, internationalization is its inseparable part, because it proposes opportunities for receiving innovative language, formation of intercultural thinking, and openness to communications and research at the international level. In this context, it is an adaptive feature of higher education, which is implemented through openness to the demands of the modern global world [14].

In the scientific literature, there are several popular approaches to the assessment of the role of internationalization in higher education. Thus, it may be considered as the following:

- One of the elements of a complex system of higher education management, which contains the innovative component and penetrates all sub-systems of universities, reorienting and redirecting them at international cooperation [16];
- The process of training specialists for work in the conditions of the global economy. According to this, competencies, which are connected with mobility, ability to work in a multicultural environment, and ability to create and manage international communication channels acquire larger relevance [12];
- The factor of ensuring the competitiveness of the university through its involvement in the world educational space, which is most fully interpreted through global rankings of higher education establishments [5].

The first large-scale attempt at the internationalization of Russian higher education was made in the early 2000s when Russian universities tried to integrate into the world's educational space. This allowed for a significant increase in the level of mobility of students and lecturers and the formation of more complex forms of cooperation with foreign universities. However, the comprehensive internationalization was implemented only by several Russian universities, including Lomonosov Moscow State University, RUDN University, National Research University "Higher School of Economics," etc. Government programs for internationalization—projects "Export of education" and "Priority—2030" gave additional stimulus to Russian universities. This ensured government support and scaling of internationalization at all universities of Russia for an increase in the share of non-resource exports, which is combined with the development of scientific potential, development of intellectual capital, and resolution of demographic problems [16]. The main tasks of the projects were the improvement of the attractiveness of Russian universities and conditions of foreign students' stay in Russia, the development of the brand of Russian higher education, and the improvement of the financial results of universities' activities and the entire system of the country's higher education [5].

It should be noted that the internationalization of higher education is a complex process that can be interpreted in different manners from the positions of its participants. Thus, exporters of educational services can consider it as an element of foreign economic and cultural expansion, a tool of "soft power', which ensures the achievement of not only economic but also cultural and political goals. Importers of higher education services can consider internationalization as a tool for raising the intellectual capital of the country, improving the level of education quality, and joining the global educational and innovative space. Another focus of attention of exporters of higher education services is the search, training, and involvement of talents to solve national or global problems.

In this context, Russia belongs to exporters and importers of higher education services at the same time. On the one hand, Russian students study in universities in the USA and Europe, receive access to the leading knowledge in the sphere of technology, and acquire cultural values of the Western world. On the other hand, Russian universities accept students from the countries of the CIS, South-East Asia, and Africa, ensuring their cultural and information expansion.

A specific feature of such dual status in modern conditions is that the inclusion of Russian students into the Western educational environment is decreasing, and the influence on the educational space of developing countries is set onto the similar expansion of Western universities. Besides, it is necessary to heed the increase in China's impact on the global environment of higher education. In such conditions, competition in the sphere of internationalization of higher education grows constantly.

Against the background of these processes, it should be noted that the active implementation of internationalization in the system of higher education in Russia was connected with a range of problems. The main of these problems are connected with intercultural differences, the problem of insufficient commercialization of Russian universities, differences between internationalization and localization of higher education, and the necessity of realization based on the principles of sustainable development. The problems of Russia's higher education in the context of large-scale internationalization are shown in Table 1.

One of the relevant problems of Russian universities in the context of their internationalization is institutional inconsistency. According to it, the conceptual basis of internationalization in its modern understanding has been formed by Western universities. It involves certain transformation of the Russian system of higher education, which is not complete yet. Thus, the preservation of the trend of internationalization of Russian universities puts them in a difficult situation. Possibilities of development higher education in modern Russia in the context of the large-scale process of internationalization of universities are shown in Fig. 1.

The main institutional problem in this context is the unpreparedness of Russian universities for comprehensive entrepreneurial activities in the form of autonomous subjects of entrepreneurial activities.

Besides, the problems of Russian universities in these conditions are closely connected with the necessity of raising their scientific & pedagogical and information & methodological potential, deepening horizontal ties between the institutions of

Table 1 The key problems of higher education in Russia in the conditions of large-scale internationalization of universities

Problem	Description
Incomplete institutional transformation	Russian universities have not fully realised themselves as independent entrepreneurial subjects, which limits their autonomy and leads to dependence on government financing
A mismatch between the modern status of the university and its market basis	Universities, as market subjects, must seek methods for independent commercialization of their scientific and educational programs.
Insufficient level of development of the scientific potential of universities	One of the components of internationalization is the availability of research centers with universities, which must be tools for attraction of financing, development of innovative knowledge, and attraction of talents.
Weak horizontal ties between Russian universities	Internationalization—as a tool of decentralization and democratization—involves active interaction and cooperation between universities of different profiles for the formation of a comprehensive scientific and educational product.
The mass character of higher education with its negative influence on quality	The concept of the mass character of higher education, which formed in the late twentieth century, negatively influenced its quality through unpreparedness of the infrastructure, academic staff, and insufficient motivation of a wide range of students.
Low level of personnel and methodological readiness of universities for internationalization	Insufficient number of highly qualified academic staff with knowledge of foreign languages, lack of educational and methodological materials
Low level of digital literacy	Limited use of the digital potential of internationalization of Russian universities due to the insufficient level of digital literacy of lecturers and students.

Source Made by the authors based on [5, 10, 12, 15, 16]

higher education of different profiles, solving the problem of low motivation for obtaining higher education, and ensure development by the principles of sustainability.

Another direction for these problems is the limitations of Russian universities in the comprehensive use of digital technologies and remote and online education for improvement and expansion of international cooperation in the context of internationalization.

4 Internationalization of Russian Universities: Problems and Solutions

The large-scale internationalization of Russian universities is a complex process, within which important problems arise. They restrain the development of higher education and do not allow achieving the goals of internationalization. Most of these problems are of non-system character and thus require additional research and discussion in the expert and scientific circles. Thus, the negative influence of the mass character of higher education on its quality has not been fully proven. Such

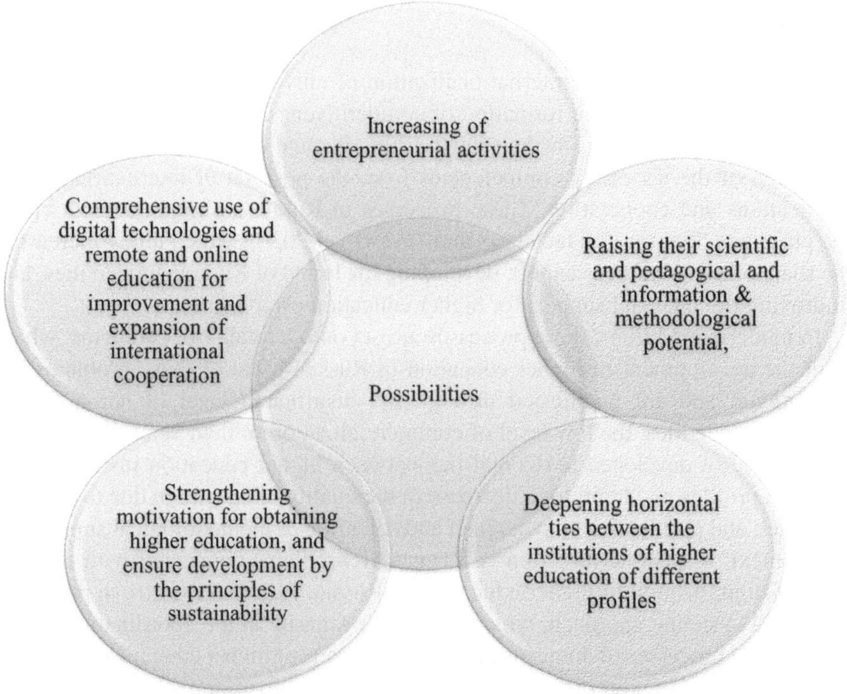

Fig. 1 Possibilities of development higher education in modern Russia in the context of the large-scale process of internationalization of universities. *Source* Developed by the authors

conclusions are mainly based on empirical studies or have small representativeness. To confirm or disprove the presence of a negative correlation between the quality of higher education and its mass character, including due to internationalization, it is necessary to conduct deeper scientific research based on a more representative sample.

One of the goals of the internationalization of universities is the effective transformation of higher education through the adaptation and modernization of national educational systems based on parity and equality. Adherence to these principles in the conditions of modern internationalization is often not ensured. The study and discussion of this problem are an important scientific task, the resolution of which will allow ensuring real mutual enrichment of the systems of higher education in different countries, which interact through academic exchange, scientific and methodological cooperation, etc.

However, the most relevant—in the modern conditions—problem of internationalization of higher education in Russian universities is a decrease in cooperation with universities from the USA and Europe. In this context, it is important to determine the real influence of such a situation on the system of higher education in Russia and develop and substantiate the compensation mechanisms for reorienting the international cooperation of Russian universities.

5 Conclusions

The large-scale process of internationalization of universities is an important process, which is aimed at the formation of competitiveness of the Russian market of higher education, support for the growth of intellectual capital of the country, and formation of the necessary competencies from the position of international communications and cooperation. These processes in Russia are implemented within the programs "Export of education" and "Priority—2030," which allow increasing the share of non-resources export, developing the brand of Russian universities, and improving the financial support for higher education establishments.

In modern conditions, these processes are set onto a totality of problems, which limit the development of higher education in Russia. Most of these problems are institutional and are manifested through the insufficient level of autonomy of Russian universities, the low level of commercialization of their scientific product, the absence of developed horizontal ties between higher education institutions of different profiles, the trend for a decrease in the quality of education due to its mass character, and problems with personnel and scientific & methodological support. In this context, a separate direction is formed by problems connected with the low level of digitalization of higher education in Russia. Resolution of these problems requires a systemic approach, which is aimed primarily at the coordination of the institutional processes of domestic and global markets of higher education, improvement of the level of entrepreneurial activity and autonomy of universities, and resolution of current problems regarding reorientation of the directions for internationalization in the current conditions.

References

1. Crowley-Vigneau, A., Kalyuzhnova, Y., & Baykov, A. (2024). World-class universities cut off from the West: Russian higher education and the reversal of the internationalization norm? *Higher Education Quarterly, 78*(3), 709–729. https://doi.org/10.1111/hequ.12481
2. Dixon, R. (2024). Assessing the internationalization of higher education: A case study of three universities in the Kentucky-Ohio Region. *Journal of Higher Education Policy and Leadership Studies, 5*(3), 153–166. https://doi.org/10.61186/johepal.5.3.153
3. Ghasab, A. A., Heydari, M., & Arab, M. (2025). Foresight for Internationalization of Medical Sciences Universities in Iran in the 2040 horizon. *European Journal of Futures Research, 13*(1), 2. https://doi.org/10.1186/s40309-024-00245-1
4. Labraña, J., Henríquez, A., LaTorre, P., Puyol, F., Gómez, M. R., & López, N. (2024). New public management and internationalization in Latin American top-tier universities: A multiple case study. *Journal of Studies in International Education, 28*(4), 568–587. https://doi.org/10.1177/10283153231221652
5. Riazantseva, I. V., Soboleva, E. L., & Matveeva, I. V. (2022). University internationalization as a key aspect of training competitive specialists. *Tvorchestvo i sovremennost* [Creativity and Modernity], *2*(17), 51–59. Retrieved from https://cyberleninka.ru/article/n/internatsionalizatsiya-vuza-kak-klyuchevoy-aspekt-podgotovki-konkurentosposobnyh-spetsialistov (Accessed 22 December 2023).

6. Yang, X., & Li, T. (2024). Impacts of different internationalization modes on the global competence of Chinese university students. *ECNU Review of Education, 7*(4), 945–970. https://doi.org/10.1177/20965311241256363
7. Altbach, P. G., & Knight, J. (2007). The internationalization of higher education: Motivations and realities. *Journal of Studies in International Education, 1*, 290–305. https://doi.org/10.1177/1028315307303542
8. De Wit, H. (2012). The history of internationalization of higher education. In D. K. Deardorff, H. de Wit, & J. D. Heyl (Eds.), *The SAGE handbook of international higher education* (pp. 43–60). SAGE. https://doi.org/10.4135/9781452218397.n3
9. Knight, J. (1997). Internationalization remodeled: Delnition, approaches, and rationales. *Journal of Studies in International Education, 1*(8), 5–31. https://doi.org/10.1177/102831530326O832
10. Smaliakou, D. (2021). Institutional conditions of internationalization of higher education: Mass higher school. *The Education and Science Journal, 23*(5), 11–37. https://doi.org/10.17853/1994-5639-2021-5-11-37
11. Beliaeva, V. S., & Frolova, E. D. (2021). Differentiation of universities as a feature of the internationalization of the modern stage of higher education. *International Scientific Research Journal, 8–4*(110), 110–116. Retrieved from https://cyberleninka.ru/article/n/differentsiatsiya-universitetov-kak-osobennost-internatsionalizatsii-sovremennogo-etapa-vysshego-obrazovaniya (Accessed 23 December 2023).
12. Prokhorov, A. V. (2015). Problems of internationalization of modern educational space. *Nauka. Obshchestvo. Gosudarstvo* [The Science. Society. State], *3*(3), 16. Retrieved from https://cyberleninka.ru/article/n/problemy-internatsionalizatsii-sovremennogo-obrazovatelnogo-prostranstva (Accessed 23 December 2023).
13. Andriushina, E. V., & Riabinina, A. M. (2023). Consequences of the internationalization of higher education in the modern world: Theory and practice. *Vlast* [The Power], *31*(2), 24–29. Retrieved from https://cyberleninka.ru/article/n/posledstviya-internatsionalizatsii-vysshego-obrazovaniya-v-sovremennom-mire-teoriya-i-praktiki (Accessed 24 December 2023).
14. Yvchen, W., & Bazarova, T. S. (2023). Internationalization of education is a leading trend in the modern world. *Bulletin of Buryat State University. Education. Personality. Society, 1*, 13–18. https://doi.org/10.18101/2307-3330-2023-1-13-18
15. Nikitenko, E. V. (2023). Internationalization of higher education in Russia: In search of ways of development. *Vysshee obrazovanie v Rossii* [Higher Education in Russia], *32*(5), 125–137. https://doi.org/10.31992/0869-3617-2023-32-5-125-137
16. Riabinina, A. M. (2020). Internationalization of the higher education system in the era of scientific and technological revolution. *Gosudarstvennoe upravlenie. Elektronnyi vestnik* [Public Administration. Electronic Bulletin], *81*, 297–321. https://doi.org/10.24411/2070-1381-2020-10090

Image Communications of Universities: Current Technologies

Yulia N. Zemskaya, Evgeniya A. Kuznetsova, and Anna V. Osmolovskaia

Abstract The research comprehensively examines the image communications of a higher education institution. The research aims to investigate the influence of the teacher's image as a communicator on the institution's image. The research objectives are (1) to identify the components of the concept of a positive image of a contemporary university, (2) to explore media and non-media technologies of institutional image communications, and (3) to study the impact of the teacher's image on the image of a contemporary university. The research presents a typology of communications in a higher education institution involving the participation of a teacher. These communications were divided into four groups: (1) personal communications that involve the absence of digital intermediaries; (2) Intranet communications (e.g., broadcasting lectures via intra-university channels); (3) Internet communications (e.g., a teacher's participation in communications on social networks, posting their publications on websites, etc.); and (4) media communications (e.g., the appearance of a teacher on a television show as an expert). As a result of the survey-based research, the authors established that the teacher's image influences the university's image among the student audience; the information conveyed by the teacher instills trust in the students, and the way this information is conveyed facilitates the learning process. The research hypothesis, which posited that the teacher as a communicator impacts the image of the higher education institution, was fully confirmed.

Keywords Image · Image communications · Image of the university · Image of the teacher · Image perception

---

Y. N. Zemskaya · A. V. Osmolovskaia (✉)
RUDN University, Moscow, Russia
e-mail: zemskaya_yun@pfur.ru; glagoleva_av@pfur.ru

E. A. Kuznetsova
Higher School of Economics, Moscow, Russia
e-mail: evgeniya.kuznetsova@hse.ru

B. S. Sergi et al. (eds.), *University 4.0 and Educational Technology for Sustainable Development*, Education in the Asia-Pacific Region: Issues, Concerns and Prospects 75, https://doi.org/10.1007/978-981-96-7376-6_18

JEL Codes M31 · M37 · M39 · L14 · I23 · P360

1 Introduction

Higher education institutions are responsible for advancing science and cultivating highly skilled professionals with up-to-date tools and deep knowledge in specific fields. They often serve as intellectual and cultural hubs within cities and urban areas. A contemporary university must be dynamic and innovative, establish effective communication with various audiences, and contribute to maintaining and developing its positive image [1].

Research on the image of universities has been actively developing in recent decades in Russian and international academia [2]. This is largely due to a reevaluation of the role of universities in shaping future professionals and increasing competition among educational institutions and fields of study [3]. Educational institutions are objectively compelled to follow societal trends, leading to the emergence of new faculties and areas of study and the introduction of new disciplines in the curriculum; interest from applicants and students in traditional leaders wanes [4]. To maintain the effectiveness of a university, continuous monitoring of various aspects of its activities is necessary [5].

The relevance of this research is also determined by the change in the format of educational services delivery after the COVID-19 pandemic and the need to assess the potential impact of this factor on the perception of the educational structure's image [6]. During the COVID-19 pandemic, the previous direct interaction between the teacher and the group of students became impossible; it was replaced by mediated interaction based on the use of digital technologies. This sparked a public debate about whether universities are needed as intermediaries in providing knowledge, or if knowledge can be obtained directly through modern electronic technologies. It also raised questions about the necessity of a teacher, their role, and the qualities they should possess to inspire trust in the knowledge they disseminate to modern students who have access to unlimited information content. In other words, the image of universities has become more dependent than ever on the image of the teachers. The situation of external constraints in our country has also led to a certain transformation of the traditional concept of providing educational services.

The research aims to investigate the influence of the image of a teacher as a communicator on a higher education institution's image.

The specified goal is achieved by consistent solutions to the following interrelated research tasks:

- To identify the components of the concept of a positive image of a contemporary university;
- To explore media and non-media technologies of institutional image communications;

- To study the impact of the teacher's image on the image of a contemporary university.

2 Approaches to Identify the Image of Higher Education Institutions

To understand the phenomenon of image, it is necessary to consider the individuality of human perception.

Since the research on the image of higher education institutions has been going on for decades in Russia and abroad, it is possible to identify not only a certain commonality of methodological approaches but also some methodological regularities despite the diversity of scientific works.

The formation of a university's brand image should have a comprehensive strategic nature [7]. Part of such a strategy includes research on the image communication of universities in terms of the effectiveness of the technologies used.

In general, research interest is focused on studying knowledge about the university, associations arising regarding its qualitative characteristics, information in the media, correspondence of expectations to the actual state of the educational structure, satisfaction of students and staff, positioning in relation to competing universities, knowledge exchange, academic connections, and contacts [8–11].

Regional features of the university and the organization of events within the university (e.g., anniversary celebrations) are also important from the perspective of research interest [12]. The image is influenced by the assessment of research activities, indicators of graduate employment, and academic mobility.

The nomination influences the image of the social object. Already in the name—higher education institution—there are premises for its positive perception by the audience: a university is traditionally associated with high quality of services provided [13], which gives it a certain level of trust. Simultaneously, there is a constant gradation of universities based on rating assessments. Thus, for a specific educational structure, a positive image is important not only in itself but also in comparison to competitors among other higher education institutions [14].

Recently, the view of higher education institutions as an enterprise has become increasingly widespread [7, 15]. Particularly, M. Probst [16] considers the university from the perspective of its servile function, that is, as an entity offering comprehensive services not limited to education. Based on this position, a positive image of a university should align with the parameters of an effective organization.

Media communications play a crucial role in shaping and managing the university's image. From the standpoint of contributing to maintaining a positive university image, media communication channels can be divided into managed channels, which include university periodicals and websites, and unmanaged channels, which encompass all other media and require continuous PR interaction.

Researchers regularly study the perception of the university's image, which is formed in the media and the Internet [17–20]. Scholars highlight the influence of cross-platform capability, multichannel communication, multimedia content, and regular content updates on the positive brand image of a higher education institution in the media.

In addition to media channels, there are other communication channels and technologies that contribute to managing the image of an educational institution. In this research, the authors view teachers as communicators who, by actively participating in various university communications, exert influence on the perception of the university's image. Accordingly, the research hypothesis posits that a teacher, as a communicator, impacts the image of the higher education institution.

The authors categorize university communications involving faculty members into the following four groups:

- Personal communications involving the absence of digital intermediaries;
- Intranet communications (e.g., broadcasting a lecture via intra-university channels);
- Internet communications (e.g., a teacher's participation in communications on social networks, posting their publications on websites, etc.);
- Media communications (e.g., the appearance of a teacher on a television show as an expert).

3 Empirical Research of the Image of the Moscow Universities

As demonstrated by the analysis of practices in higher education institution image research, the standard methods include a survey and associative experiment to determine spontaneous associations. To achieve the purpose of this research, the authors applied the survey method.

To develop a methodology for researching the image of a higher education institution, a significant stage is the selection of respondents.

It should be noted that there is no single approach to respondent sampling in the research of the higher education institution's image. The division of the image of higher education institutions into internal (formed among the internal audience: staff, students, etc.) and external (formed among the so-called external audiences: the population of the city where the university is located, journalists, etc.) is fundamental for many Russian and foreign studies.

The following target groups are also distinguished [12]: residents of the community where the educational institution is located, politicians and entrepreneurs, university staff (engaged in research and educational activities and not), students, and graduates [21]. Separate studies are devoted to the audience of university stakeholders: survey instruments oriented to them are being developed and tested [22].

It should be noted that this list does not include the group of potential consumers of educational services—university entrants. Despite the importance of the opinions of this group, it should be recognized that its allocation is conditional because it is challenging to determine which university the applicant will eventually enter. Despite this, the interest of researchers is also directed to this group. For example, they include H. Habicht [22], who calls them potential learners.

The choice of respondents for this research was conditioned by the research aim and the audience's specifics. The respondents of this research were students of 3–4 years of study at Moscow universities. Such a sample was conditioned by the fact that students of initial courses, in the researchers' opinion, are less likely to have work experience in their specialty.

The respondents were divided into two groups:

1. Students who, in addition to studying at university, have work experience in their future specialty;
2. Students who do not have this experience.

The distribution of students into these two groups is determined by the fact that students who already have work experience in their specialty, as a rule, can evaluate the class material from the point of view of its applied value and are more informed about alternative sources of obtaining professionally demanded information.

A total of 200 students were interviewed, of which 100 belonged to the first group and 100 to the second. The study was of an exploratory nature, and its results were to contribute to the formulation of hypotheses for the next research stage.

The research period is August–September 2023.

The range of research questions concerned the following topics:

- "Whether the teacher's image affects the university's image";
- "Whether the information communicated by the teacher is trustworthy;
- "Whether the knowledge the teacher communicates to the audience is unique."

Unique knowledge was understood as knowledge that cannot be freely available, as well as the method and form of delivering this knowledge to the audience.

Scaling was used to distribute the answers. The respondents were asked to choose a response option in the following range:

- Very important;
- Important;
- May be important;
- Not really important;
- Not important.

For example, the distribution of respondents' answers to the question about how important the image of a particular teacher is for the formation of a positive image of the university is presented in Fig. 1.

Responses to the question about students' perceptions of the importance of non-media communications with faculty were distributed as follows (Fig. 2):

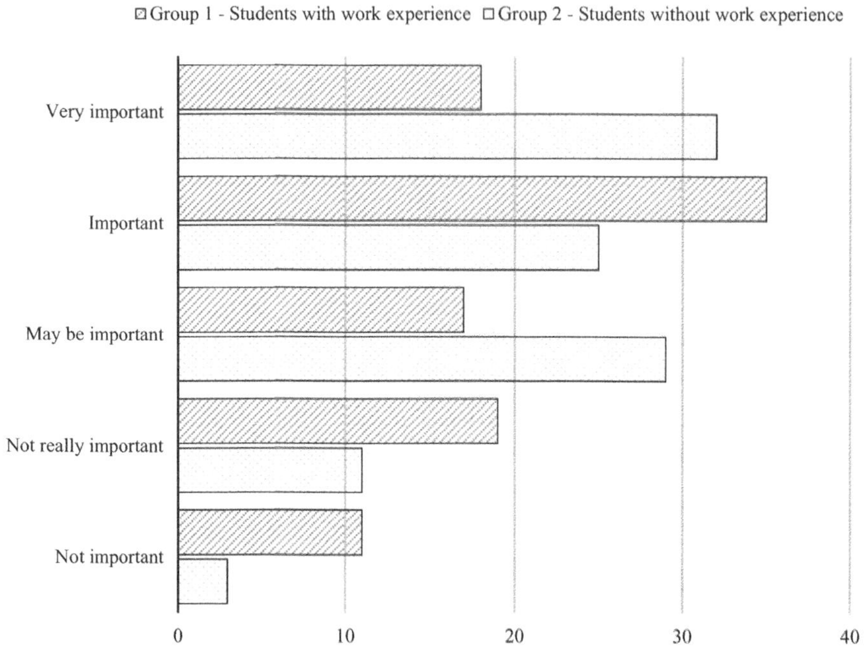

Fig. 1 Distribution of respondents' answers to the question about the influence of the teacher's image on the university's image. *Source* Developed by the authors

Despite significant differences in the answers of different groups, analyzed in a separate stage of the study, the research results show that, according to the respondents' estimates, the teacher's image and communication with him or her influence the university's image among the student audience. The information communicated by the teacher causes trust among students; the way of delivering this information facilitates the learning process.

Thus, the research hypothesis is fully confirmed.

4 The Influence of the Teacher's Image as a Communicator on the University's Image

The novelty of this research lies in the study of the influence of the teacher's image as a communicator on the university's image. Considering current technologies of image communication of universities, it should be noted that the teacher is an active and one of the most influential participants in university communication aimed at the student audience. Accordingly, a teacher needs to master and apply advanced teaching technologies that have educational and image character. If we consider

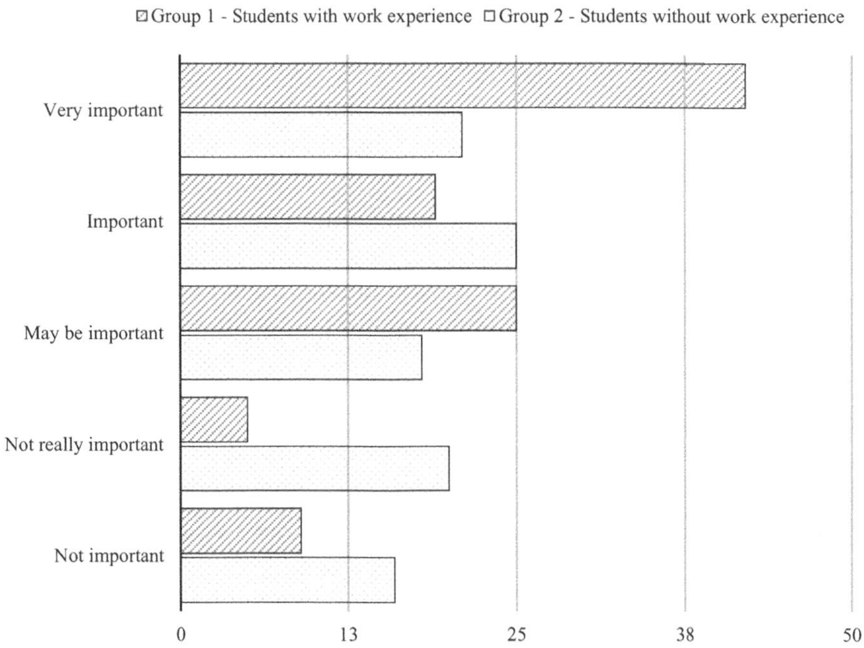

Fig. 2 Distribution of respondents' answers to the question about the importance of non-media communications with the teacher. *Source* Developed by the authors

such technologies as the global university, the teacher's image forms a positive image of the educational project.

The sampling of the student audience was carried out in such a way as to give an opportunity to assess the teacher's image to students undergoing theoretical training and students who already have experience of professional activity. This division of respondents made it possible to obtain objective results.

Simultaneously, the authors cannot claim that the teacher's image affects the perception of the university's image among people who are not students (e.g., external audience). To determine the possibilities of the influence of the teacher's image on the university's image among external audiences, it is necessary to conduct separate independent research.

5 Conclusion

Thus, the conducted research allowed the authors to identify and evaluate the influence of the teacher's image on the image of higher education institutions. During the research, the authors studied the components of the concept of a positive image of contemporary higher education institutions, considered media and non-media technologies of image communications of higher education institutions, and used a survey to conduct the research on the influence of teacher's image on the image of contemporary higher education institution.

References

1. Mendoza-Villafaina, J., & López-Mosquera, N. (2024). Educational experience, university satisfaction and institutional reputation: Implications for university sustainability. *International Journal of Management Education, 22*(3), 101013. https://doi.org/10.1016/j.ijme.2024.101013
2. Muslim, A.B., Imperiani, E., Musthafa, B., Farlian, T., & Francisco, A. (2025). Enhancing international student motivations in Indonesian universities: Building academic reputation in the Global South. *Globalisation, Societies and Education*. https://doi.org/10.1080/14767724.2024.2439420
3. Phan, T. C., Ngo, V. M., Nguyen, T. P., Saefullah, K., & Doran, D. (2024). University social responsibility and teachers' satisfaction: The mediating role of reputation and image. *Journal of Social Studies Education Research, 15*(1), 57–90.
4. Romiani, Y., Ghoraishi Khorasgani, M. S., & Norollahee, S. (2024). Higher education managers' perception of university reputation components: A model for Middle Eastern countries. *International Journal of Educational Management, 38*(2), 525–548. https://doi.org/10.1108/IJEM-09-2023-0433
5. Soysal, Y. N., Baltaru, R. D., & Cebolla-Boado, H. (2024). Meritocracy or reputation? The role of rankings in the sorting of international students across universities. *Globalisation, Societies and Education, 22*(2), 252–263. https://doi.org/10.1080/14767724.2022.2070131
6. Sun, Z. (2024). Reactive management of university reputation during the COVID crisis: Governance and the control of public narratives in Chinese higher education. *Higher Education Policy*. https://doi.org/10.1057/s41307-023-00338-9
7. Stensaker, B. (2005). *Strategy, identity and branding – Re-inventing Higher Education Institutions*. City University.
8. Alessandri, S. W., Yang, S.-U., & Kinsey, D. F. (2006). An integrative approach to university visual identity and reputation. *Corporate Reputation Review, 9*(4), 258–270. https://doi.org/10.1057/palgrave.crr.1550033
9. Arpan, L. M., Raney, A. A., & Zivnuska, S. (2003). A cognitive approach to understanding university image. *Corporate Communications: An International Journal, 8*(2), 97–113. https://doi.org/10.1108/13563280310475535
10. Baier, G., & Zocher, D. (2008). Imageanalyse der Westsächsischen Hochschule Zwickau. *Methoden und Ergebnisse einer Unternehmensbefragung*. Camppus 3, Magazin der WHZ, 2. Jg., N1, März.
11. Ivy, J. (2001). Higher education institution image: A correspondence analysis approach. *The International Journal of Educational Management, 15*(6), 276–282. https://doi.org/10.1108/09513540110401484
12. Fischer, J. (2008). *Das Image der Universität Leipzig: Ergebnisse von Erhebungen 2007*. Amt für Statistik und Wahlen.
13. Guhn, M. (2007). *Hochschulfinanzierung und Hochschulqualität. Eine interdisziplinäre Analyse zum Verhältnis beider Bereiche der Hochschulökonomie* (Dissertation). Universität St. Gallen.
14. Benkenstein, M., Bastian, A., & Pflieger, D. (2003). *Marketing-Management für Bildungsinstitutionen. Aus- und Aufbau langfristiger Wettbewerbsvorteile*. Deutscher Universitätsverlag.
15. Hermann, W. (2008). Unternehmen Universität. In I. Hausladen (hrsg.v.), *Management am Plus der Zeit: Strategien, Konzepte und Methoden* (pp. 33–38).

16. Probst, M. (2008). *Vom Eindruck zur Meinung: Der Zusammenhang zwischen Image und Reputation bei komplexen Dienstleistungen* (Dissertation). Cuvillier Verlag.
17. Glagoleva, A. V., Kuznetsova, E. A., & Zemskaya, Y. N. (2019). Brand communications of leading Russian universities in social media (on the example of RUDN University). In A. D. Krivonosov (Ed.), *Branding as a communicative technology of the 21st century: Proceedings of the V International Scientific and Practical Conference* (pp. 177–180). Saint Petersburg State University of Economics.
18. Glagoleva, A. V., Zemskaya, Y. N., & Kuznetsova, E. A. (2018). System of brand communications of the global university. In A. D. Krivonosov (Ed.), *Branding as a communicative technology of the 21st century: Proceedings of the V International Scientific and Practical Conference* (pp. 71–75). Saint Petersburg State University of Economics.
19. Glagoleva, A. V., Zemskaya, Y. N., & Kuznetsova, E. A. (2021). From the corporate culture of national university to the global university: The case of the RUDN University. *RUDN Journal of Economics, 29*(1), 64–75. https://doi.org/10.22363/2313-2329-2021-29-1-64-75
20. Siebenhaar, K. (2007). *"Master your Service!" Die Universität als Dienstleister. Band 12: Public Relations und Kommunikationsmanagement*. Vistas Verlag.
21. Schneider-Sliwa, R., Erismann, C., Griebel, C., Hoffmann, A., Saalfrank, C., & Wieland, B. (2010). *Attraktivität und Image der Universität Basel aus Sicht von Studierenden* (Vol. Band 33). Schwabe Verlag.
22. Habicht, H. (2009). *Universität und Image: Entwicklung und Erprobung eines stakeholderorientierten Erhebungsinstitutionen*. Gabler.

Features of Influence Marketing Technologies in the Field of the Promotion of Higher Education Institution

Nina V. Trubnikova and Viktoria V. Shlenskaya

Abstract The research aims to identify the image of a blogger that best aligns with the objectives of promoting Russian universities. The authors employ a comparative analysis of influencers based on several characteristics (platform, blog topic, self-positioning of the blogger, etc.). Practical materials for the research were selected from the communication campaign of the Faculty of Life Science at the Moscow Institute of Psychoanalysis (MIP) in the blogosphere for 2023. The analysis results in a spectrum of influencer image characteristics that are crucial to consider when planning a communication campaign for university promotion. In this case, it is most effective to collaborate with bloggers whose image reflects intellectuality and specialization in a particular field. The blogger's interest in science and involvement in educational projects will have a positive impact. The research systematizes the key components of the influencer's image, providing recommendations for selecting bloggers for university promotion. Using materials from MIP's communication campaign as an example, the authors identified the influencer's image as the most relevant for university promotion. The data obtained from the research can assist communication professionals in building more effective audience engagement when promoting higher education services in the blogosphere, considering several new aspects.

Keywords Blogger image · Influencer marketing · Blogosphere · Higher education institutions · Educational services

JEL Code M210

N. V. Trubnikova · V. V. Shlenskaya (✉)
RUDN University, Moscow, Russia
e-mail: trubnikova-nv@rudn.ru

B. S. Sergi et al. (eds.), *University 4.0 and Educational Technology for Sustainable Development*, Education in the Asia-Pacific Region: Issues, Concerns and Prospects 75, https://doi.org/10.1007/978-981-96-7376-6_19

1 Introduction

Nowadays, influencer marketing is one of the most popular ways to promote products in the digital environment. However, it is diverse and multifaceted. Its specificity is determined by the object of promotion: goods, services, or ideas. Are there significant differences between marketing goods and services? How does this manifest in the digital environment? The specificity of services makes it possible to consider independent patterns in the field of influencer marketing for this category.

Nowadays, the interpretation of influencer marketing as a promotion tool is well-developed in scientific literature. There is a consolidated understanding of influencer marketing in the professional environment. Russian experts and scholars often address this topic in their works, including the works of D. Lemeshevskiy and L. R. Grunin [1], and O. S. Logunova and P. A. Lebedev [2], N. Zhiltsova [3]. Nevertheless, influencer marketing in the field of services is still a relatively underexplored subject. The specific properties of services and the complexity of their study contribute to this. Services involve duration and a significant human element, which complicates the control and evaluation of the effectiveness of marketing actions. Moreover, services are often intangible until the point of acquisition. Services are variable in terms of quality and cannot be stored, as indicated by P. Kotler [4]. Selecting influencers for promoting services is almost always a task of increased complexity. While effective promotion of goods is possible through various forms of digital advertising with minimal personal involvement, services often require an individual with an established personal brand who can explain the service and, through their authority, build trust with users.

Most advertising formats in the blogosphere involve the influencer's personal participation in using the product or service. Bloggers showcase positive results from using the product, explaining its properties based on their experience. This "activity-based" approach is often necessary for promoting services. Advertising services require more sophisticated mechanics than, for example, tagging a brand's account in a photo. Bloggers may be invited to places where services are provided (e.g., beauty salons or clinics), on tours (for advertising hotels, airlines, or travel agencies), as suggested by Azarova et al. [5], or to offline events. During offline events, influencers prepare material and later share it on their social media pages for followers to comprehend the entire service.

Educational services stand out among services, having a commercial and a significant social component. Such comprehensive services require more complex mechanics and tools. The result and the actual process of learning are almost impossible to demonstrate in a short advertising material. According to E. V. Baldin [6], a detailed narrative by the influencer about all the features and advantages is appropriate for advertising educational services; the advertising campaign should be lengthy. In addition to classic advertising formats, collaborations are possible, such as creating educational projects in collaboration with influential individuals, where the blogger is involved in the educational process and even provides an educational service, acting as a teacher. This option is suggested for university advertising by

E. Moshtag [7]. Gundarin et al. propose inviting influencers to offline events at the university and covering them on the blogger's page [8].

In the context of influencer marketing, education deserves closer examination. In 2020, due to quarantine, users' interest in educational platforms, especially those with robust digital support, increased by 51% (according to an analysis by the BloggerBase company, reflected on the Cossa website) [9]. During this time, users began to pay increased attention to and trust influencer advertising of educational services as they found time for self-development (according to the HypeFactory agency, reflected on the Adindex website) [10].

The education sector has segments with distinct characteristics, depending on whether it involves higher education, supplementary education, digital education, corporate training, etc. [11]. These segments have significant differences [12]. Thus, the tools and platforms in influencer marketing are chosen according to this specificity [13].

It is important to note that the education sector is not entirely business services; the cultural component is also crucial. On the one hand, services (sometimes paid) are provided [14]. On the other hand, there is a social and educational subtext because education also involves people's integration into social communities [15]. This is why choosing an influencer to promote educational services has its peculiarities. It is necessary to consider the intellectual capacity of the influencer and how they can share and promote the cultural elements of education.

To promote educational services, Kostin [16] recommends choosing the following types of blogs: educational blogs, creative blogs (culture, art, book reviews, etc.), popular science blogs, self-development blogs, interviewers' and journalists' blogs, and blogs about careers and business, and mass culture blogs.

As already mentioned, the education sector encompasses several diverse segments. For instance, over the last two decades, there has been rapid development in online courses, which can serve as a supplement and, sometimes, an alternative to traditional education, as emphasized by R. Grewal et al. [17]. When promoting online courses, it is recommended to choose influencers with high professional competencies in areas relevant to the course's topic. Additionally, the influencer should share the values of the educational organization offering the courses [18].

Promoting higher education through collaboration with influencers bears similarities to promoting supplementary education programs. However, collaboration with influencers is more challenging due to its specificity—wider selection scale, generally higher cost of educational services, and the presence of two target audiences: prospective students and their parents. Within the promotion of universities, it should be considered that the majority of applicants are representatives of the digital generation who actively use social networks. Social networks are also a place for communication for parents who are familiar with the digital environment, albeit with a more demanding approach to the information quality of the content. For this reason, advertising universities through influencers currently has great potential, as indicated by I. Gabelaia [19]. According to Vasilieva [20], individuals of this generation see bloggers as role models for behavior and lifestyle, which they transmit, including through advertising messages. Research on influencer marketing and the blogosphere shows that bloggers now influence not only the sales domain but are also capable of transforming many social areas [21].

However, the cultural aspect inherent in the process of education also plays a crucial role in promoting higher education. When applying influencer marketing to advertise universities, it is necessary to choose bloggers who will be suitable for prospective students, current students, and the university environment. The main research question should be formulated: "What should be the image of the blogger to meet the goals of promoting Russian universities?"

When choosing an influencer to promote a university, the blogger's personality must be considered. In general, we can distinguish celebrities, industry professionals, and amateurs among influencers [22]. It is important for students, applicants, and, in some cases, their friends and relatives, to perceive the blogger as an opinion leader, trust their point of view, and follow their recommendations. Moreover, universities can collaborate not only with influencers who act as experts in a specific field but also with well-known alums, students, or faculty members [23].

According to Kovaleva [24], the blogger's image is reflected in their positioning on social networks. We can agree with the significance of several components proposed by Kovaleva:

1. Profile description. In this section, an influencer typically indicates the focus of their blog and briefly describes their activities. When choosing an influencer for university advertising, one can refer to this section to understand the area in which they specialize. For instance, bloggers may specify that they conduct lectures; such information speaks to their expertise.
2. Content. By examining recent publications, one can conclude the blogger's views and values. For the promotion of educational services (courses, training, and universities), it is expedient to collaborate with bloggers dedicated to self-improvement, reflection, and the dissemination of informative materials.
3. Tone of voice (TOV). The way in which an influencer communicates with their audience is reflected in the TOV. It should correspond to the university's communication style. For higher education institutions, a neutral tone is advisable, with a slightly more informal tone acceptable on social networks. The same applies to bloggers; they can interact with subscribers neutrally, using humor and slang in moderation. Such a communication style will be relevant for university advertising because it is oriented toward a young audience.

2 Communication Campaign of the Faculty of Life Science of the Moscow Institute of Psychoanalysis

The characteristics of a blogger's image proposed by Kovaleva [24] and the classification of influencers by Zniva et al. [22] were used in the research. Kostin explored the topics of blogs in the context of promoting educational services [16].

The following details are provided about the research conducted by the authors through a comparative analysis of bloggers. The practical materials for the research were data from the communication campaign of a Russian university in 2023.

Currently, several educational institutions in Russia implement promotion campaigns involving the dissemination of knowledge through bloggers. The promotion of the Moscow Institute of Psychoanalysis (MIP) through collaboration with influencers can be examined more closely. Influencer marketing became an experimental direction. As a result, it confirmed the previously hypothesized effectiveness of promoting this university through bloggers. In the summer of 2023, the university implemented a communication campaign to promote the new Faculty of Life Science, which offers professions in the field of neurobiology and bioecology (according to the Moscow Institute of Psychoanalysis website) [25]. In total, 15 bloggers were involved in the campaign. Additionally, publications were posted in 14 thematic Telegram channels; other tools were also used.

The communication campaign in the blogosphere included two flights (in June and July), which represented relatively independent parts of the campaign and were implemented on several social platforms: Telegram, Instagram, and YouTube. Thanks to the collaboration with influencers, over 49,000 views, 530 website visits, and 25 applications were achieved.

Expert bloggers from different fields were selected for the campaign; the themes of their blogs more or less corresponded to the programs of the Faculty of Life Science: ecology, urbanism, psychology, neuropsychology, biology, etc. In Telegram, bloggers published a post and an image. For Instagram, a format of 3–5 video stories and the provision of promo codes for a discount were chosen. For YouTube, advertising integration for 1.5–2.5 min was implemented (with a promo code). The influencers selected for collaboration and key data about their blogs are reflected in Table 1.

Posting with most of the bloggers was effective and helped to get applications for training. The results are reflected in Table 2.

3 Components of the Blogger's Image That Are Most Crucial for University Promotion

The most successful publication came from the blogger Egor Rafikov (Telegram channel "Captain Grantov"), garnering 6298 views, 207 website visits, and 11 applications for the faculty. The influencer shares announcements of grants, competitions, and educational digests for NGOs and activists. The content of this Telegram channel is informative; the audience is keen on learning and self-development. Egor Rafikov attracts funds for social projects through grant competitions, delivers lectures, and collaborates with universities (according to the conference website "White Nights of Fundraising – 2023") [26]. Undoubtedly, the blogger's image aligns with the university's promotional objectives. This alignment is the key to

Table 1 Bloggers selected for the communication campaign of the Faculty of Life Science of the Moscow Institute of Psychoanalysis

Name	Platform	Audience	Topic	Publication date
Urbanism as the meaning of life	Telegram	6300	Urban studies	June 9, 2023 July 25, 2023
Cerebrum	Telegram	24,000	Neuropsychology	June 7, 2023 July 31, 2023
No fanaticism	Telegram	17,000	Ecology	June 6, 2023
Alexey Radchenko	Telegram	3800	Urban studies	July 11, 2023
Captain Grantov	Telegram	37,000	Activists	July 15, 2023
Ecologist Zhora Kavanosyan	Telegram	28,700	Ecology	June 19, 2023
Black about neuronics	Telegram	24,000	Neural networks	June 23, 2023
@katrin_ecotrue	Instagram	145,000	Ecology	June 9, 2023
@chisto.men	Instagram	19,000	Ecology	June 14, 2023
@daryapilipak	Instagram	7000	Urban studies	June 13, 2023
@victoria.miliary	Instagram	12,500	Psychology	July 12, 2023
@eco_diana	Instagram	5300	Ecology	July 13, 2023
@ecologic_blog	Instagram	18,100	Ecology	July 19, 2023 July 27, 2023
@olga_viatore.m	Instagram	15,400	Biology	July 21, 2023
Alexander Goprov	YouTube	50,500	Urban studies	July 27, 2023

Source Developed by the authors

Table 2 Applications received through cooperation with bloggers within the framework of the communication campaign of the Faculty of Life Science of the Moscow Institute of Psychoanalysis

Name	Platform	Audience	Topic	Number of applications
Captain Grantov	Telegram	37,000	Activists	11
@ecologic_blog	Instagram	18,100	Ecology	4
Cerebrum	Telegram	24,000	Neuropsychology	4
Urbanism as the meaning of life	Telegram	6300	Urban studies	2
No fanaticism	Telegram	17,000	Ecology	1
@katrin_ecotrue	Instagram	145,000	Ecology	1
@victoria.miliary	Instagram	12,500	Psychology	1
Alexander Goprov	YouTube	50,500	Urban studies	1

Source Developed by the authors

success, confirming the research hypothesis regarding the significance of image-related aspects in choosing an influencer.

Additionally, the placement of posts with Instagram blogger Svetlana Kovalevskaya (@ecologic_blog) proved quite effective. Two video stories about the Faculty of Life Science were posted on her account. Despite a relatively small number of views in the first and second mentions (1722 and 1341 views for each story), 32 followers of the blogger transitioned to the website; 4 enrollment applications

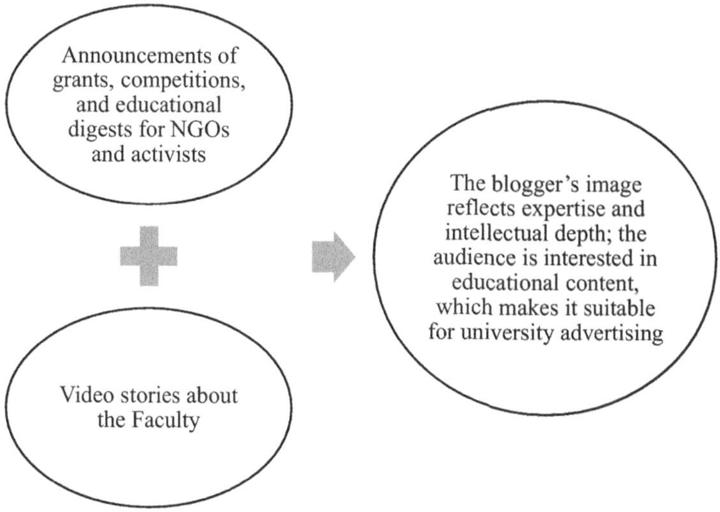

Fig. 1 Components of the blogger's image that are most crucial for university promotion. *Source* Developed by the authors

were received. These data indicate an active and engaged audience for the influencer. Svetlana Kovalevskaya is an ecologist and award winner. The blogger also gives lectures and appears on television (according to Svetlana's Instagram account) [27]. Thus, the blogger's image reflects expertise and intellectual depth; the audience is interested in educational content, which makes it suitable for university advertising.

Currently, the implementation of communication campaigns by higher education institutions in the blogosphere is unexplored within the academic community. An analysis of sources shows that authors approach this issue primarily from a theoretical standpoint and focus predominantly on examining collaboration formats with influencers.

By scrutinizing the recent campaign of the Moscow Institute of Psychoanalysis in the blogosphere, the authors identified components of the blogger's image that are most crucial for university promotion (Fig. 1).

The research findings are interpreted in a manner that suggests they could hold value within the context of promoting any educational institution in Russia through influencer marketing, particularly those providing higher education services.

4 Conclusion

The analysis of the campaign to promote the Moscow Institute of Psychoanalysis in the blogosphere provides the following conclusions:

1. Currently, influencer marketing is an effective tool for promoting universities in the digital environment, yet not all Russian universities have embraced it fully. Advertising through bloggers enables communication with a young audience accustomed to trusting influencers' opinions, positively assessing the brand presence and active promotion on social media. It is crucial for university leadership to comprehend the essence of the ongoing shifts in media trends, where there is a noticeable shift in user interest from traditional media to blogging. Previous methods of communicative impact are rapidly losing their effectiveness; entities managing budgets for university promotion must align with these new realities.
2. The acquisition of 25 applications resulted directly from posting content through bloggers. Notably, influencer publications proved more effective compared to posts in thematic Telegram channels, each of which generated an average of 1–3 applications. This reaffirms the importance of clarification for service promotion, emphasizing the significance of the personal, human factor. In such cases, the audience knows who specifically is recommending education at the university, listens to the blogger, and trusts their judgment.
3. The optimal strategy for a university promotion campaign in the blogosphere involves engaging bloggers whose image reflects intellectual depth and expertise. A positive impact stems from a blogger's genuine interest in science and involvement in the field of education. Additionally, the influencer should maintain an active and engaging blog, remaining relevant to the interests of the young audience.

References

1. Lemeshevskiy, D., & Grunin, L. R. (2021). Assessment of market size for the Instagram ad market in Russia. *Monitoring of Public Opinion: Economic and Social Changes Journal, 5*, 185–191. https://doi.org/10.14515/monitoring.2021.5.1987
2. Logunova, O. S., & Lebedev, P. A. (2021). Digital celebrities: Transformation of theoretical approaches. *Monitoring of Public Opinion: Economic and Social Changes Journal, 5*, 12–31. https://doi.org/10.14515/monitoring.2021.5.1992
3. Zhiltsova, O. N. (2019). Collaboration and influencer marketing as an alternative tool for low-budget promotion. *Journal Marketing MBA. Marketing Management Firms, 10*(4), 106–119.
4. Kotler, P. (2007). *Marketing essentials (Transl. from English)*. Williams Publishing. (Original work published 1980).
5. Azarova, S. P., Balova, S. L., Bezdenezhnykh, I. A., Goryacheva, M. M. V., Zhiltsova, O. N., Zhiltsov, D. A., et al. (2021). *Development of service marketing in the digital environment: Theory and methodology*. Prometej.
6. Baldin, E. V. (2022). Influencer marketing in educational projects: Trends and prospects. In A. A. Abaev, A. G. Golova, L. A. Chasovskaya, & V. N. Akulinin (Eds.), *Education 4.0: Competition, competencies, communication, and creativity: Collection of scientific papers on the materials of the VI International scientific and practical conference* (pp. 54–60). Russian State University for the Humanities.

7. Moshtag, E. (2023). *The online teaching entrepreneur: A guide to starting and growing your online teaching business*. Qualia Cycle.
8. Gundarin, M. V., Oleshko, P. A., & Hassan, Z. (2022). University communication strategies in modern conditions. *PR and Advertising in a Changing World: A Regional Aspect, 26–27*, 16–21.
9. Cossa. (2020, August 13). *"Influencer Marketing. Online education market": BloggerBase published the key findings of a closed study*. Retrieved September 22, 2023, from https://www.cossa.ru/news/272342/
10. Adindex. (2020, July 3). *How the pandemic reshaped influencer marketing*. Retrieved September 22, 2023, from https://adindex.ru/publication/opinion/internet/2020/07/3/283387.phtml
11. Phillips, C. H., & Jones, S. J. (2024). Strategic and tactical marketing strategies for regional public universities to address the enrollment cliff. *Journal of Higher Education*. https://doi.org/10.1080/00221546.2024.2378640
12. Truong, T. H., Lin, B. -W., & Tung, C. -P. (2024). Strategic legislation for the promotion of university–industry collaborations: A case study of Taiwan. *Journal of Technology Transfer*. https://doi.org/10.1007/s10961-024-10110-9
13. Wagué, O. (2024). Higher education and the approach to promoting gender equality in Mauritania: Quantitative and qualitative analysis. *Revista Espanola de Educacion Comparada, 45*, 46–63. https://doi.org/10.5944/reec.45.2024.39804
14. Zarubina, V., Zarubin, M., Yessenkulova, Z., Salimbayeva, R., & Satbaeva, G. (2024). Digital transformation of the promotion of educational services of Kazakhstani universities. *Journal of Innovation and Entrepreneurship, 13*(1), 3. https://doi.org/10.1186/s13731-023-00355-3
15. Zhao, Y. (2024). Intelligent promotion strategies for the integration of traditional culture and ideological and political education in higher education. *Applied Mathematics and Nonlinear Sciences, 9*(1), 20243304. https://doi.org/10.2478/amns-2024-3304
16. Kostin, D. M. (2023). Blogger as a tool for building personal brand of a persona and company reputation: Trends and prospects. *Scientific Journal, 1*(66), 18–27.
17. Grewal, R., Meyer, R., & Mittal, V. (2022). Education and marketing: Decision making, spending, and consumption. *Journal of Marketing Research, 59*(1), 1–10. https://doi.org/10.1177/00222437211068091
18. Heights Platform. (n.d.). *How to use influencer marketing to promote an online course*. Retrieved September 25, 2023, from https://www.heightsplatform.com/blog/how-to-use-influencer-marketing-to-promote-an-online-course
19. Gabelaia, I. (2020). Do not be terrified to pivot: Adaptation of influencer marketing on student enrollment management. In *Proceedings of the 6th International Scientific Symposium "Economics, Business & Finance"; 3rd International Multidisciplinary Academic Conference (IMAC)* (pp. 32–40). IRIS – ALKONA.
20. Vasilieva, O. A. (2021). Opinion leaders as an effective channel for shaping consumer preferences in Generation Z. *Trends in the Development of Science and Education, 75-2*, 154–157. https://doi.org/10.18411/lj-07-2021-69
21. Zakharkin, R. A., & Argylov, N. A. (2021). Influencers as media significant others: Modern trends of secondary socialization. *Vlast' [The Authority], 29*(6), 27–35. https://doi.org/10.31171/vlast.v29i6.8674
22. Zniva, R., Weitzl, W. J., & Lindmoser, C. (2023). Be constantly different! How to manage influencer authenticity. *Electronic Commerce Research, 23*, 1485–1514. https://doi.org/10.1007/s10660-022-09653-6
23. Shchetinina, N. N., Kidinov, A. V., & Gundarin, M. V. (2020). Promoting higher education services for Generation Z applicants. *Science Prospects, 12*(135), 303–307.
24. Kovaleva, A. (2023). *Opinion leaders. How to work with influencers: From proposal letters to successful collaborations*. Bombora.
25. Moscow Institute of Psychoanalysis. (n.d.). *Faculty of life science*. Retrieved September 27, 2023, from https://inpsycho.ru/faculty/8

26. White Nights of Fundraising – 2023. (2023). *Egor Rafikov*. Retrieved September 27, 2023, from https://fr.ngokitchen.ru/rafikov
27. Kovalevskaya, S. [@ecologic_blog] (n.d.). *Ecologist, ecospeaker, ecoscientist, and nutritionist*. Instagram. Retrieved September 27, 2023, from https://instagram.com/ecologic_blog

Study of the Impact of Reputation of Higher Education Institution on Its Ranking

Natalya A. Aleksandrova and Julia A. Aleksandrova

Abstract The research explores the definition of the reputation of a higher educational institution (HEI) as an integral concept and evaluates the role of the reputational responsibility of universities in enhancing the competitiveness of the educational institution, influencing its position in domestic and international rankings. The research pays particular attention to the structure of the university's reputation as a multifaceted concept, highlighting Russian and international approaches to shaping reputation as a factor in increasing the university's competitiveness in contemporary conditions. The "Priority 2030" program and the requirements for the reputational profiles of universities participating in this program are examined, considering the digitization of Russian society and education. During the research, the authors propose an original methodology for assessing the reputation of a university based on the key requirements of "Priority 2030," present the results of the testing of this methodology, and provide reputational profiles for three leading Russian universities on which the assessment is conducted. The expert assessment data for the three leading universities in Russia were collected in 2022, considering contemporary trends in digital education development and the requirements of the "Priority 2030" program. The methodology and approach disclosed in this research can enrich the toolkit for assessing the reputation of a university and building a reputational profile. It also complements the criteria and evaluation methods by incorporating digital components, which are currently absent in most classical assessment approaches. The research may be valuable for education administrators, students, graduate students, and anyone interested in the topic of assessing a university's reputational profile with the aim of enhancing its competitiveness in domestic and international rankings.

N. A. Aleksandrova
RUDN University, Moscow, Russia

The Presidential Academy, RANEPA, Moscow, Russia

J. A. Aleksandrova (✉)
The Presidential Academy, RANEPA, Moscow, Russia

B. S. Sergi et al. (eds.), *University 4.0 and Educational Technology for Sustainable Development*, Education in the Asia-Pacific Region: Issues, Concerns and Prospects 75, https://doi.org/10.1007/978-981-96-7376-6_20

Keywords Higher education institution · University · Reputation · Image · Rating of university · Competitiveness of university

JEL Codes I21 · I23 · I25 · O21

1 Introduction

The reputation of a university is an assessment of its educational and scientific activities, as well as the quality of student and graduate training. Thus, the reputation of a higher education institution (HEI) should be understood as a certain degree of public recognition based on its achievements, serving as an external evaluation of the institution's success over a specific period from the perspective of the public, students, faculty, and other stakeholders.

It is worth noting that a university may have differentiated reputations in different markets or concerning various stakeholder groups [1]. For instance, the university may establish one reputation in the educational services market, another in the market of scientific research and publications, and a third with employers. It is essential to highlight that reputation may also vary depending on the specialty and region. Therefore, each university is unique, possessing its entirely distinctive reputational profile [2].

The assessment of a university's reputation has been the focus of many international and Russian scholars for many years. Yu. O. Rezakova [1] and A. V. Zakharova, E. Yu. Karmalova, and A. D. Krivonosov [2] dedicate considerable attention to the formation of a university's reputation as an element of the university's marketing management process, including in the digital environment. A. V. Protsyuk, G. M. Tokhmetova, A. V. Glagoleva, Y. N. Zemskaya, and E. A. Kuznetsova delve into evaluating communication channels of the university and the impact of its business reputation on competitiveness, especially in the context of its presence on social networks.

The academic and public reputation of the university, priorities, methods, and tools for its management are presented in the works of A. L. Abaev, A. G. Golova, and T. B. Serebrovskaya. Yu. A. Arefkin and E. E. Kharlamov focus on rankings as a tool for assessing the business reputation of a university.

In line with most definitions presented in the works of the aforementioned authors [3, 4], the reputation of a university is formed as an integral concept from a multitude of factors, such as the quality of education, scientific achievements, international activity, graduate employment, and several others.

2 The Assessment of a University's Reputation in Special Rankings

The assessment of a university's reputation can be relevant for various purposes, including attracting students and faculty, appealing to investors, obtaining government grants, enhancing the institution's image, and improving its position in domestic and international rankings based on various criteria [5, 6].

A ranking is the outcome of a university's effectiveness in managing various reputational components of the institution [7, 8]. It represents a list of the best universities or educational programs of universities in a specific field, arranged in ascending or descending order [9]. Rankings are a sought-after and effective tool for shaping public opinion.

In 2001, the Ministry of Education and Science of the Russian Federation initiated the development of a university ranking methodology through Order "On the ranking of higher education institutions" (February 26, 2001 No. 631). One of the most notable projects in this niche is recognized as "DreVO," presenting a league table of Russian universities compiled by the public organization "Business Russia" [10]. University rankings are also compiled by the All-Russian public organization "Business Russia," the publishing house "Kommersant," the independent agency "Reuters," the Higher School of Economics, RIA Novosti, and several other public organizations. Most public organizations and agencies representing university rankings use the following types of ranking classifications [11]:

- Rankings with the assignment of a single final score are used to rank universities overall. During the ranking compilation, a set of indicators is compiled, each of which is assigned its weight, considering which a unified final assessment of the quality is calculated for a given university.
- Rankings of universities by specific disciplines, study programs, or individual subjects. Such rankings are prepared by "Perspectives" (Poland), "Stern" (Germany), "Business Week," "Financial Times" (UK), and others.
- Rankings with a combined approach. In this case, there are unique ranking or table methods that cannot be unified into a single type.

In addition to Russian developments, methods, and approaches, there are several notable international university rankings and evaluation systems. The most significant international rankings of higher education institutions are as follows [12]:

1. QS World University Rankings evaluates universities based on academic reputation, employer reputation, and the citation index of scientific publications;
2. Times Higher Education World University Rankings considers various factors, including educational activities, scientific research, internationalization, and citation of scientific articles;
3. Academic Ranking of World Universities (ARWU) assesses universities based on indicators such as the number of graduates and staff who have received the Nobel Prize, citation index, and the number of publications per staff member;

4. U.S. News & World Report Best Global University Rankings ranks universities based on their educational activities, scientific research, employer reputation, and student reviews;
5. Webometrics Ranking of World Universities evaluates universities based on the quantity and quality of their websites, as well as the citation count of their publications.

Moreover, in today's conditions, a key challenge in Russia and abroad is the development of methods and approaches to compile rankings of the most developed and active universities that respond to the challenges of the digital economy. The reputation of a university in the digital economy, in addition to the classical factors, is closely related to its innovative projects, the use of digital technologies in education, infrastructure, scientific research, and collaboration with other organizations and enterprises in the digital economy [13].

In response to this challenge, the Ministry of Science and Higher Education of the Russian Federation created the "Priority 2030" program aimed at enhancing the competitiveness of Russian universities and ensuring their inclusion in global rankings, considering the digitization of the economy and the introduction of new evaluation parameters in rankings. Universities implementing the "Priority 2030" program in Russia have a positive reputation among internal and external audiences and achieve the highest rating scores according to many domestic and international rankings. These universities are the Lomonosov Moscow State University (MSU), St. Petersburg State University (SPBU), National Research University Higher School of Economics (HSE), ITMO University, National University of Science and Technology MISIS, Moscow Institute of Physics and Technology (MIPT), Moscow Engineering Physics Institute (MEPhI), Peoples' Friendship University of Russia (RUDN), Pirogov Russian National Research Medical University, I. M. Sechenov First Moscow State Medical University (Sechenov University), Voronezh State Medical University named N. N. Burdenko, Moscow Aviation Institute (MAI), Bauman Moscow State Technical University (BMSTU), and others.

3 Indicators and Authors' Methodology for Assessing the University's Reputation

Considering the requirements of the main provisions of the "Priority 2030" program and the key trends in the digitalization of education, the authors developed a proprietary expert approach to assessing the reputation profile of universities. The methodology was tested using three leading universities in Russia: Peoples' Friendship University of Russia (RUDN), Russian Presidential Academy of National Economy and Public Administration (RANEPA), and Lomonosov Moscow State University (MSU). The method included the following ten blocks of evaluation criteria (each of which contained 3–10 indicators):

1. Quality of educational programs with a digital component is evaluated based on exam results, feedback from students and university faculty, and input from key employers. The assessment also considers the digital component, including the use of digital technologies in the educational process (e.g., electronic courses, online conferences, and virtual laboratories), the use of distance learning systems and online communication, and the creation and use of digital platforms for interaction between students, faculty, and employers.
2. Scientific achievements and research activity of the university are assessed based on the number of publications, patents, awards, participation in international conferences, and other achievements within the university.
3. Reputation and competencies (including digital competencies) of faculty are assessed based on the experience, qualifications, and achievements of faculty members. The evaluation of digital competencies of faculty is based on their proficiency in digital skills, including knowledge and use of various programs and applications in the educational process, the ability to apply digital technologies to create and distribute educational materials, skills in working with information systems, and more.
4. Employer reviews are evaluated based on feedback regarding graduates and their success in the job market.
5. Key competencies and digital competencies of students and graduates use various methods such as testing, project completion, participation in research, article writing, conference participation, etc. Indicators for assessing digital competencies of students and graduates include the ability to work with spreadsheets, knowledge of databases, programming skills (knowledge of programming languages), Internet skills (ability to find and analyze information, use search engines and social networks for communication and data exchange), use of social networks, and digital safety (knowledge of basic information security, ability to protect personal and other people's data).
6. University infrastructure can be evaluated based on parameters such as the presence of new educational buildings that meet safety standards, dormitories, cafeterias, sports facilities, libraries, computer labs, digital research laboratories, campuses, and classrooms.
7. Innovations in the university can be assessed based on the adoption of artificial intelligence for data analysis and decision support, integration of augmented and virtual reality in education, use of blockchain technologies, development and implementation of innovative methods for assessing student knowledge, creation of digital platforms for interaction between students, faculty, and employers, use of IoT technologies to optimize campus infrastructure management, development of distributed ledger technologies for information accounting and storage, and integration of the university into the international scientific and educational space using digital technologies.
8. International cooperation and collaboration are evaluated based on the assessment of the presence and number of faculties and institutes in the university leading international programs and collaborations, international recognition of the university (QS, THE, and other rankings), cooperation between the

university's research laboratories and partners, and the presence of international exchange and student training programs.

9. Reviews and employment of graduates assess satisfaction with the received specialty, employment in the field, satisfaction with work, career growth, and employer reviews.
10. Promotion of the university in the digital environment includes such indicators as the quality of content and its regular updates, relevance of information, use of social networks, interaction with the audience through chats and messengers, SEO optimization of the website, handling electronic reviews from students, potential students, and employers, active online presence on leading platforms in the Russian-speaking Internet and YouTube, availability of electronic educational courses on platforms such as "Open Education," "Coursera," and others, and high Internet metric indicators (conversions, etc.).

The proposed methodology for assessing the university's reputation with the use of highlighted indicators is presented in Fig. 1.

The methodology for assessing the university's reputation, devised by the authors, encompassed the following stages:

1. Definition of goals and assessment tasks: evaluating the impact of the university's reputation on its rankings;
2. Data collection: gathering data involved analyzing educational programs, scientific achievements, feedback from students, faculty, employers, and other indicators based on the aforementioned ten criteria blocks;
3. Expert assessment and data analysis: data were analyzed, compared, and juxtaposed to determine the reputation of the evaluated top three universities according to the presented ten criteria blocks and indicators. During this stage, a 360-degree expert assessment was conducted involving various audiences (faculty, students, graduates, key employers, department heads, and university management, as well as independent experts). A total of six independent groups were formed, comprising 330 individuals, with 55 individuals in each group. The preliminary stage included forming a candidate list based on specific criteria (e.g., experience in the field of education, scientific achievements, publications, etc.),

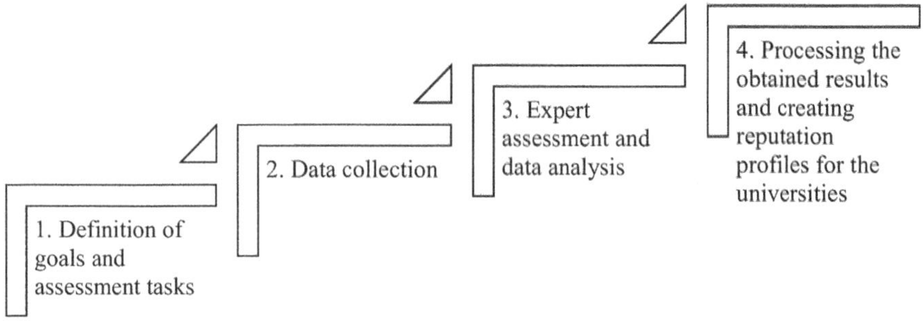

Fig. 1 Authors' methodology for assessing the university's reputation. *Source* Compiled by the authors

assessing the competence of experts, and developing an anonymous questionnaire. Subsequent stages involved direct expert evaluation, summarization of results, and calculation of the consistency of expert opinions in each subgroup based on Kendall's coefficient of concordance;
4. Processing the obtained results and creating reputation profiles for the universities.

4 Assessment of the Reputation of the Leading Russian Universities

As a result of processing the obtained data, the authors concluded that the sample was relevant and there was a high consistency of opinions within each of the six groups (Kendall's coefficient of concordance in each of them was higher than 0.85) and, thus, that the obtained data had sufficiently high reliability.

Let us present the main results obtained in Table 1.

Figure 2 presents the reputation profiles based on the results of the assessment of the three HEIs. Thus, for the profiles of the three HEIs, independent experts, graduates, employers, and teachers noted shortcomings in the quality of educational programs (not all programs implement digital competencies), shortcomings in innovation and infrastructure of HEIs (many buildings, research laboratories, and equipment are recognized outdated), shortcomings in the promotion of HEIs in the digital environment. Many people expressed their wishes to develop the digital competencies of teachers and students.

5 Conclusion

Overall, it is essential to note that the majority of Russian universities need to develop and manage a reputation strategy that would enable them to achieve the following:

- To enhance the quality of educational programs and scientific research;
- To attract highly qualified faculty and researchers;
- To strengthen connections with employers and other organizations;
- To foster international relations and collaboration;
- To work on improving infrastructure and learning conditions;
- To create opportunities for the professional development of students and faculty;
- To support student engagement in scientific and community activities;
- To inform the public about the university's achievements and its contribution to the development of science and education.

Table 1 Expert assessment of the reputation of three leading Russian universities

Factor	Factor weight, wi (determined expertly)	MSU (score in points)	RUDN (score in points)	RANEPA (score in points)	Weighted Assessment, MSU	Weighted Assessment, RUDN	Weighted Assessment, RANEPA
Quality of educational programs	0.1	8	7	9	0.8	0.7	0.9
Scientific activity of the university	0.2	10	7	9	2	0.7	0.9
Faculty reputation	0.1	10	8	9	1	0.8	0.9
Employer feedback	0.05	10	7	8	0.5	0.7	0.8
Competencies of students or graduates	0.1	8	7	7	0.8	0.7	0.7
HEI infrastructure	0.05	7	7	9	0.35	0.7	0.9
Innovations at the HEI	0.2	6	6	8	1.2	0.6	0.8
International cooperation	0.05	10	9	9	0.5	0.9	0.9
Feedback and employment of graduates	0.1	10	10	9	1	1	0.9
Promotion of the HEI in the digital environment	0.05	8	7	9	0.4	0.7	0.9
Total points	1	87	75	86	8.55	7.5	8.6

Source Calculated and compiled by the authors

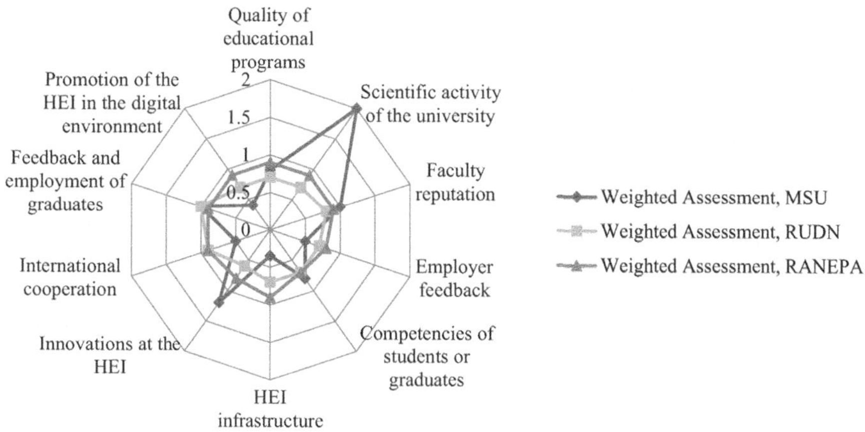

Fig. 2 Reputation profiles of MSU, RUDN, and RANEPA. *Source* Calculated and constructed by the authors

In the presence and implementation of an effective reputation strategy addressing the challenges of the digital economy, many leading Russian universities will be able to significantly enhance their image and enter Russian and international rankings.

References

1. Rezakova, Y. O. (2020). Formation of HEI reputation as an element of HEI marketing management process. In *Radio electronics, electrical engineering, and power engineering: Abstracts of reports* (p. 537). RADUGA.
2. Zakharova, A. V., Karmalova, E. Y., & Krivonosov, A. D. (2022). Crisis communications in the digital environment and reputation of the university. *Russian School of Public Relations, 24*, 59–72.
3. Qazi, A., & Al-Mhdawi, M. K. S. (2025). Benchmarking higher education excellence: insights from QS rankings. *Benchmarking*. https://doi.org/10.1108/BIJ-03-2024-0195
4. Sällberg, H., & Folino, E. (2024). The relative importance of distance education challenges to instructors in higher education—A ranking-type Delphi study. *Education and Information Technologies, 29*(11), 13495–13522. https://doi.org/10.1007/s10639-023-12345-9
5. Shih, Y. -H. (2024). Higher education for sustainable development in Taiwan: An analysis of universities listed in the top 500 of the QS World University Rankings 2024. *Frontiers in Education, 9*, 1421813. https://doi.org/10.3389/feduc.2024.1421813.
6. Smith, M. D., & Colpitts, B. D. F. (2025). Japan's pursuit of meritocracy, cosmopolitanism, and global rankings in higher education: A Bourdieusian interpretation. *Asia Pacific Journal of Education, 45*(1), 178–193. https://doi.org/10.1080/02188791.2022.2086102
7. Thomas, S. (2025). Does higher ranking ensure higher student satisfaction: Evidence from higher education institutions in India. *Journal of Applied Research in Higher Education, 17*(1), 320–333. https://doi.org/10.1108/JARHE-08-2023-0358
8. Tóth, B., Motahari-Nezhad, H., Horseman, N., Berek, L., Kovács, L., Hölgyesi, Á., et al. (2024). Ranking resilience: Assessing the impact of scientific performance and the expansion

of the Times Higher Education Word University Rankings on the position of Czech, Hungarian, Polish, and Slovak universities. *Scientometrics, 129*(3), 1739–1770. https://doi.org/10.1007/s11192-023-04920-1
9. Prokhorov, A. V. (2020). Reputation of a higher education institution in the context of its social media presence. *Memoirs of Novgorod State University, 4*(29), 3. https://doi.org/10.34680/2411-7951.2020.4(29).3
10. Protsuk, A. V. (2016). Evaluation of the communication channels of a higher education institution that affect its reputation. *Theory and Practice of Modern Science, 6–2*(12), 67–70.
11. Glagoleva, A. V., Zemskaya, Y. N., Kuznecova, E. A., & Aleshina, I. Y. (2021). Communicative aspect of university reputation assessment (on the case of RUDN University). *Scientific Research and Development. Modern Communication Studies, 10*(6), 75–82.
12. Serebrovskaya, T. B. (2020). Strengthening the university's reputation in the context of modernizing the education system. *Problems of Theory and Practice of Management, 5*, 88–99.
13. Arefkina, Y. A., & Kharlamova, E. E. (2016). Rating as a tool for assessing the business reputation of a university. *Management Business Power, 1*(10), 102–104.

Features of Promoting the Image of Higher Educational Institutions in the Media Space

Irina Yu Glinskaya and Olga N. Popova

Abstract The research reflects the results of an analysis of the influence exerted by the image of higher education institutions on social networks within the framework of promoting educational services. The relevance of the research is driven by the widespread use of digital technologies and the Internet, which have made social networks one of the key tools for managing the image of universities. An important role is played by the constantly growing dependence of the younger generation on the Internet and social networks. Active acquisition and dissemination of information by youth through social networks hold separate significance. The conducted research reflects various aspects of the concept of image, with these aspects considered in the context of higher education institutions. An analysis of image management of educational organizations was carried out to enhance the competitiveness of higher education institutions through social networks. The authors identified the main mechanisms of image management through contemporary information channels represented by social networks. The authors also grouped the stages of the development of the situation with social networks and closely related digital channels in contemporary Russia. The research explores the main factors and characteristics of the influence exerted by the activities of organizations on social networks on their image. The research highlights the prospects and strengths of image management for educational institutions through social networks.

Keywords Higher educational institutions · Image of the organization · Management of the image · Image of higher educational institutions · Social media

JEL Code M3

I. Y. Glinskaya (✉) · O. N. Popova
RUDN University, Moscow, Russia

B. S. Sergi et al. (eds.), *University 4.0 and Educational Technology for Sustainable Development*, Education in the Asia-Pacific Region: Issues, Concerns and Prospects 75, https://doi.org/10.1007/978-981-96-7376-6_21

1 Introduction

The relevance of this research is prompted by the widespread use of digital technologies and the Internet in the world, transforming social networks into one of the key tools for managing the image of higher educational institutions.

In this regard, the main goal of the research is to analyze the practical role that played by the image of universities in promoting their educational services through social networks.

To achieve the set research goal, the authors identified the following main tasks:

- To analyze the term image in the context of higher educational institutions;
- To determine the role and place of managing the image of universities through digital information channels, especially social networks, within the framework of increasing competitiveness;
- To analyze the strengths and main current prospects of managing the image of universities through social networks.

In writing the research paper, the authors used the works of the following authors, addressing the issues of organizational image and its formation: D. A. Bykov, E. V. Vylegzhanina, S. N. Komarova, A. A. Malkovich, A. A. Mukhina, L. A. Nedykhalov, S. V. Sviridovich, L. M. Semenova, K. A. Tarasevich, and others.

2 Various Approaches to the Definition of the University's Image and Ways of Its Formation

The market for educational services is one of the most dynamic and rapidly growing sectors of the world's economy. The number of educational institutions offering various programs increases annually. Competition in the market becomes more intense. In such conditions, creating and maintaining an image becomes one of the key tasks for higher educational institutions.

Image is the established perception of a company, product, or service in people's minds. It is formed based on many factors, including the quality of service, level of professionalism, compliance with customer expectations, and many others. For higher education institutions, the image plays a crucial role in attracting students and ensuring the loyalty of graduates. Therefore, universities actively use social networks to shape their image [1–6].

For many, the term "image of the organization" is associated with a certain averaged perception of the organization. In other words, it is a kind of general representation. There is no universally accepted definition of the organizational image. Let us consider the approaches of some authors (Table 1).

Some authors define the image of the organization as a purposeful process of creating a specific image. Others believe that the image forms spontaneously. It is fair to assert that the image is managed by the organization itself or by other entities.

Table 1 Approaches to defining the image of an organization by some authors

Author	Definition of the term "image of the organization"
A. B. Borisov	Reputation, company prestige, product quality, communication with customers, adherence to its commitments with precision [7]
S. Yu. Golovin	An emotionally colored image of the organization, having a stereotype [8]
R. R. Gorchakova	An image formed and developed by the organization, corresponding to its norms and values [9]
G. Dowling	A set of beliefs and feelings that a person develops about the organization [10]
S. I. Ozhegov	The result of the organization's reflection on the individual's consciousness [11]
A. Yu. Panasyuk	Opinions about the organization held by a population group based on the image formed in the minds of these people as a result of direct contact with the organization or information about the organization from others [12]
M. L. Razu	Specially designed image in the interests of the organization based on the specifics of its functioning, endogenous regularities, properties, merits, qualities, and characteristics. It is intentionally introduced into the consciousness (or subconscious) of the target audience, aligns with the organization's expectations, and serves as the foundation for distinguishing features of the organization from others [13]
M. V. Tomilova	Holistic perception as a comprehensive understanding and evaluation of the organization by various groups in society. It is formed based on stored data in the memory of these groups about various aspects of the company's activities [14]

Source Compiled by the authors based on the works of the mentioned authors

Let us compare the methods of assessing the formation of the university's image by different authors.

According to A. V. Skripkin, we can talk about the following key elements of the strategy for creating the university's reputational capital:

(a) Conducting an analysis of the external environment of the educational institution;
(b) Identifying the target audience, which is crucial for image formation;
(c) Identifying and analyzing existing internal resources;
(d) Determining key tasks necessary for image formation;
(e) Identifying the motivation of the participants in the process;
(f) Determining the current ratio of all main components of the university's image (educational, scientific, social, and artistic-creative);
(g) Highlighting key principles of image formation;
(h) Identifying technologies and methods suitable for managing all elements of the educational institution's image in a specific situation;
(i) Monitoring the results determined by the corresponding image [15].

T. N. Shcherbakova distinguishes between permanent and variable components of the positive image of a university. According to her, the faculty plays a central role in the image of an educational institution. Under the influence of the teachings and worldview of the faculty, students largely form their own worldview, master the methodology of acquiring scientific knowledge, and learn to form their judgments and critically evaluate the information received. This ultimately shapes the image of the educational institution.

S. D. Reznik suggests determining the following factors that influence the increase in the reputational responsibility of an educational institution:

- Fulfillment of obligations (payment of taxes and fees);
- Employment data of the university's graduates;
- A comprehensive system of continuous training and employment of graduates;
- Personal awareness of the responsibility of teachers for the learning outcomes;
- Scientific achievements of the university (its staff);
- Features of the personnel policy of the educational institution, the presence of social guarantees;
- Features of competitive struggle;
- Features of enhancing the competitiveness of students in the labor market;
- Conducting extracurricular activities with students;
- Development and implementation of an adequate strategy for the educational institution;
- Results of management decisions made by the university administration;
- System of university management, the quality of educational services;
- Consideration of the interests of the target audience.

F. V. Malakhova believes that the strategy of image management includes the following two components:

- Formation, consolidation, and maintenance of the desired reputation;
- Reputation protection or crisis management actions.

F. V. Malakhova suggests certain rules for successful overcoming in the event of a crisis. Their essence is as follows:

1. Publicly admit guilt or point out the guilty party in the crisis situation;
2. Demonstrate existing concern about what happened to the public;
3. Properly interpret the severity of the crisis;
4. Use various types of research and analysis to better understand how to react in each specific crisis situation;
5. Engage third parties with such capabilities to spread the necessary information in support and defense of the organization;
6. Actively interact with the media, which is an important channel for disseminating information;
7. React promptly and timely to any crisis situation;
8. Provide responses and comments in any crisis situation in the simplest and most concise manner possible;
9. Use the tactic of continuously spreading positive information about the organization;
10. Use digital technologies and the Internet as a channel for disseminating information, with the corporate website playing a leading role;

11. Have a clear action plan in case of various crisis situations;
12. Timely inform the staff and the public about the resolution of the crisis situation [16].

Having considered various approaches to the definition of the organization's image and ways of its formation, it is proposed to treat this concept as an average conscious image of the organization among a certain group of people formed under the influence of several factors dependent and independent of the organization, as well as resort to the use of social networks to form a positive reputational capital of the university.

3 The Modern Experience of Using Social Networks for Image-building of a University

Periodic evaluation of an organization's image makes it possible to draw conclusions about the dynamics of changes in public opinion regarding the organization. In other words, the results of the current assessment are compared to the previous one (assuming the use of the same assessment method).

An important characteristic is the duration of the sustainability of the image. On the one hand, it allows us to speak about stability and is a consequence of this stability. On the other hand, it indicates the difficulty of improving the image and the low effectiveness of corresponding efforts [17].

One related characteristic is the functional effectiveness of the image, which is understood as the ability of the organizational image to influence the achievement of goals and tasks, corresponding to the organization's capabilities.

Communicative efficiency is also highlighted as the ability of the company's image to influence the target audience through communication tools. Simultaneously, an assessment is made of the quantity of information transmitted and received by the organization, the quality of this information, and the peculiarities of its interpretation by the target audience.

The effectiveness of the image can be determined with general criteria that are independent of the organization's characteristics and the nature of its activities and with specific criteria that consider the unique features of the organization.

Each university has its individual value characteristics, highlighted as competitive elements. For some universities, it is the ranking and status; for others, it is student life, high scholarships, or support for learners. These elements are identified as competitive advantages. In most Russian universities, the educational process goes beyond the framework of direct education and includes such forms of extracurricular activities as KVN [Club of the Funny and Inventive], volunteering, and joint celebration of special days. Student councils are formed to engage students in university life (actively involving them in the social environment) and express the opinions and preferences of the students [18].

Currently, the information flow is shifting towards online information and actively expanding. To shape the desired image of an organization, including higher educational institutions, it is necessary to consider and influence the digital landscape.

In our opinion, the formation of an image in the digital environment is carried out directly (through the creation of information sources in the digital space (official websites, participation in social networks and digital platforms such as Telegram, VKontakte, Odnoklassniki, WhatsApp, YouTube, etc.)) and through reactions to external information flows (e.g., interviews with independent media). The directions for shaping the image of a university in the digital environment are illustrated in Fig. 1.

Practically all universities in Russia have official websites where information about the university is posted, including photos of licenses and accreditations, contacts, group enrollment vacancies, specialties, schedules, and the faculty. The official website holds particular importance in light of the development of distance learning technologies. It serves as a crucial communication tool for the university, significantly shaping its image. The website allows the university to showcase information that serves as a competitive advantage, highlight the institution's strengths, and influence the perception of the university by the target audience.

The design of the website complies with the requirements outlined in the relevant regulatory act: the Order of the Federal Service for Supervision in Education and Science "On approving the requirements for the structure of the official website of an educational organization on the Internet and the format for presenting information on it" (May 29, 2014 No. 785) [19].

When working with the website, it is essential to timely post and update information, ensure a user-friendly format, present the most important information in an

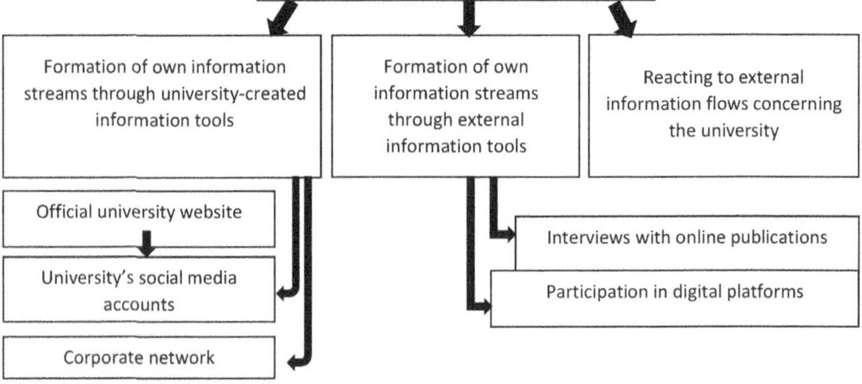

Fig. 1 Directions for shaping the image of a university in the digital environment. *Source* Compiled by the authors

accessible manner, facilitate user feedback, and monitor the informativeness, openness, and technical reliability of the website.

As for the internal corporate network, this primarily refers to Intranet—an internal corporate network built on Internet technologies. The network serves as an internal corporate web portal, mainly created for the organization, storage, and processing of intracorporate data. The Intranet is accessible only within the organization's local network, including remote branches. Users access the portal pages through a web browser, allowing people with basic computer skills to use intranet systems [20].

Social networks are built on the formation of the so-called virtual communities, providing users with the opportunity to express their opinions and desires on the Internet. Advertising on social networks makes it possible to connect the appropriate audience (based on needs and values) with companies whose products satisfy those needs. When selling organizations join social channels on websites, customers begin to interact with the companies directly. This interaction can be more individual and private for customers. Social networks operate similarly to electronic word-of-mouth.

The development of social networks in Russia can be delineated into several stages (Table 2).

Since 2022, Instagram and Facebook have been blocked for the Russian audience. YouTube enforces a "ban" policy for a pro-Russian stance. This has led to an increase in users on Telegram, VKontakte, Odnoklassniki, and Rutube.

One of the main tasks of higher educational institutions on social networks is to attract new students. To achieve this, they employ various marketing strategies, such as advertising campaigns, contests, promotions, etc. Additionally, many universities and colleges create their pages on social networks, where they publish information about their educational programs, achievements, and news.

This trend is driven by the high growth rate of Internet users over the past 20 years. As of 2022, this number reached around 129.80 million people [21].

Table 2 Stages of social network development in Russia

Stage	Brief description
2006–2009	The emergence of the social network VKontakte and Russian-language Facebook; the emergence of branded societies: the sending of spam messages: invitations to formed groups
2010–2014	The emergence of Instagram and Telegram; the creation of mobile applications for social networks and the development of targeted advertising; companies realize the importance of their reputation and image in social networks
2015–2021	The appearance of special "Buy," "Order," and "Goods" buttons in social networks as advertisers' tools; the emergence of bots in advertising (in dialog boxes); increased role of opinion leaders; the emergence of the VKontakte exchange; improved settings of targeted advertising; decrease in the effectiveness of traditional advertising methods
Since 2022	Blocking of social networks Instagram and Facebook in Russia; removal of some pro-Russian content on YouTube; active promotion of video hosting Rutube; growth of users of social networks VKontakte, Odnoklassniki, and Telegram.

Source Compiled by the authors

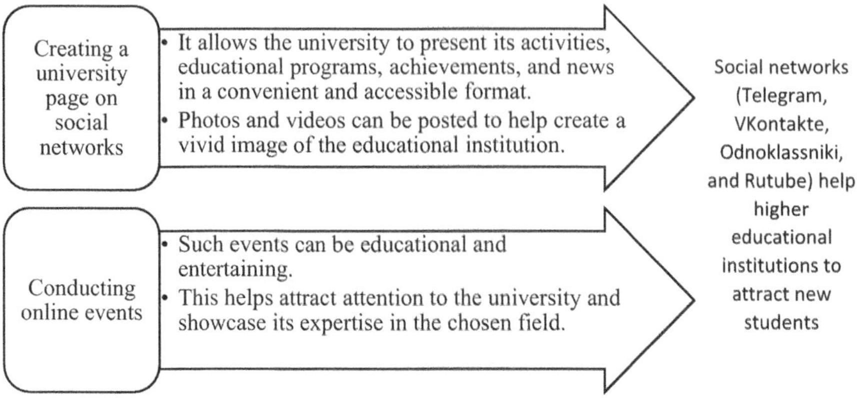

Fig. 2 The main ways to use social networks for image-building of a university. *Source* Compiled by the authors

High growth rates are also evident in social networks. For example, the total number of social media users in Russia at the beginning of this year accounted for 72.70% of the country's total population (Fig. 2).

One of the main ways to use social networks for image-building is by creating a university page on these platforms. This page allows the university to present its activities, educational programs, achievements, and news in a convenient and accessible format. Photos and videos can be posted on the university's page to help create a vivid image of the educational institution.

Another way to use social networks for image-building is to conduct online events. Such events can be educational and entertaining. For example, a university can host a webinar on a topic related to its specialization or organize an online competition for students. This helps attract attention to the university and showcase its expertise in the chosen field.

4 Conclusion

Currently, the information flow is shifting towards online information and actively expanding. To shape the desired image of an organization, including higher educational institutions, it is necessary to consider and influence the information field.

The formation of the image in the digital environment is carried out directly (through the creation of proprietary information sources in the digital space: official websites, participation in social networks and platforms such as Telegram, VKontakte, Odnoklassniki, WhatsApp, Rutube, etc.) and by reacting to external information flows (e.g., interviews with independent media).

The primary factor determining the image of an educational institution is the quality of education. This situation is directly related to the fact that the quality of education influences further employment—the main reason for obtaining higher education.

The most popular information sources where university rankings can be found include rating agencies, notably the RAEX-100 ranking of the best universities in Russia, the Subject Ranking by RAEX, and the Forbes ranking of Russian universities.

Practically all universities in Russia have official websites, where information about the university, photos of licenses and accreditation, contacts, vacancies for enrollment, specialties, schedules, and the teaching staff are posted. The official website holds special significance in connection with the development of distance technologies. The official website of a university is a crucial communication tool. It significantly shapes the university's image. Website makes it possible to place information that serves as a competitive advantage for the university, showcasing its strengths to influence the perception of the university by the target audience.

It is noted that social networks are a powerful tool for influencing the image of a higher educational institution. In this regard, monitoring indicators of the information flow on social networks and evaluating the effectiveness of investments in promoting the university play a crucial role.

References

1. Suárez-Perdomo, A., Garcés-Delgado, Y., & Arvelo-Rosales, C. N. (2024). Academic procrastination and inappropriate use of social networks among university students. *Studies in Higher Education*. https://doi.org/10.1080/03075079.2024.2373190
2. Valecillos Vázquez, C. A., & López-Navas, C. (2025). The problematic use of social networks in university students: New perspectives and study needs. *European Public and Social Innovation Review, 10*, 1–13. https://doi.org/10.31637/epsir-2025-717
3. Villalustre-Martínez, L., & Cueli, M. (2024). Size and intensity of use in digital social networks: How it is related with personal and social skills perceived by university students? *Revista Complutense de Educacion, 35*(2), 227–237. https://doi.org/10.5209/rced.83129
4. Vylegzhanina, E. V., & Sherstobitov, M. A. (2019). Assessment of the business reputation of the organization. *International Journal of Humanities and Natural Sciences, 2-2*, 69–72. https://doi.org/10.24411/2500-1000-2019-10574
5. Wang, J., Wang, N., Liu, Y., & Zhou, Z. (2025). Experiential avoidance, depression, and difficulty identifying emotions in social network site addiction among Chinese university students: A moderated mediation model. *Behaviour and Information Technology*. https://doi.org/10.1080/0144929X.2025.2455406
6. Weerasinghe, R. N., & Gunawardena, K. S. L. (2025). Unveiling university networks aligning with SDGs: A social network analysis of the national innovation system of Sri Lanka. *Journal of Lifestyle and SDG's Review, 5*(1), e02967. https://doi.org/10.47172/2965-730X.SDGsReview.v5.n01.pe02967
7. Borisov, A. B. (2021). New product: Reasons for failures and key success factors. *Journal of Economics, Entrepreneurship and Law, 11*(8), 2027–2044.
8. Golovin, S. Y. (1997). *Dictionary of practical psychologist*. Harvest.
9. Gorchakova, R. R. (2012). Peculiarities of corporate image formation. *University Proceedings Volga Region Social Sciences, 2*(22), 185–192.

10. Dowling, G. (2003). *Creating corporate reputations: Identity, image, and performance* (A. V. Kuznetsova, Transl. from English). Image Contact Consulting Group; Infra-M. (Original work published 2001).
11. Ozhegov, S. I., & Shvedova, N. Y. (2006). *Explanatory dictionary of the Russian language: 80000 words and phraseological expressions* (4th ed.). A Temp.
12. Panasyuk, A. Y. (2007). *Image formation: Strategy, psychotechnologies, and psychotechniques*. Omega-L.
13. Razu, M. L., Bronnikova, T. M., Razu, B. M., Titov, S. A., & Yakutin, Y. V. (2012). *Project management: Fundamentals of project management*. KnoRus.
14. Tomilova, M. V. (1998). *Model of organization's image*. St. Petersburg, Russia.
15. Kim, S. A. (2022). *Marketing: Textbook* (2nd ed.). Dashkov and Co.
16. Komarova, S. N., & Prokopets, T. N. (2020). Development of process of forming the image of organization and method of its estimation. *Vestnik of Rostov State Economic University (RINH), 1*(65), 95–99.
17. Glushchenko, V. M., & Tarasova, S. V. (2023). State policy of The Russian Federation in the field of reforming higher education. *Modern Pedagogical Education, 4*, 18–23.
18. Izakova, N. B., & Kapustina, L. M. (2021). Marketing research of the brand awareness impact on the real estate agencies competitiveness. *Practical Marketing, 9*(295), 11–17. https://doi.org/10.24412/2071-3762-2021-9295-11-17
19. Kyshtymova, E. S. (2021). Formation of psychological media competence of specialists in the field of advertising and public relations. *Baikal Research Journal, 12*(4), 102–108.
20. Lagutina, E. A. (2022). Factors determining the choice of the instrument of communication influence. *Problems of Management Theory and Practice, 9*, 90–98.
21. Malkevich, A. A. (2021). *Organizing and conducting public relations campaigns: Textbook for universities* (2nd ed.). Urait.

The Product Approach in Marketing Control: The Practice of Implementation in Educational Institutions of Higher Education

Svetlana A. Kucheryavenko

Abstract The article discusses the practice of using a product approach in the framework of marketing control based on the materials of the National Research University "BelSU". The analysis of sources on this topic makes it possible to conclude that at present the educational market is represented by aggressive competition of educational institutions of higher education. This requires an assessment of the effectiveness of the marketing policy chosen by the university, including those aimed at identifying the needs and expectations of consumers. The literature provides a complete description of the methodology of marketing control for commercial organizations. However, there is no descriptive practice of implementing marketing control in educational organizations. These factors have led to the emergence of a research request for the practice of applying the product approach in the framework of marketing control. As part of the study, the practice of monitoring consumer satisfaction of the National Research University "BelSU" for 2010–2022 is considered on the example of the category "students of complementary vocational education"; an attempt is made to assess the effectiveness of using the product approach in marketing control; basic indicators are determined taking into account the specifics of educational activities for the group of respondents "students of additional vocational education"; the influence of the product approach in marketing control is revealed for the sustainable development of the university.

Keywords Marketing · Marketing control · Marketing activities · Monitoring of customer satisfaction · Education market · Labor market · Additional vocational education

JEL Code M31

S. A. Kucheryavenko (✉)
Belgorod State National Research University, Belgorod, Russia
e-mail: Kucheryavenko_s@bsuedu.ru

B. S. Sergi et al. (eds.), *University 4.0 and Educational Technology for Sustainable Development*, Education in the Asia-Pacific Region: Issues, Concerns and Prospects 75, https://doi.org/10.1007/978-981-96-7376-6_22

1 Introduction

The analysis of publications on the research problem allows us to conclude that at today the educational market is represented by aggressive competition of educational institutions of higher education. This requires an assessment of the effectiveness of the marketing policy chosen by the university, including those aimed at identifying the needs and expectations of consumers [1–3].

The methodology for the formation and development of a marketing control system in commercial organizations is described quite fully in marketing textbooks, however, there is no described practice of implementing marketing control in educational organizations, which requires a detailed interpretation in the application of the methodology taking into account the specifics of the activity.

Inconsistency or lack of methodological approaches in the organization of strategic marketing activities of the university create certain difficulties in the implementation of the function of marketing control as an assessment of the effectiveness of marketing strategy. Taking into account the specifics of the educational market, the fundamental factor affecting the competitiveness of the educational program is the quality of education [4, 5].

In order to achieve sustainable success in the university's activities, there is an objective need to develop appropriate tools that contribute to the formation, effective functioning and development of marketing in educational organizations and its individual functions, in particular marketing control.

The research aims to identify the main conditions for the application of the product approach used in the framework of marketing control in educational institutions of higher education on the example of a group of respondents who are "students of additional vocational education."

The research objectives are to consider the practice of using the product approach in the framework of marketing control based on the materials of the National Research University "BelSU"; to determine the basic indicators, taking into account the specifics of educational activities for the group of respondents - "students of additional vocational education"; to identify the impact of the use of the product approach on the sustainable development of the university.

2 Marketing Control Tools in Educational Institutions of Higher Education

In the context of the transformation of the national system of higher education, the use of marketing control tools in educational institutions of higher education is becoming one of the competitive advantages [6–8]. Marketing control indicators provide an opportunity to evaluate the effectiveness of the chosen marketing policy, as well as to strike a balance between the educational organization, the microenvironment and the macroenvironment [9–11].

The research data were the annual reports on the monitoring of consumer satisfaction of the National Research University "BelSU" for 2010–2022 [12], as well as the implemented roadmaps for improving consumer satisfaction. The research methods were content analysis of scientific articles on the research problem in the Russian scientific electronic library eLibrary.Ru and marketing analysis of the educational market. This article is a continuation of the author's research in the field of formation of the methodology of the organization of strategic marketing activities of universities based on the concept of marketing control.

3 Implementation of the Product Approach in Marketing Control in Educational Institutions of Higher Education in Russia

One of the tools of marketing control within the framework of the product approach is the monitoring of customer satisfaction of the university. As part of the study, the practice of monitoring the satisfaction of consumers of the National Research University "BelSU" for 2010–2022 is considered on the example of the category "students of additional vocational education," an attempt is made to assess the effectiveness of using the product approach in the framework of marketing control [12].

Thus, 85% of respondents are currently satisfied with the schedule of study (Fig. 1). It should be noted that over the past 4 years, the level of satisfaction has increased slightly (from 84.3% in 2019 to the already mentioned 85% in 2022).

The majority of students were satisfied with the possibility of choosing additional professional programs in 2022 (Fig. 2). In comparison with the survey data of 2013, the level of satisfaction with this aspect of the educational process has more than doubled (from 40% to 87.1%).

In 2022, the majority of students on programs of additional vocational education are satisfied with the quality of teaching (89.8%), and this indicator has been highly

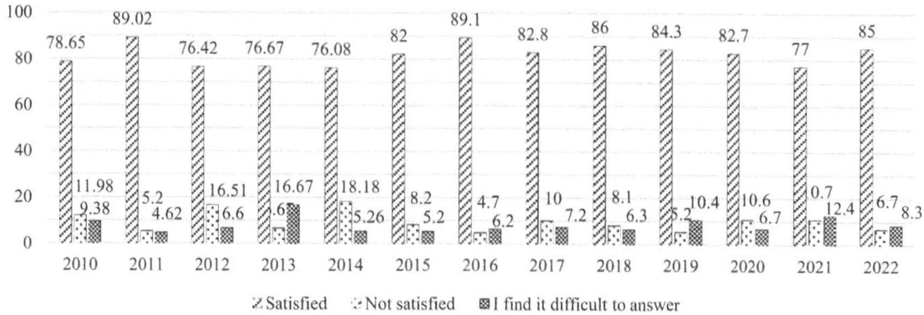

Fig. 1 Satisfaction with the schedule of study of the students of additional vocational education. *Source* Compiled by the author on the basis of [12]

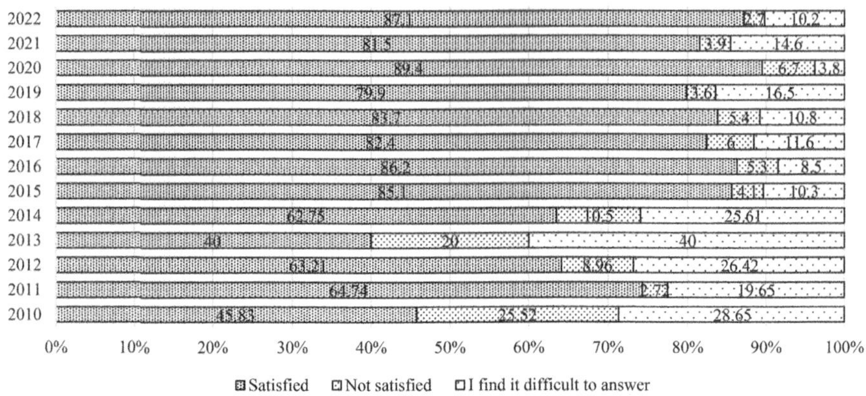

Fig. 2 Students' satisfaction with the possibility of choosing additional professional programs. *Source* Compiled by the author on the basis of [2]

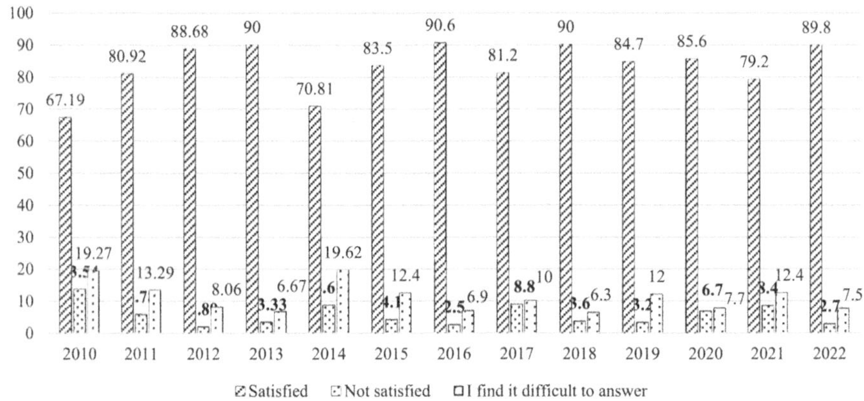

Fig. 3 Students' satisfaction with the quality of teaching. *Source* Compiled by the author on the basis of [12]

appreciated by the consumer of additional education services over the past 3 years (Fig. 3).

The majority of respondents highly appreciated the provision of the educational process with the classroom fund (Fig. 4). 82% of respondents are currently satisfied with the cleanliness and equipment of classrooms (79.1% in 2019, 80.8% in 2020, 73.6% in 2021).

The vast majority of respondents in 2022 positively assessed the level of relations of the university administration and teachers to them (86.9%), which is 2.6% more than last year (Figs. 5 and 6). Such results may show positive dynamics of the indicator. However, the dissatisfaction rate has also increased slightly over the past year (from 1.1% in 2021 to 3.8% in 2022).

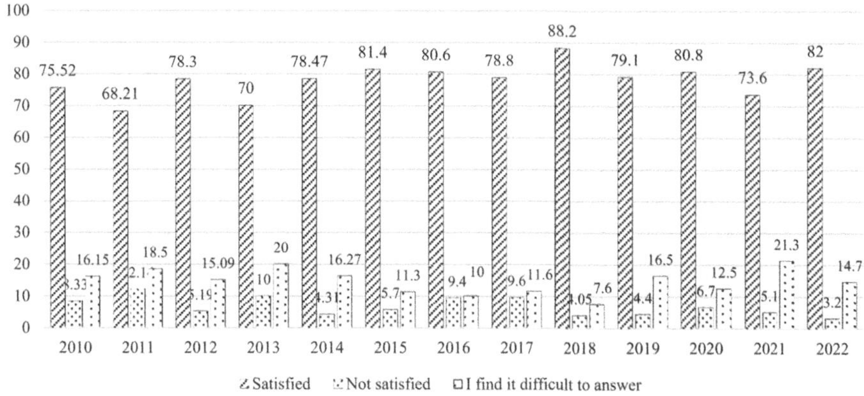

Fig. 4 Students' satisfaction with the equipment and cleanliness of the classrooms. *Source* Compiled by the author on the basis of [12]

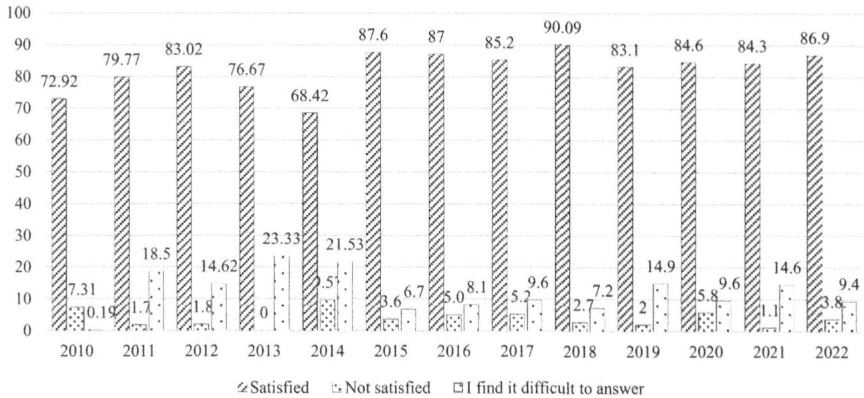

Fig. 5 Students' satisfaction with the attitude of teachers, staff, administration of the institute (faculty, center). *Source* Compiled by the author on the basis of [12]

In general, the quality of the organization of additional vocational education at the National Research University "BelSU" is assessed by students as satisfactory—88.7%. Moreover, this indicator has been showing positive dynamics since 2020.

4 Conclusion

Monitoring of consumer satisfaction of the National Research University "BelSU," within the framework of systematic marketing control, makes it possible to conclude about the identified positive dynamics over the past 10 years. In this regard, it

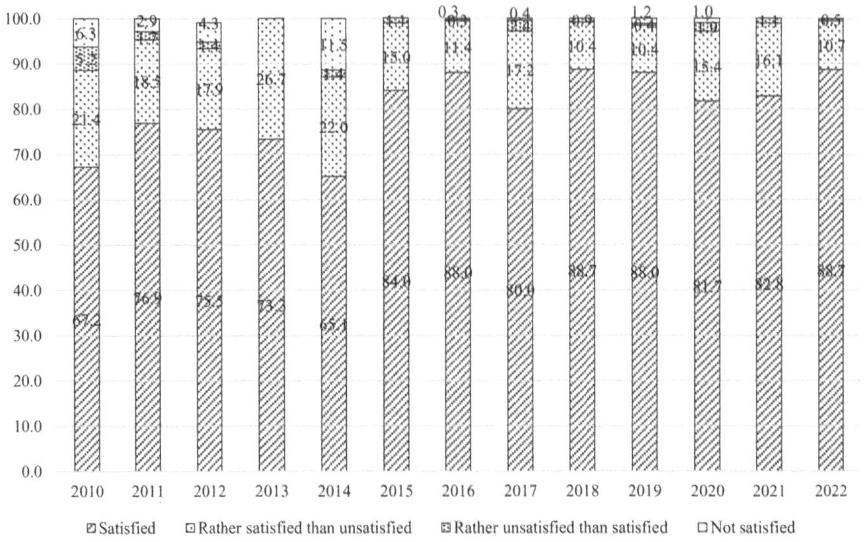

Fig. 6 Students' satisfaction with the quality of the organization of additional vocational education at the NRU "BelSU" as a whole. *Source* Compiled by the author on the basis of [12]

can be noted the effectiveness of the consumer satisfaction monitoring, as well as the ongoing activities of the annual roadmap for improving consumer satisfaction at the university level and at the level of individual institutes, which are developed based on the results of marketing control. Recommendations and suggestions for improving individual problematic indicators are received in the form of memos from heads of structural divisions or suggestions for improvement from teachers, students and employees. The results of the annual monitoring of customer satisfaction of the university are reviewed by the Academic Council.

To achieve sustainable success, the marketing policy of the university should be based on an understanding of present and future needs and be be accomplished through constant interaction with the consumer. Therefore, marketing control should be systematic and include monitoring of the customer satisfaction of the university.

The use of the product approach is seen as part of monitoring of consumer satisfaction. In order to obtain objective information in educational institutions of higher education, it is necessary to identify and examine the following main categories of consumers: students of all areas of training and specialties, postgraduates, doctoral students, students of vocational courses, employees, teachers, employers and parents of students of secondary vocational education. The final stage of this procedure should be the development of measures aimed at increasing customer satisfaction of the university. It is important to place constantly available links to questionnaires for each group of respondents on the official website of the university in order to provide an opportunity to receive feedback.

The system of basic indicators of customer satisfaction should include:

- satisfaction with the schedule of studies, the content of studies, the possibilities of choosing a specialization and variable disciplines;
- satisfaction with the quality of teaching, organization of practices;
- satisfaction with the possibilities of obtaining additional education at the university;
- satisfaction with the availability of educational, and methodical literature, the availability of libraries, the availability of electronic information sources;
- satisfaction with the conditions and quality of nutrition, the opportunity to receive primary health care in a polyclinic, the opportunity to receive social support measures;
- satisfaction with the opportunity to participate in scholarship programs with research work, opportunities for creative self-expression, recognition of successes and achievements, the attitude of teachers, staff, faculty and university administration, relationships with the curator, relationships in the student team.

The conducted research has proved a close relationship between the indicators of customer satisfaction and the achievement of sustainable success of an educational institution of higher education.

References

1. Chernyatin, S. V. (2019). The concept of marketing corporate education services based on evaluation of customer satisfaction. In *Collection of scientific papers of the International Scientific and Technical Symposium "Economic mechanisms and management technologies of industrial development" of International Kosygin Forum "Modern problems of engineering sciences"* (pp. 244–247). Federal State Budgetary Educational Institution of Higher Education "Kosygin Russian State University (Technologies. Design. Art)".
2. Kuznetsova, N. V. (2021). Competitiveness of an educational organization: On the issue of determining development criteria at the present stage. In *Proceedings of the VI All-Russian (national) Scientific and Practical Conference "Organization management, accounting and economic analysis: Issues, problems and prospects of development"* (pp. 78–84). Magnitogorsk.
3. Nadtochiy, Y. B. (2021). The quality of teaching and the quality of training sessions as a competitive advantage of an educational organization. *Self-Government, 4*(126), 521–524.
4. Kolesnikova, A. A., Kolesnikov, O. L., Sinitsky, A. I., & Shishkova, Y. S. (2020). Assessment of student satisfaction in the conditions of distance education. *Modern Problems of Science and Education, 4*, 71. https://doi.org/10.17513/spno.30068
5. Tuchina, O. R., Shaposhnikova, T. L., & Egorova, A. Y. (2021). Diagnostics of the stability of the educational environment based on the assessment of its risks. *Pedagogical Journal, 11*(1–1), 461–467. https://doi.org/10.34670/AR.2021.40.64.059
6. Fritz, A. M., & Smith, A. M. (2024). Marketing higher education on YouTube: A content analysis of college promotional videos. *Journal of Marketing for Higher Education.* https://doi.org/10.1080/08841241.2024.2400088
7. Gómez-Bayona, L., Valencia-Arias, A., Orozco-Toro, J. A., Tabares-Penagos, A., & Moreno-López, G. (2024). Importance of relationship marketing in higher education management: The perspective of university teachers. *Cogent Education, 11*(1), 2332858. https://doi.org/10.1080/2331186X.2024.2332858

8. Liu, L., Aw, E. C. -X., & Cham, T. -H. (2025). Driving brand advocacy in higher education marketing: An integrative perspective on China students studying overseas. *Asia Pacific Business Review*. https://doi.org/10.1080/13602381.2025.2452857
9. Maldonado-López, B., Ledesma-Chaves, P., & Gil-Cordero, E. (2024). Personality, loneliness and the metaverse: Exploring their interaction for higher education in marketing. *Journal of International Education in Business*. https://doi.org/10.1108/JIEB-07-2024-0085
10. Negm, E. (2024). Recognizing the impact of value-belief-norm theory on pro-environmental behaviors of higher education students: Considering aspects for social-marketing applications. *International Journal of Sustainability in Higher Education, 25*(2), 289–305. https://doi.org/10.1108/IJSHE-04-2023-0135
11. Tomasella, B., Akbar, B., Lawson, A., Howarth, R., & Bedford, R. (2024). Embedding the sustainable development goals into higher education institutions' marketing curriculum. *Journal of Marketing Education, 46*(2), 155–174. https://doi.org/10.1177/02734753241231182
12. Belgorod State National Research University (NRU "BelSU"). *Quality management documents in the field of educational services for 2010–22*. Retrieved from https://bsuedu.ru/bsu/info/officialdocs/sections.php?ID=160 (Accessed 19 February 2023).

Connection with the Labor Market as a Manifestation of the Inclusiveness of Universities 4.0 in Russia and Central Asia

Preface to the Fourth Part of the Book

This fourth part of this book addresses the organizational and managerial aspects of training personnel at Universities 4.0 in Russia and Central Asia. The significance of the chapter by Lu Bin and Nurbubu A. Asipova, titled "Survey on Culture Knowledge System in Senior High School English Textbooks," for developing the topic of this part lies in its establishment of a new philosophy of teaching English in senior high schools through the systematization of national cultural knowledge.

The chapter "Collaboration for Supporting Creative Activities" by Tatiana N. Ivanova, Maria P. Pavlova, Nikita Yu. Gulyaev and Sergey N. Shirobokov supports the topic of this part by proposing a combination of creative and innovative initiatives to enhance the quality of personnel training at Universities 4.0.

The key conclusion of the chapter "Training IT Specialists for Cooperative Organizations in Contemporary Conditions" by Natalya V. Golyaeva, Nadezda V. Akamova, and Olga I. Nemikina is that high-quality training of IT specialists at Universities 4.0 is essential, given their high demand in the labor markets of Russia and Central Asia.

The contribution of the chapter "Training Green Personnel in Higher Education Institutions for Cooperatives from the Standpoint of Pedagogy and Psychology" by Irina I. Kondrashkina, Dmitriy N. Dudoladov, Marina I. Bespalova, Anna S. Timonina, and Arina R. Pugacheva lies in its development of theoretical foundations and the identification of promising teaching methods for organizing independent work in the training of green personnel at Universities 4.0 in Russia and Central Asia.

In the chapter "Impact of Neural Networks on the Labor Market" by Tatiana M. Lebedintseva, Andrey V. Nikitin, Anna M. Vasilieva, Olga Yu. Dmitrieva, and Valery A. Filippov, the requirements for personnel training at Universities 4.0 are reexamined in light of shifts in the Russian labor market – where traditional professions are being displaced by artificial intelligence and neural networks – thus advancing the topic of this part.

Also noteworthy are the proposals in the chapter "Improving the Organizational Management Structure of a Research Organization Based on a Project Approach" by Ludmila V. Marabaeva, Tatyana E. Shilkina, and Mariya V. Myagkova, which suggest transforming the organizational structure of Universities 4.0 in Russia and Central Asia to optimize their management processes and research activities.

The chapter "Professional Training of Green Personnel in Higher Educational Institutions through Information and Communication Technologies" by Dmitry A. Nazarenko, Maryana V. Arkhipova, Victoria B. Gorbunova, Ekaterina A. Redkina, and Vladimir-Alexander T. Arkhipov further develops this topic by proposing scientific and practical approaches to training green personnel at Universities 4.0 using information and communication technologies.

Supporting the topic of this part, the chapter "Preparation of Ecologically Oriented Personnel for Collaboration Through Cooperation with Green Non-Governmental Organizations" by Alexander S. Pechatkin, Natalya G. Tarakanova, and Tatyana Yu. Pyatkina presents an important new scientific conclusion: ecologically oriented training of specialists at Universities 4.0 for cooperative business structures initiates a qualitative transformation of personal values, fosters life-affirming attitudes, and deepens the understanding of the interconnectedness between humans and their environment.

Thus, each of the eight chapters in this part contributes uniquely to the topic, exploring various practical organizational and managerial mechanisms for training green and creative personnel, as well as IT specialists, and offering practical recommendations to enhance these mechanisms in the operations of University 4.0 in Russia and Central Asia.

Survey on Culture Knowledge System in Senior High School English Textbooks

Lu Bin and Nurbubu A. Asipova

Abstract The research studies attitudes toward selecting a cultural knowledge system conducted among Chinese high school English teachers. The research aims to reveal teachers' attitudes towards the cultural knowledge system in English textbooks for senior high schools and strengthen the reference base with the addition of materials about Chinese culture. The research focuses on finding answers to several issues: determine the main views on English textbook reform, find out the attitude towards the knowledge system of the national culture of Chinese English language teachers in high school, and determine suggestions for improving pedagogical competence in designing and thinking about constructing a knowledge system in English textbooks. To test teachers' attitudes toward building a cultural knowledge system, the authors surveyed teachers in 11 schools in Henan Province and conducted in-depth interviews with nine English teachers out of 230 respondents. The research provided several results. First, most English language teachers in senior secondary schools agree with constructing a knowledge system on national culture. Second, they suggest paying attention to a more complete reflection of the knowledge system about Chinese culture, which will increase the effectiveness of teaching national culture. Third, English teachers in senior high schools consider it necessary to systematize knowledge about national culture and improve students' basic English literacy.

Keywords Senior high school · English textbooks · Culture knowledge · National culture · Attitude survey

JEL Code Z19

L. Bin
School of Foreign Languages, Luoyang Normal University, Luoyang, China

N. A. Asipova (✉)
Kyrgyz National University Named After J. Balasagyn, Bishkek, Kyrgyzstan
e-mail: nurbubu.asipova@knu.kg

B. S. Sergi et al. (eds.), *University 4.0 and Educational Technology for Sustainable Development*, Education in the Asia-Pacific Region: Issues, Concerns and Prospects 75, https://doi.org/10.1007/978-981-96-7376-6_23

1 Introduction

As one of the primary substrates of knowledge creation [1, pp. 127–151], the construction of cultural knowledge system in English textbooks covers the following aspects. The first aspect is the diversity characteristic of cultural knowledge, especially the content of local culture and national cultural knowledge [2] in English textbooks, which should be diverse and focused. That is, the construction of knowledge should cover the cultural knowledge of different countries, ethnic groups, and regions so that students can understand a wider range of cultural knowledge and clearly compare the cultural orientation of their culture. The second aspect is the practicality of cultural knowledge in the cognition construction process of students' worldview and knowledge view [3, pp. 241–258]. The culture knowledge in English textbooks should be practical, which can help students better understand the English language and international culture, as well as improve their English communication skills based on the diverse presentation of national cultures and comparative study on intercultural competence development [4]. The third aspect is the depth and breadth of cultural knowledge. Cultural knowledge in English textbooks should have a certain depth and breadth, covering basic cultural knowledge and aiming to conduct an in-depth study of cultural aspects of ideology and philosophy so that students can fully understand English and culture [5]. Hence, constructing the cultural knowledge system of English textbooks is a complex and important task, which requires English teachers to have more competence and teaching skills to lead the students to explore and practice their cultural literacy and English proficiency continuously.

This research aims to reveal teachers' attitudes towards the cultural knowledge system in English textbooks for senior high schools and strengthen the reference base with the addition of materials about Chinese culture.

2 Methodological Foundations for Constructing English Language Textbooks

The methodological premises of the research are based on the following assumptions. First, from a pedagogical effectiveness perspective, this research focuses on teachers' attitudes toward the representation of cultural content in English textbooks and their ability to work with textbooks to enhance students' reading interests and aesthetics [6, 7]. Second, from the point of view of professional competence, the focus is placed on teaching and assessing students' knowledge of the knowledge system of national cultures, which serve as the basis for the formation of patriotism and national cultural identity [8]. Therefore, this research is designed to identify English teachers' readiness to reconsider the development and preparation of basic knowledge about students' cultural cognition development [9, 10].

The theoretical basis of this research is the theory of constructivism, which makes it possible to understand the nature of the creation, construction, and development of any activity, especially the constructive activity of teachers [11], who, by the nature of their activities, have to search for and develop teaching materials constantly.

As a novel cognitive theory, constructivism has been regarded as a revolution in modern educational psychology [12]. Constructive learning theory has been stated as "a revolution taking place in contemporary educational psychology" [13]. It is developed based on the principles set forth by J. Piaget, L. S. Vygotsky, and other scientists, who assert that knowledge is acquired through active cognitive processes and that learning involves the construction of meaning by the cognitive agent [14], who, being one of the main founders of the constructivism theory, emphasizes the formation of the psychological function of the subject, and his theory can be said to be an internalization theory or an internalized process of individual construction. However, it is incomplete because individuals cannot spontaneously coordinate mental functions, which are constantly internalized [1 , pp. 110–122]. Therefore, this research aims to enrich the theory of constructing a cultural knowledge system and the effectiveness of its interaction during the period of maximizing the use of English textbooks. Analyzing the significance of English as an international language (EAL) and the percentage of references to the cultures of developed leading countries, A. A. Budnikova notes, "In all analyzed textbooks, the percentage of references to countries and cultures of the external and expanding circles is more than 50." Simultaneously, the number of countries mentioned once or twice is 27.8% [15 , p. 65]. This mainly concerns less developed countries.

V. Dagar considers the need to adapt constructivism for pedagogical purposes. He believes that this approach helps the teacher "to decide the type of pedagogy he or she will follow to create a constructivist classroom. When the teacher structures situations for their students then they are restricting the learners to carry out their mental actions" [16].

One of the important methodological foundations for constructing English language textbooks is learning the system of scientific knowledge of indigenous peoples (IKS). Analyzing this aspect, J. Tharakan notes, "At its most elemental level, IKS can be considered the foundation upon which local communities make determinations about local issues. These decisions pertain to various areas of endeavor, including the use, conservation, and management of water and other resource, agriculture, health care issues, as well as providing information and public outreach and education within a local community" [17]. A. Lareau focuses on the interest of young people of different social backgrounds (particularly the middle class and workers) in the study of cultural knowledge and key aspects in social life [18].

Another group of authors led by Morteza Hosseinzadeh reveals the representation of cultural content in ELT textbooks in primary schools in Iran through comparative analysis [19].

The study of the problems of covering knowledge about culture in English textbooks in senior high schools in China was accompanied by an interest in the Russian language in sexual and gender-based violence (SGBV) countries, where this language performs the function of intercultural interaction. In 2023, according to the global competitiveness index, the Russian language took fifth place and retained its position compared to 2022, despite a decrease in the number of speakers [20].

Based on the above premises, this research discusses the following two issues:

1. What is the main attitude towards senior high school English teachers for textbook compilation reform on the national culture knowledge system?
2. What are the suggestions for improving teaching design competence and reflection of knowledge system construction in English textbooks?

The next section describes the main research methods. This study is a survey that was conducted and lasted 3 months, from September 2022 to December 2022. The three-month study intends to test the effectiveness and quality of English language teaching and learning in the new culture knowledge system mode. The teaching experiment involves 11 senior high schools in Henan Province, China. The main methods are as follows.

The authors use in-depth interviews to gain a deep understanding and test the willingness of culture knowledge construction of the overall English teachers in 11 senior high schools in Henan province of China to understand the cultural knowledge system mode and interaction activities between senior high school English teachers and textbooks.

This survey also uses the questionnaire to understand the application willingness of senior high school English teachers and their attitude towards culture knowledge system construction in English textbooks. The culture knowledge system of the questionnaire involves three dimensions: (1) "culture knowledge contents," (2) "knowledge cognition construction," and (3) "acceptance willingness." It also includes teachers' views and reflections on teaching practice to understand the current situation of senior high school English culture construction and the acceptance of the culture knowledge system.

3 Problems and Prospects of Development of the Culture Knowledge System in Senior High School English Textbooks

This section describes the data analysis and survey results on attitudes toward readiness to apply the cultural knowledge construction framework to high school English language teaching. To obtain more realistic and extensive data, the authors used the Internet to distribute English teacher questionnaires in Henan province, China. The participants are surveyed by questionnaires. The total sample includes 238 people, of which eight invalid questionnaires were excluded. Thus, 230 valid questionnaires were obtained, with a return rate of 96.6%. Table 1 summarizes the main information about the English language teachers surveyed.

Table 1 Basic information about the research sample

Items	Contents	Frequency	Percentage
Sex	A. Male	52	22.6%
	B. Female	178	77.4%
Age	A. Over 50 years	21	9.1%
	B. 41–50 years	48	20.9%
	C. 31–40 years	43	18.7%
	D. 20–30 years	118	51.3%
Teaching experience	A. Over 15 years	34	14.8%
	B. 10–15 years	34	14.8%
	C. 5–10 years	26	11.3%
	D. 1–5 years	136	59.1%
Education	A. College	4	1.7%
	B. Bachelor	125	54.3%
	C. Master	94	40.9%
	D. Doctor	7	3.0%
Total		230	

Source Developed by the authors

From the sample of 230 teachers, 52 people are men, and 178 are women. In terms of age, young teachers aged between 21 and 30 accounted for half of the total number of teachers surveyed (51.3%). The number of teachers aged between 31 and 40 is 18.7%. The number of teachers aged between 41 and 50 is 20.9%, which is the lower percentage of the total participants, indicating that the new teaching reform on the construction of a cultural knowledge system is likely to be well received.

This section mainly investigates English teachers' attitudes towards the cultural knowledge system construction in their current English classroom teaching and interaction effectiveness with English textbooks. There are 15 questions in the English teachers' questionnaire. Questions 1–6 are basic investigations of the cultural characteristics of English textbooks. Questions 7–15 are the main attitude survey toward the contents of the construction of the cultural knowledge system in English textbooks (Table 2).

As can be seen from Table 2, answering question 7 ("Do you think that most English teachers should have a high level of ability to organize the construction of cultural knowledge system?"), 59.5% of teachers think that English teachers should have a high level of ability to organize cultural teaching activities. Answering question 8 ("Do you think the culture knowledge system activity for classroom interaction can help improve English teaching?"), over 75.2% of teachers felt that the culture knowledge system for cognition development was useful in improving teachers' ability to organize culture teaching design. The answers to question 11 ("In your classroom, do you ask more open-ended questions than closed-ended questions when teaching culture subjects?") revealed that 29.5% of English teachers asked mainly open-ended questions, which is lower than the number of teachers who asked mainly closed-ended questions. According to question 15 ("I will be

Table 2 English teachers' attitude towards the construction of culture knowledge system

	Strongly agree	Agree	Not sure	Disagree	Strongly disagree
Q7	35 (15.2%)	102 (44.3%)	56 (24.3%)	36 (15.6.9%)	1 (0.4%)
Q8	44 (19.1%)	129 (56.1%)	30 (11.7%)	27 (11.7%)	0 (0.00%)
Q9	56 (24.3%)	93 (40.4%)	42 (18.2%)	38 (16.5%)	1 (0.4%)
Q10	18 (7.8%)	41 (17.8%)	85 (36.9%)	80 (34.7%)	4 (1.7%)
Q11	23 (10.0%)	45 (19.5%)	53 (23.0%)	108 (46.9%	1 (0.4%)
Q12	27 (11.7%)	131(57.0%)	44 (19.1%)	28 (12.1%)	0 (0.0%)
Q13	17 (7.4%)	60 (26.1%)	79 (34.3%)	72 (31.3%)	2 (0.9%)
Q14	12 (5.2%)	43 (18.6%)	96 (41.7%)	74 (32.1%)	5 (2.1%)
Q15	56 (24.3%)	114 (49.5%)	23 (10.0%)	33 (14.3%)	4 (1.7%)

Source Developed by the authors

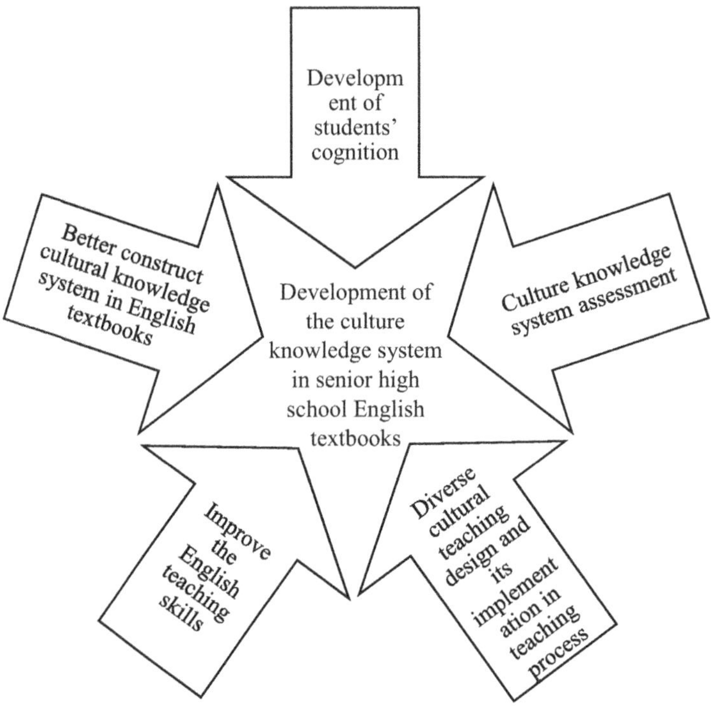

Fig. 1 Prospects of development of the culture knowledge system in senior high school English textbooks. *Source* Developed by the authors

likely to accept the new textbooks that improve the construction of culture knowledge system"), 73.8% of English teachers agreed they would welcome the new compilation version of English textbooks with culture knowledge system design.

The analysis of the above data shows that during the construction of a culture knowledge system, the development of students' cognition is not given much attention; culture knowledge system assessment is relatively single (Fig. 1).

Therefore, most English teachers are acceptable to use the new perspective on culture knowledge system construction to improve their teaching design competence on culture knowledge system construction. They all agree that teachers should pay attention to diverse cultural teaching design and its implementation in their teaching process, improve their English teaching skills, and better construct cultural knowledge system in English textbooks.

4 Conclusion

Thus, the analysis of teachers' responses confirms the research hypothesis. From the survey of teachers' attitudes given above, it is clear that teachers and students, to some extent, recognize the importance of the presence of elements of Chinese culture in English textbooks. English teachers realize the educational function of students' learning elements of national culture and the necessity of introducing students to traditional Chinese culture in English and developing patriotic consciousness and intercultural communication skills. English teachers use the method of comparing Chinese and foreign cultures in the teaching process, thus forming students' ideas about the peculiarities of national cultures. Students are more at the stage of interest in learning the elements of Chinese culture. They lack the understanding and motivation to study the elements of Chinese culture actively.

Based on the above, the authors propose paying attention to some suggestions. For example, teachers should raise awareness of the interpenetration of Chinese and other cultures when teaching English, develop students' cultural confidence, and raise awareness of the active spread of Chinese culture to foreign countries. Teachers should guide students to deeply study the cultures of other countries with a more positive attitude and an objective view. The time has come to work on the reform of the content on cultural knowledge in English textbooks and use digital resources and materials about the diversity of cultures in senior high schools in the process of teaching English.

References

1. Fernando, S., & Marikar, F. (2017). Constructivist teaching/learning theory and participatory teaching methods. *Journal of Curriculum and Teaching, 6*(1), 110–122. https://doi.org/10.5430/jct.v6n1p110
2. Venaik, S., & Brewer, P. (2016). National culture dimensions: The perpetuation of cultural ignorance. *Management Learning, 47*(5), 563–589. https://doi.org/10.1177/1350507616629356
3. Khaksar, S. M. S., Chu, M. -T., Rozario, S., & Slade, B. (2023). Knowledge-based dynamic capabilities and knowledge worker productivity in professional service firms the moderating role of organizational culture. *Knowledge Management Research & Practice, 21*(2), 241–258. https://doi.org/10.1080/14778238.2020.1794992.

4. Ayub Khan, M., & Smith Law, L. (2018). The role of national cultures in shaping the corporate management cultures: A three-country theoretical analysis. In J. Vveinhardt (Ed.), *Organizational culture* (pp. 1–25). IntechOpen. https://doi.org/10.5772/intechopen.78051
5. Slavin, R. E. (1994). *Education psychology: Theory and practice*. Allyn and Bacon.
6. Wang, Y., Xin, Y., & Chen, L. (2024). Navigating the emotional landscape: Insights into resilience, engagement, and burnout among Chinese high school English as a foreign language learners. *Learning and Motivation, 86*, 101978. https://doi.org/10.1016/j.lmot.2024.101978
7. Wei, Y., Sulaiman, N. A., & Ismail, H. H. (2024). Overcoming academic writing challenges: English language learning strategies for Chinese international high school students in blended learning environments. *Theory and Practice in Language Studies, 14*(10), 3285–3294. https://doi.org/10.17507/tpls.1410.32
8. Wu, X. (2024). Stories of names: Exploring Chinese high school language-other-than-English learners' foreign name(s) usage and its role in multilingual identity construction. *Journal of Multilingual and Multicultural Development, 45*(10), 4415–4429. https://doi.org/10.1080/01434632.2023.2170392
9. Saito, S., & Turner, M. (2024). Language diversity in "monolingual" Japan: Language awareness among high school teachers of English. *Journal of Language, Identity and Education*. https://doi.org/10.1080/15348458.2024.2385837
10. Xie, S. (2024). An analysis of the cognitive demands of senior high school english as a foreign language textbooks in China. *SAGE Open, 14*(3). https://doi.org/10.1177/21582440241280457
11. Ebrahimi, R., Adli, F., & Mehran, G. (2023). The role of academic culture on the knowledge creation from perspectives of experts of the higher educational system. *Research and Planning in Higher Education, 21*(1), 127–151.
12. Taber, K. S. (2019). Constructivism in education: Interpretations and criticisms from science education. In IRMA (Ed.), *Early childhood development: Concepts, methodologies, tools, and applications* (pp. 312–342). IGI Global. https://doi.org/10.4018/978-1-5225-7507-8.ch015
13. Zhukova, I. N., Lebedko, M. G., Proshina, Z. G., & Yuzefovich, N. G. (2015). *Glossary of intercultural communication terms* (2nd ed.). Flinta; Nauka.
14. Mercer, N. (2016). Culture, context and the construction of knowledge in the classroom. In P. Light & G. Butterworth (Eds.), *Context and cognition* (pp. 28–46). Routledge.
15. Budnikova, A. A. (2021). Analysis of textbooks of English as an international language for universities: Goal setting, content, principles. Proceedings of Volgograd State Pedagogical University, 7(160), 62–69. Retrieved July 20, 2023, from https://cyberleninka.ru/article/n/analiz-uchebnikov-angliyskogo-yazyka-kak-mezhdunarodnogo-dlya-vuzov-tselepolaganie-soderzhanie-printsipy/viewer
16. Dagar, V., & Yadav, A. (2016). Constructivism: A paradigm for teaching and learning. *Arts and Social Sciences Journal, 7*(4), 1–4. Retrieved October 12, 2023, from https://www.hilarispublisher.com/open-access/constructivism-a-paradigm-for-teaching-and-learning-2151-6200-1000200.pdf
17. Tharakan, J. (2017). Indigenous knowledge systems for appropriate technology development. In P. Venkatesan (Ed.), *Indigenous people* (pp. 124–133). IntechOpen. https://doi.org/10.5772/intechopen.69889
18. Lareau, A. (2015). Cultural knowledge and social inequality. *American Sociological Review, 80*(1), 1–27. https://doi.org/10.1177/0003122414565814
19. Hosseinzadeh, M., Heidari, F., & Choubsaz, Y. (2022). A comparative analysis of the cultural contents and elements in international and localized ELT textbooks. *International Journal of Society, Culture & Language, 10*(1), 109–124. https://doi.org/10.22034/ijscl.2021.246790
20. Press in Education. (2023, May 24). *Index of the position of the Russian language in the world—2023: Fifth in competitiveness, second in Internet content.* Retrieved October 8, 2023, from https://edupressa.vm.ru/gazeta/plody-prosvescheniya/indeks-polozheniya-russkogo-yazyka-v-mir/

Collaboration for Supporting Creative Activities

Tatiana N. Ivanova, Maria P. Pavlova, Nikita Yu Gulyaev, and Sergey N. Shirobokov

Abstract The authors examine innovative activities within the context of the initial stage of forming an innovative space or innovative infrastructure from the perspective of in-depth sociological analysis of various features of progressive society development and the socio-economic system. The authors focus on creativity, which subsequently transforms into innovative activity. The research presents a combination of collaboration to support creative and innovative activities, which will facilitate the progressive development and stimulation of creativity for its transformation into innovative activity and organize an effective mechanism for implementing creativity in society. The research presents a series of sociological studies. Within the framework of the first study, the authors revealed preferences for reading interests among men and women. The result of the study is the importance and value of books and reading for the residents of Russia. According to the second sociological study conducted among residents of Togliatti, the authors found that respondents do not attach great importance to reading. Nevertheless, they prefer paper media. The third sociological study focused on determining the use of information and communication tools by teachers of different generations. The study showed that the Internet has pros and cons, which students receiving higher education need to be aware of.

Keywords Innovation management · Social trigger · Creativity · Innovation activities · Sociological study · Respondents · Books · Reading · Higher education · Internet · Information

JEL Codes J6 · J60 · Z1 · Z13

T. N. Ivanova (✉) · M. P. Pavlova · N. Y. Gulyaev · S. N. Shirobokov
Togliatti State University, Tolyatti, Russia
e-mail: IvanovaT2005@tltsu.ru; s.shirobokov@tltsu.ru

B. S. Sergi et al. (eds.), *University 4.0 and Educational Technology for Sustainable Development*, Education in the Asia-Pacific Region: Issues, Concerns and Prospects 75, https://doi.org/10.1007/978-981-96-7376-6_24

1 Introduction

Research on innovative activities largely relies on the prevailing stereotype that innovative activity is a commercial type of activity, combining processes of commercializing the results of intellectual activity and obtaining monopolistic profits. Innovative activity is now seen as the beginning of forming an innovative space or innovative infrastructure. It is declared as the main driving force of the country's modernization processes, ensuring the growth of scientific and technical potential and serving as the primary state instrument for economic development according to an innovative scenario. The effectiveness of these processes is indeed evident and tangible [1].

In this research, the authors propose to consider innovative activity as the result of processes aimed at its emergence, maintenance, and development in a progressive society, in the socio-economic system.

Creative activity is primary; it provokes the emergence of creativity, which subsequently transforms into innovative activity [2].

Creative activity is aimed at creating new images and actions. Our imagination, experience, memory, surrounding society, and emotions are involved [3]. We cannot influence biological and genetic factors, but we can change and control the social factors that influence creativity, which directly depend on everyday reality.

2 Collaboration for Supporting Creative and Innovative Activities

Research on scientific works in the field of supporting creative and innovative activities [4–9] made it possible to identify three main types: pedagogical, psychological, and economic.

Thus, collaboration for supporting creative and innovative activities constitutes the following combination:

1. Pedagogical support manifests itself in preschool, school, and student age. At this age, children are freer in their fantasies and more willingly accept their reality [10]. The works of T. -A. Ribot demonstrated research where imagination and emotions are inversely related to the development of thinking. Therefore, in mature age, with each passing year, creative imagination fades, emotions acquire a controllable nature to consciousness, and creativity is more likely to become a chance occurrence in the prose of practical everyday life [11].
2. Psychological support aims to form the adaptive potential of the creative individual necessary for maintaining one's integrity (resilience and self-efficacy). This is a complex process that affects the personal capabilities of the individual in new and changing conditions of the social environment [12].

3. Economic support develops in the following directions: government financing, public-private investments, private investments, and household funds. There are many works and research in this area. Apparently, developing innovative commercial activities aimed at obtaining monopolistic profits and maintaining the product's competitiveness requires significant investments to organize the production process or its modernization to conduct activities for entering the market. In this case, the creativity does not require capital investment.
4. As a result of social support, creativity transforms into innovative commercial activity. This type of support acts as an adapter for the interaction of the creative individual with economic entities to introduce creative activity results into economic circulation and with society to organize the process of accustoming to innovations. The binary transformation of the result of creative activity accompanies the interaction of the social trigger and the creative individual. First, the result of creative activity is transformed into innovation as a commodity. Next, the organized process of accustoming turns innovation into a sustainable social practice among the population, which will subsequently form a new everyday reality.

3 ICT Tools for Acquiring Knowledge in the Higher Education

Thus, in mature age, with each passing year, creative imagination fades, emotions acquire a controllable nature to consciousness, and creativity is more likely to become a chance occurrence in the prose of practical everyday life [3]. To develop innovative commercial activities aimed at obtaining monopolistic profits and maintaining the product's competitiveness, significant investments are necessary to organize the production process or its modernization and enter the market. The interaction of the social trigger and the creative individual is accompanied by the binary transformation of the result of creative activity: first, the result of creative activity is transformed into innovation as a commodity.

Let us consider a series of sociological studies. The authors conducted a sociological study on reading interests.

There are insignificant differences in the preferences of men and women in choosing books. The male population does not have specific preferences in literature. They read everything: foreign classics (26%), the latest and classical Russian literature (22% and 20%, respectively), and contemporary foreign literature (18%). Women choose contemporary foreign literature (30%) and foreign and Russian classics (28%). The contemporary Russian literature is further down the list. Russian writers remain the same and practically in the same places, whereas foreign ones have all changed.

Genre interests also diverged. Both categories placed novels in the first place (32% of men and 56% of women). Then, the preferences of the groups diverged.

The male population chose science fiction (20%) and novellas (16%). Women prefer poetry (12%) and detective stories (10%). Men have no interest in poetry and mysticism (2% each), while women have no interest in drama (2%) and thrillers (0%).

It is worth noting that Russians with secondary education are interested in contemporary foreign and Russian literature (28.6%). Individuals with higher education prefer foreign and Russian classics (32% and 26.7%).

There are no significant differences in genres across different age groups and educational backgrounds.

Examining interests by cities, it can be observed that residents of Moscow and St. Petersburg prioritize novels (60.7%), followed by science fiction and novellas (17.9%) and philosophy (14.3%). Residents of regional capitals show similar preferences: they prefer novels (38.7%), followed by novellas (16.1%) and science fiction (12.9%). Residents of provincial towns also prioritize novels (36.6%), followed by science fiction and psychology (12.2%) and detective stories, poetry, and children's literature (9.8%).

Thus, books and reading are important and valuable to the residents of Russia.

The end of the twentieth century marked a period of active integration of information technologies into all areas of society. Humanity has never developed so rapidly before.

Essentially, people remained the same as they were before. We still contemplate the meaning of life, good and evil, and love and mercy. Based on the best moral concepts, Russian classical literature has always helped find answers to these questions. By making us think and empathize with the characters of its works, it enriches the inner world of humans, raises the spiritual level, and provides a moral compass in life.

Social and cultural changes in society affect the reading audience. The demands and preferences of contemporary readers, as well as channels and forms of information delivery, are changing. Mass culture becomes the sole source of spiritual experiences for many. The younger generation, prioritizing information in compressed, mainly electronic form, grows up estranged from the rich heritage of Russian literature. Reading ceases to be a source of spiritual and moral development for individuals, and books cease to be teachers and textbooks of life [13].

The object of sociological research is the residents of the Togliatti urban district aged 18 to 50 years. Data collection methods include surveys conducted on the city streets and in the institute during school hours. Data processing was carried out using the SPSS program.

During the survey, 150 people aged 18 to 50 were surveyed. Most respondents are employed (74%), 16% work and study, and only 7% are students. There are also respondents who neither work nor study, comprising only 3%.

According to 62% of respondents, their financial status is average, with a slight difference. In terms of sampling by gender, 68% of female respondents consider themselves to be in families with average income, and 53% of male respondents share this view.

Only 2% of the respondents "barely get by"; almost none deny themselves anything (3%). In the authors' opinion, these numbers cannot be considered because the residents of Togliatti may have different concepts of income within their families, and they may not be aware of all the nuances. Moreover, they may exaggerate or underestimate their financial status.

Based on the research, regardless of students' place of residence, reading occupies a similar place in their lives. Nevertheless, there are differences in percentages because the place and role of reading in different cities of Russia are diverse.

Comparing the results of this research with others allows the authors to identify similarities and differences. Preferences for fiction literature among readers in two universities of the Omsk region are distributed as follows: 71% and 49% read contemporary foreign literature, 48% and 24% read contemporary Russian literature, 39% and 11% read Russian classical literature, and 14% and 11% red the literature of Russian abroad. According to this data, each city and university has its own interests and preferences regarding reading fiction literature.

The authors asked respondents what shapes fiction literature. Of all respondents, 55% believe that literature shapes articulate speech, 51% believe it fosters erudition and a broad outlook, 9% responded that it distracts from lively and direct communication, and only 1% stated that it encourages a sedentary lifestyle.

There are no significant differences across categories. This indicates that perceptions of fiction literature among the residents of Togliatti are consistent regardless of gender, age, or education.

According to the results of this research, most respondents purchase books or borrow them from friends.

Regarding preferred information carriers, most students prefer paper-based media (76%); 33% opt for electronic books, 23% prefer books on computers, and only 8% choose books on disks. From this, it can be inferred that respondents are accustomed to traditional reading, specifically reading on paper-based media.

In this research, the following hypotheses were formulated:

1. Residents of the Tolyatti municipal district do not prioritize reading. They only read books during their study or work processes. This hypothesis was not confirmed because only 8% of the respondents read many books before exams;
2. Elderly Tolyatti residents tend to read more poetry, novels, detective stories, and science fiction, while younger ones prefer adventures, mysteries, thrillers, or have no particular preferences. The hypothesis was confirmed for older residents and not confirmed for the younger generation;
3. Young people read more often than adults. This hypothesis was not fully confirmed. Comparing age categories, residents aged 41–50 read more "for the soul" (53%), followed by those aged 31–40 (33%), and 18–30 (20%);
4. Tolyatti residents do not attach great importance to reading. This hypothesis was confirmed: 44% of respondents can find other activities besides reading;
5. "Light reading" for leisure and scientific literature for education dominate in terms of reading preferences. The hypothesis was partially confirmed: the motives for turning to reading for relaxation are dominant only among the male

population. Thus, motives for turning to books are "entertainment," "leisure," and "emotional motives" (18%);
6. Friends and family are the most important among the factors determining reading preferences. This hypothesis was fully confirmed;
7. Tolyatti residents prefer reading electronic books over paper books. This hypothesis was not confirmed because most students prefer paper media (76%); only 33% prefer electronic books;
8. Tolyatti residents do not listen to audiobooks. This hypothesis was not confirmed: 8% of respondents prefer books on disks.

Additionally, the authors organized and conducted a sociological study on the use of information and communication technologies by teachers of different generations.

During sociological analysis, the authors primarily relied on generational characteristics, which helped identify differences in using information and communication technologies by teachers of different generations.

During the research, the authors surveyed ten respondents. All individuals underwent training at the School of Young Teachers of the university. At the time of the study, two of the respondents were no longer working as teachers but still represented an interest in our research topic.

The interview program consisted of four sets of questions covering respondents' attitudes towards their work, their ideas, professional perspectives, self-assessment of young teachers according to several criteria, etc. In this regard, the authors decided to ask respondents about their attitude towards integrating ICT tools into the educational process and what advantages and disadvantages they could identify (Fig. 1).

It turned out that all respondents have a positive attitude towards using information and communication technologies in the educational process, explaining this by several reasons. First, teachers can visually present lecture material. Second, the work of teachers and students in searching for information was significantly simplified, providing access to electronic textbooks and scientific articles. Third, education has become more mobile with the emergence of distance learning. Fourth, when independently selecting information, students intensely develop individual thinking.

However, respondents also highlighted several drawbacks. Half of the respondents stated that with the implementation of information tools, live communication between teachers and students had been replaced by computerized acts of communication, making it difficult to form a proper opinion about students, their abilities, and aspirations. Additionally, the material students find on the Internet often does not meet the qualitative requirements. When preparing for practical classes, students download essays, bringing them to classes where they familiarize themselves with the content for the first time. As noted by one of the respondents, in the current situation, "topics for essays should be given that are not so easy to find, or students should be given tasks for conducting real research instead of essays. Students' individual thinking is formed only through constant analysis. However, almost all

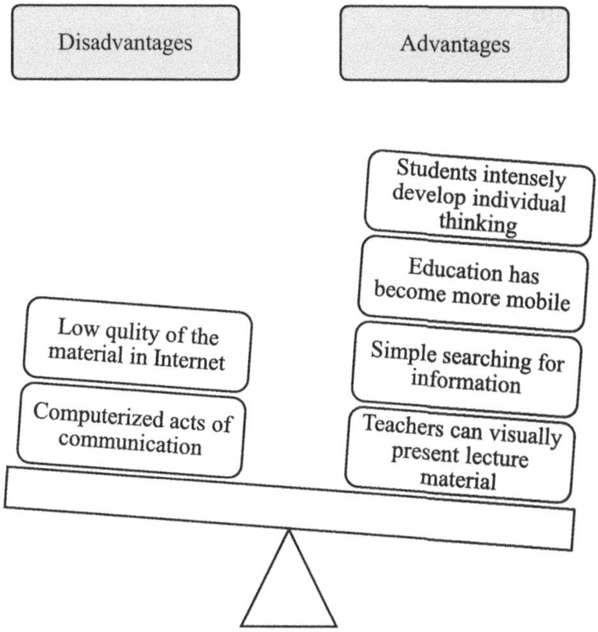

Fig. 1 Respondents' attitude towards integrating ICT tools into the educational process in universities. *Source* Created by the authors

respondents confidently stated that the Internet is a powerful tool for acquiring valuable knowledge in the educational process. However, two respondents noted that the Internet becomes a valuable resource only if used correctly. One respondent also emphasized the importance of "the abilities and possibilities of the person who uses Internet resources, their motivation to acquire knowledge. For example, a person should realize the need for a deeper study of the material, know exactly what he or she is looking for and how to set search parameters, where to look for information, etc." According to one of the respondents, data taken from the Internet needs to be thoroughly filtered because it contains a lot of "chaff." Additionally, one respondent commented that the Internet is the reason why the education system needs to be fundamentally revised.

Thus, respondents' attitudes toward using the Internet in education are generally positive. However, it is also necessary to remember the significant drawbacks that have emerged in student learning with the advent of the Internet.

4 Conclusion

The proposed collaboration will make it possible to establish a progressive process of development and stimulation of creative activity for its transformation into innovative endeavors and facilitate the organization of an effective mechanism for implementing creativity in society.

Thus, innovative activity is considered as a result—the outcome of processes aimed at its emergence, maintenance, and development in a progressive community and socio-economic system.

Creative activity is primary. It provokes the emergence of creativity, which subsequently transforms into innovative activity aimed at creating new forms and actions.

References

1. Baiserkeeva, A. E. (2016). The process of human activity development. *Current Problems of Humanities and Natural Sciences, 9*(1), 177–179.
2. Makarova, T. A. (2008). Creativity as a mode and conditions of activity. *Analytics of Cultural Studies, 1*(10), 56–60.
3. Vygotsky, L. S. (2020). *Imagination and creativity in childhood*. Publishing House "Perspektiva".
4. Suárez-Brito, P., Vázquez-Parra, J. C., Alonso-Galicia, P. E., & Cruz-Sandoval, M. A. (2025). Creative thinking as a relevant element for industry 4.0 professionals. A study in Mexican University Students. *Journal of Latinos and Education, 24*(1), 48–61. https://doi.org/10.1080/15348431.2024.2355146
5. Urban, M., Děchtěrenko, F., Lukavský, J., Hrabalova, V., Svacha, F., Brom, C., et al. (2024). ChatGPT improves creative problem-solving performance in university students: An experimental study. *Computers and Education, 215*, 105031. https://doi.org/10.1016/j.compedu.2024.105031
6. Yan, W. -J., & Li, K. -R. (2023). Sustainable cultural innovation practice: Heritage education in universities and creative inheritance of intangible cultural heritage craft. *Sustainability, 15*(2), 1194. https://doi.org/10.3390/su15021194.
7. Zatkova, T. S., Tokovska, M., & Solcova, J. (2024). What psychosocial support do first year university students need? A mixed creative methods study. *International Journal of Emotional Education, 16*(1), 140–153. https://doi.org/10.56300/AXLK4968
8. Zhang, R., & Wang, H. -P. (2024). The impact of Chinese university music teachers' teaching beliefs on creative teaching behaviors: The mediating role of technological acceptance. *Frontiers in Education, 9*, 1404541. https://doi.org/10.3389/feduc.2024.1404541.
9. Zhang, R., & Wang, H. -P. (2024). The impact of teaching beliefs on creative teaching behavior among Chinese university educators: The mediating roles of technology acceptance and organizational innovation climate. *International Journal of Technologies in Learning, 32*(1), 159–188. https://doi.org/10.18848/2327-0144/CGP/v32i01/159-188.
10. Rybakova, E. A. (2013). Educational support for preschool child in creative activities. *Humanities, Social-Economic and Social Sciences, 3*, 74–81.
11. Ribot, T. -A. (2019). *Essay on the creative imagination* (Transl. from French) (2nd ed.). Lenand. (Original work published 1900).

12. Anthropova, A. Y. (2012). Psychological support of creative and research activities of students. *Innovative Development of Vocational Education, 2*(02), 46–49.
13. Mansurova, S. E., Smirnova, E. M., & Stupnikova, L. A. (1999). Reading fiction in the structure of students' leisure activities. *Herald of Omsk University, 3*, 97–101.

Training IT Specialists for Cooperative Organizations in Contemporary Conditions

Natalya V. Golyaeva, Nadezda V. Akamova, and Olga I. Nemikina

Abstract The development of information and communication technologies and the formation of a unified information space bring huge potential to the development of society. Specialists in the field of digital economy are increasingly in demand in the labor market. One of the main national tasks related to achieving universal digital literacy is facing all levels of the education system. In today's conditions, cooperative organizations require highly qualified personnel who possess a fundamental base of specialized knowledge and skills, are ready to solve non-standard tasks, and strive to improve their qualifications constantly. Responsible training of IT specialists for cooperative organizations is becoming one of the main factors of competitiveness and development of enterprises. The research discusses approaches and methods of training IT specialists for the economic sector and describes the problems of the formation and development of IT education. The authors analyzed the quality of training and compliance with the competence requirements of young IT specialists for cooperative organizations based on a survey of employers. The research objective is to assess the professional, soft, and special competencies necessary for the work of IT specialists in cooperative organizations based on the survey results. It is determined that the formation of competence in the area of consumer cooperation among university graduates requires the creation of conditions and methods of training, such as the definition of a specialized training program, needs analysis, the program's content, the introduction of practical experience, and feedback from employers and representatives of the cooperative sector. High-quality training of IT specialists in universities is especially in demand, given the growing demand for IT professionals.

Keywords Cooperative organizations · IT specialists · Education · Information technology · Automation · Competencies · Teaching methods

N. V. Golyaeva (✉) · N. V. Akamova · O. I. Nemikina
Saransk Cooperative Institute (Branch) of the Russian University of Cooperation, Saransk, Russia

B. S. Sergi et al. (eds.), *University 4.0 and Educational Technology for Sustainable Development*, Education in the Asia-Pacific Region: Issues, Concerns and Prospects 75, https://doi.org/10.1007/978-981-96-7376-6_25

JEL Codes A22 · M15 · M53 · I21 · I25 · Q55

1 Introduction

The growth of digitization and automation in various industries makes IT specialists one of the most sought-after professional groups in the labor market. This is associated with a high demand for specialists in this field and their unique skills. The country's leadership has set the task of achieving technological independence, requiring many highly qualified personnel, primarily IT specialists, who perform various tasks in almost all fields of activity. Currently, the economic sector of Russia is experiencing a shortage of personnel in IT directions.

To address this issue, Russia is implementing a series of projects and programs. According to the project "Personnel for the digital economy," the number of budget places in digital directions is increasing from 80 thousand in 2021 to 120 thousand in 2024. However, the economy has shown that this number of specialists is insufficient. Therefore, the budget places for faculties related to information technology were increased to 160 thousand in 2022.

With the increasing demand for IT specialists, competition among candidates is also growing. In today's conditions, cooperative organizations require highly qualified personnel with a fundamental knowledge base of specialized knowledge and skills, ready to solve non-standard tasks and constantly seeking to improve their qualifications.

The Russian education system set the main objectives of preparing competitive IT specialists in the "Strategy for the development of the information society in the Russian Federation for 2017–2030" [1]. The implementation of the "Roadmap for the development of the IT industry" is in an active phase. The government expects that the roadmap's implementation will provide an additional impetus to prepare specialists and contribute to the industry's development.

Overall, the university's tasks in training IT specialists include ensuring quality technical education, preparing students to meet the requirements of the contemporary labor market, and developing the personal skills necessary for a successful career in IT [2].

From May 2022 to September 2023, the authors conducted a study analyzing the quality of training and compliance with the competence requirements of young IT specialists for cooperative organizations based on employer surveys. The research objectives are as follows:

- To assess the degree of employer satisfaction with the quality of training of IT graduates in the Republic of Mordovia;
- To assess the level of competencies and skills of young graduates;
- To rank the importance of competencies and skills when employing young professionals.

2 The Key Competencies for Graduates of IT Specialties in Cooperative Organizations in Contemporary Conditions

The methodological framework of the research consisted of analyzing scientific and methodological literature on the research problem and conducting a questionnaire survey of representatives from state and non-governmental organizations in the Republic of Mordovia (52 individuals from 32 organizations).

Universities are faced with the task of producing specialists capable of proactive work, able to promptly solve current problems facing the economy and plan and forecast the future. Traditional teaching approaches, which primarily impart theoretical knowledge, are replaced by methodologies that cultivate hands-on experience in real projects. For this purpose, universities need to collaborate with employers, design an educational environment of a new format, develop educational programs in line with employer demands, establish state-of-the-art experimental laboratories, build advanced technological parks, and implement virtual reality training systems that practically help students apply the acquired theoretical knowledge.

The education system adapts teaching methods and content to the demands of contemporary society. Collaboration with consumers, professionals, and businesses is evolving, emphasizing the role of scientific research. Nowadays, universities are increasingly addressing the challenges of the real economic sector.

Thus, collaboration between universities and business representatives will ensure the preparation of highly qualified specialists. As a result of this interaction, the Association of Computer and Information Technology Enterprises (APKIT) established the Council on Professional Qualifications in the IT field and developed professional standards, which have been approved by the Ministry of Labor and Social Protection of the Russian Federation and registered with the Ministry of Justice of the Russian Federation, and the Ministry of Labor and Social Protection.

Professional standards in the field of IT include a description of the minimum level of education and qualifications required for a specific position or role in the IT sector [3]. They define the specific technical skills and competencies necessary to perform tasks in a particular position [4]. This may encompass knowledge of programming languages, operating systems, network technologies, and other IT skills [5].

One of the primary goals of any educational institution is to prepare highly qualified specialists in a specific subject area and engage students in the design and implementation of innovative technologies [6, 7]. This involvement aims to adapt students to the conditions of the real socio-industrial environment [8].

Frequently, social demands prioritize not just the transmission of knowledge but also the cultivation of individual abilities and the capacity for continuous learning. The contemporary education system should be primarily built on knowledge and innovations [9].

The use of innovative methods helps universities equip students to meet contemporary challenges in information technology, supporting their development of relevant skills and knowledge.

Today's cooperative organizations require competent IT specialists for effective operation and development. They play a crucial role in facilitating the growth and success of cooperative organizations by optimizing business processes and enhancing efficiency through information technologies.

This underscores the importance of possessing up-to-date skills and a good reputation in the industry.

The training of such specialists requires consideration of the peculiarities of the cooperative sector and the rapidly evolving IT field. The key competencies for graduates of IT specialties in cooperative organizations in contemporary conditions are as follows:

- Understanding cooperative principles: IT specialists must grasp the fundamentals of cooperative principles such as democratic governance, shared ownership, and mutual assistance to adapt their skills and solutions to the relevant context;
- Technical skills: education should encompass extensive knowledge in IT, including software development, network security, database management, and cloud computing;
- Communication skills: specialists need to develop effective communication skills to collaborate with various cooperative members, explain complex technology concepts, and solve problems;
- Industry knowledge: understanding the specifics of the cooperative sector, its goals, and needs helps IT specialists create relevant solutions;
- Ethical aspects: training should involve discussions of ethical issues in IT, such as data confidentiality and security;
- Project management: project management skills will help IT specialists efficiently plan and execute IT projects for cooperatives;
- On-demand learning: IT specialists must be prepared to learn and adapt to new technologies and cooperative requirements;
- Innovation and development: training should stimulate specialists to seek innovative solutions that can enhance the efficiency and competitiveness of cooperative organizations;
- Community support: IT specialists can participate in educating and supporting cooperative members, helping them adopt technologies;
- Legislation and regulation: IT specialists should be aware of laws and regulations governing the cooperative sector and comply with them in their work.

In these conditions, universities play a crucial role in preparing IT specialists. The important task is to provide fundamental knowledge and develop competencies in the field of cooperative activities.

3 Competencies of IT Graduates from Universities for Cooperative Organizations in Contemporary Russia

For a high-quality education, it is crucial to maintain close collaboration with the employers of graduates and continuously work with their feedback. In this research, based on the survey results, the authors evaluated the professional, soft (soft skills), and specific competencies required for work in cooperative organizations. More than half of the respondents assessed a high level of competence in ensuring user access to information systems. Competencies related to training users in using developed software products and updating and supporting information systems were considered to be at an intermediate level of formation. More than half of the respondents indicated that most professional skills are formed at a low level (Fig. 1).

According to employers, a highly developed soft skill is the ability to make independent decisions (67%). Skills such as "high self-control and a commitment to

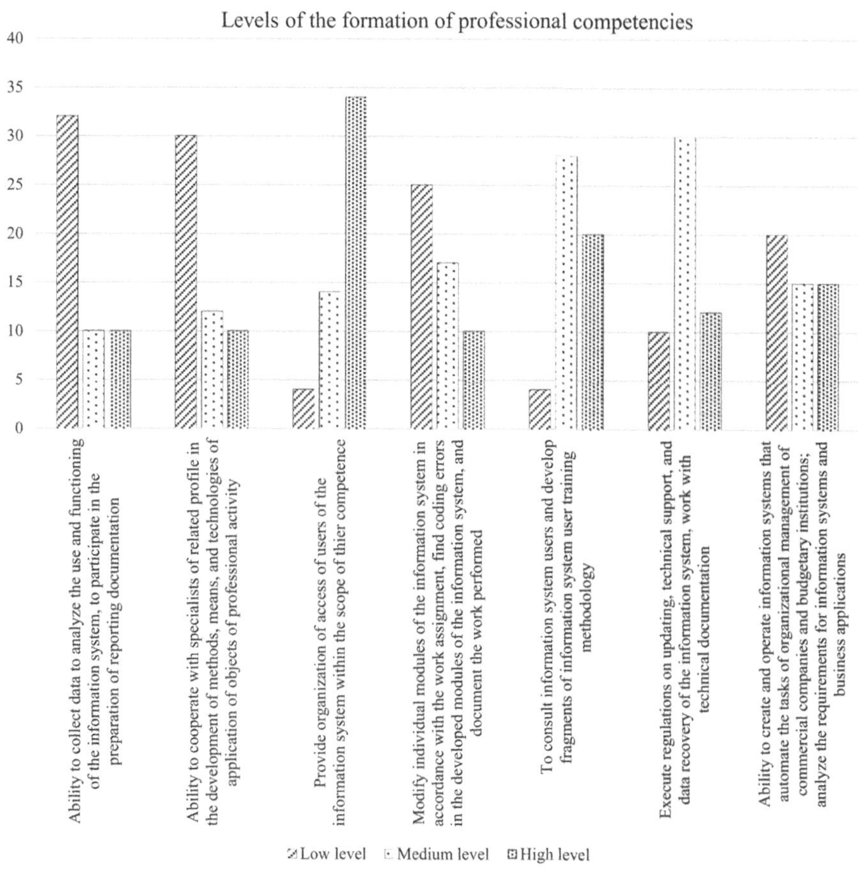

Fig. 1 Levels of the formation of professional competencies. Source: Compiled by the authors

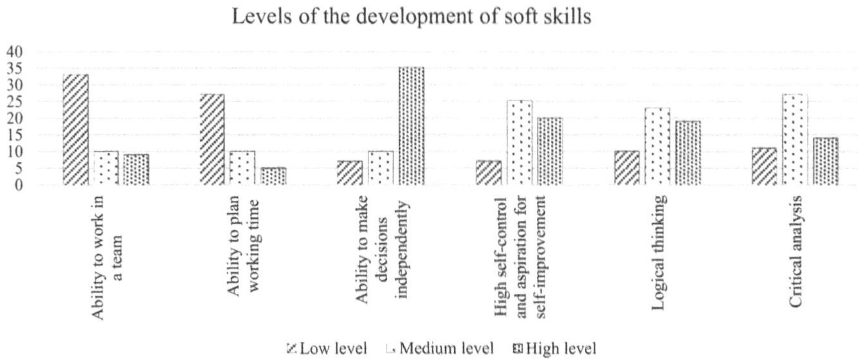

Fig. 2 Levels of the development of soft skills. Source: Developed by the authors

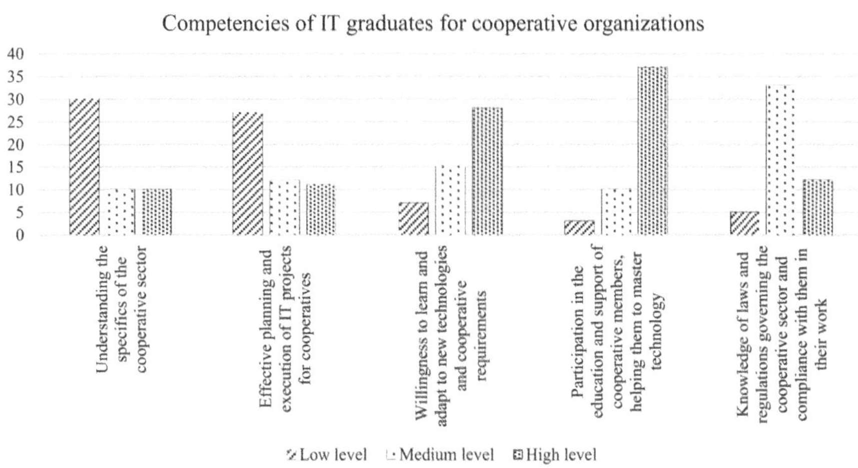

Fig. 3 Competencies of IT graduates for cooperative organizations. Source: Developed by the authors

self-improvement," "logical thinking," and "critical analysis" were considered at an intermediate level by the majority of respondents. Skills classified as either unformed or formed at a very low level include "ability to work in a team" and "ability to plan working time." The results are presented in Fig. 2.

Evaluating the competencies of graduates in IT disciplines for cooperative organizations, more than half of the respondents noted that the "understanding of the specifics of the cooperative sector" and the "effective planning and execution of IT projects for cooperatives" are at a low level. However, as a positive aspect, most of them identify a moderate level of proficiency in "willingness to learn and adapt to new technologies and requirements of cooperatives" and "participation in the education and support of cooperative members, helping them master technologies" (Fig. 3).

4 Conclusion

To develop competencies in the field of consumer cooperatives among university graduates, it is necessary to create necessary conditions and teaching methods, including the following:

- Specialized training program. Developing specialized training programs that consider the cooperative sector's peculiarities is necessary. This includes studying the principles of cooperative management, considering the specifics of cooperative organizations, and understanding their goals;
- Needs analysis. It is recommended to study and analyze the needs of cooperative organizations to determine which skills and knowledge in the field of information technology are most important for students;
- Program content. It is necessary to pay attention to the curriculum's content, including modules, case studies, and practical assignments that should cover technical and specific aspects of the cooperative sector;
- Hands-on experience. It is recommended to emphasize the importance of providing opportunities for practical work with technologies and solving tasks related to the cooperative sector.

An essential component of high-quality specialist training is establishing feedback with employers and representatives of the cooperative sector. It is necessary to integrate feedback mechanisms and assessments to continually improve the training program according to students' needs and the evolving requirements of the job market.

It is imperative to consistently adapt and update the training program in accordance with changes in technologies and the cooperative sector.

IT specialists must strive for continuous renewal of their knowledge and skills to stay abreast of the latest trends and best practices in the field of IT.

The preparation of IT specialists in universities is of utmost importance, considering the growing demand for professionals in the field of information technology. Some key aspects of this process are as follows:

- Curriculum actualization. University curricula are constantly updated and adapted to contemporary technological requirements, incorporating the latest trends, tools, and practices into the educational process;
- Practical training. In addition to theoretical education, students have access to practical sessions and projects. This includes hands-on labs, software development, working on real industry case studies, and internships in companies;
- Laboratories and equipment. Significant attention is given to providing students with equipment and software tools to experiment and develop projects effectively;
- Qualified teaching staff. Classes are conducted by teachers with relevant knowledge and experience in their respective fields. Regular updating and enhancement of the qualifications of teachers are carried out;

- Industry collaboration. The university actively collaborates with companies and organizations in the IT sector. This includes organizing lectures and workshops by industry representatives, student internships, and partnerships for joint projects;
- Communication skills training. In addition to technical skills, special emphasis is placed on developing communication skills, teamwork, problem-solving, critical thinking, and other skills that are equally important in the IT sector.

Nowadays, an increasing number of employers recognize the value of human capital and demand from job applicants not only a principle of versatility but also the possession of soft skills (personal qualities and social skills, such as teamwork, curiosity, initiative, critical thinking, the ability to solve complex problems, interact with different people, and prioritize effectively) [10].

In contemporary conditions, the preparation of IT specialists for cooperative organizations requires a comprehensive approach that considers technical and business aspects. This will help cooperatives effectively utilize IT resources to achieve their goals and proportional development.

The creation of specialized training programs for IT specialists working in cooperative organizations will assist them in better understanding the unique needs and values of this sector, thereby contributing to the development of more efficient IT solutions for cooperative organizations.

References

1. Presidential Executive Office of Russia. (2017). *Decree "On the Strategy for the development of information society in the Russian Federation for 2017–2030"* (May 9, 2017 No. 203). Moscow. Retrieved from http://static.kremlin.ru/media/acts/files/0001201705100002.pdf (Accessed 17 November 2023).
2. Díaz-García, I., Almerich, G., Suárez-Rodríguez, J. M., & Orellana, N. (2023). University students' competences in ICT: A view from the education domain. *Australasian Journal of Educational Technology, 39*(1), 106–124. https://doi.org/10.14742/ajet.6820
3. Dirckinck-Holmfeld, L., Bygholm, A., & Tabo, G. O. (2023). Transforming education through ICT: Exploring students' study practices in a resource-constrained university setting. *British Journal of Educational Technology, 54*(6), 1463–1483. https://doi.org/10.1111/bjet.13367
4. Fauth, F., & González-Martínez, J. (2023). The work environment and its influence on learning transfer in virtual in-service ICT training for teachers and university professors. *Education Sciences, 13*(4), 330. https://doi.org/10.3390/educsci13040330
5. González-Lorente, C., Martínez-Clares, P., Pérez-Cusó, J., & González-Morga, N. (2023). University tutoring and ICT: The experience of a teaching innovation in higher education. *Revista Complutense de Educacion, 34*(3), 495–506. https://doi.org/10.5209/rced.79477
6. Ngcapu, S. R., Simelane-Mnisi, S., & Mji, A. (2024). Aligning preservice teachers' experiences for ICT integration in education in the school of education at the university of technology in South Africa. *International Journal of Learning, Teaching and Educational Research, 23*(2), 203–226. https://doi.org/10.26803/ijlter.23.2.10
7. Sanhueza, M., Sandoval, L., Ormazabal, M., & Zúñiga, M. (2025). Effect of a teacher training program with ICT on university students' learning. *Contemporary. Educational Technology, 17*(1), ep556. https://doi.org/10.30935/cedtech/15745

8. Golyayev, S. S., Golyayeva, N. V., & Semikov, M. N. (2023). Immersive technologies as an innovative teaching method in the context of digitalization of education. *Journal of Secondary Vocational Education, 3*, 37–39.
9. Golyayev, S. S., & Golyayeva, N. V. (2022). Theoretical aspects of the digital transformation of the educational environment. *Journal of Secondary Vocational Education, 5*(321), 37–39.
10. Shipilov, V. (2016, May 17). *A list of soft skills and how to develop them.* Retrieved from https://www.cfin.ru/management/people/dev_val/soft-skills.shtml (Accessed 17 November 2023).

Training Green Personnel in Higher Education Institutions for Cooperatives from the Standpoint of Pedagogy and Psychology

Irina I. Kondrashkina, Dmitriy N. Dudoladov, Marina I. Bespalova, Anna S. Timonina, and Arina R. Pugacheva

Abstract The contemporary transformations occurring within the socio-cultural framework of society necessitate demands for an innovative paradigm in higher education, prompting the evolution of fundamentally novel techniques and pedagogical strategies. These methods should be focused on developing students' personalities, stimulating their creative activity, developing the ability to navigate the information environment, and forming universal skills for setting and solving problems in their professional and personal lives. Additionally, innovative approaches to organizing an educational approach are required from the point of view of the transition to an economy with a low level of environmental burden, which is known as a green economy. The research aims to determine the peculiarities of implementing green education and designing individual work of bachelor students. The authors applied methodological principles rooted in the research conducted by Russian and international experts in the domains of economics and pedagogy, along with guidance from international ecological organizations. The research object is the theoretical foundations and educational and methodological techniques for the use and functioning of individual work in the conditions of training green personnel.

Keywords Green economy · Personnel training · Vocational education · Individual work of students · Education strategy

I. I. Kondrashkina (✉) · D. N. Dudoladov · M. I. Bespalova · A. S. Timonina · A. R. Pugacheva
Saransk Cooperative Institute (Branch) of the Russian University of Cooperation, Saransk, Russia
e-mail: ikondrashkina@ruc.su; m.i.bespalova@ruc.su; a.s.timonina@ruc.su; a.r.pugacheva@ruc.su

B. S. Sergi et al. (eds.), *University 4.0 and Educational Technology for Sustainable Development*, Education in the Asia-Pacific Region: Issues, Concerns and Prospects 75, https://doi.org/10.1007/978-981-96-7376-6_26

1 Introduction

Being part of the global and educational space, today's society demands socially strong and creative individuals capable of showcasing themselves in non-standard situations and independently and effectively applying knowledge in diverse life circumstances that require immediate response. In the context of such societal developments, education and upbringing highlight the creation of conditions where students can realize their potential and demonstrate individual qualities. The goals of university education encompass ensuring the acquisition of knowledge in academic subjects and fostering gradual progress, creating success situations that instill confidence in the student's knowledge and abilities, and enabling openness to new global and cross-cultural connections.

Additionally, students must acquire skills for identifying environmental education. The normative document "Fundamentals of state policy in the field of environmental development of Russia until 2030" sets specific goals and tasks directed at higher education institutions [1]. It is essential to emphasize that environmental education has become a key platform in the education system, playing a crucial role in realizing humanitarian strategy in the twenty-first century. Russia's transition to advanced forms of socio-economic development requires fundamental changes in the worldview and behavioral characteristics of individuals.

The research employed the following methods:

- A review of books, scientific articles, and legislative acts related to the preparation of green personnel and cooperatives;
- Analysis of educational programs, methodological materials, and other documentation addressing the integration of environmental education;
- Observation of the learning process and the interaction between students, instructors, and cooperatives within the framework of environmental education;
- Examination of specific cases of successful integration of environmental training into higher education programs and their interaction with cooperatives;
- Involvement of expert opinions in the field of pedagogy and psychology to conduct assessments and analyze proposed practices in the preparation of green personnel.

2 The Main Challenges in the Training of Young Professionals in the Field of Environmental Education in Russia

The authors identified peculiarities in the education of bachelors in the field of ecology, pinpointed deficiencies in the Russian Federation's education system, analyzed the educational-methodical complex, selected scientific-methodical material, and, based on experimental work, provided recommendations for enhancing the ecological education of bachelor students to better prepare professionals.

This thesis asserts the conformity of the educational process in the Russian Federation to the key principles of humane pedagogy, where knowledge is positioned as a means of stimulating personal development. Significant opportunities in this context are presented by information and communication technologies and their direct application in the conditions of the educational process. The stated thesis confirms the initiative of former President and Prime Minister of the Russian Federation, D. A. Medvedev, to create the project "Digital school." According to him, the involvement of students and teachers in implementing this project should be ensured. This process should be continuous for all subjects of the educational process since "technologies change, and it is impossible to gain knowledge once and for all in the field of technology" [2]. The "Concept of environmental education in the system of general education" was adopted in 2022. According to this concept, "achieving a high level of environmental culture is an indicator of the development of the country's human capital, the quality of life of Russians, the national security of the country, and its competitiveness in the world" [3].

In this regard, social institutions and civic communities express the urgent need to form students with competencies, allowing them to create a wide network of communication at various levels. The latter are in demand because they activate the processes of popularization and support communities of initiative young people (including children of different age groups) aimed at solving essential environmental problems, as well as their integration into the cooperation of the Russian Federation, expanding the possibilities of additional environmental education, primarily for enterprises that use nature, providing the economy with specialists in the field of ecology and environmental protection.

The effective implementation of personnel policy in the green economy currently provides a unique opportunity to transition to a higher level of education modernization [4]. From a global perspective, such a step will further improve students' competencies, making it possible to establish contact and successfully integrate innovative ideas into the educational process [5, 6]. One should not forget about the individualization of education, which will improve the quality of education and provide additional motivation to become a specialist in the field of green employment [7].

The considered realm of scientific inquiry necessitates the clarification of the concept of green employment [8]. Let us examine the first definition proposed by the International Labor Office in Geneva: "any activity fundamentally aimed at reducing or offsetting its impact on the environment" [9 , p. 99].

Thus, we are discussing a form of labor activity oriented toward minimizing negative environmental impact, conserving natural resources, and contributing to creating a favorable environment. This definition partially aligns with the doctrine of the United Nations (UN), where jobs in the business will be held by individuals contributing to ecology, as well as in the realm of protecting workers' rights and well-being by transitioning to cleaner energy sources, addressing climate change, and preventing biodiversity loss. It is important to note that the Geneva definition of green employment resonated in the Declaration of the United Nations Conference on the Human Environment in 1972 in Stockholm: "Education in environmental matters, for the younger generation as well as adults, giving due consideration to the underprivileged, is essential in order to broaden the basis for an enlightened opinion and responsible conduct by individuals, enterprises and communities in protecting and improving the environment in its full human dimension" [10].

In our understanding, green employment is also associated with the term sustainable development, developed by the Brundtland Commission: "Development that meets the needs of the present without compromising the ability of future generations to meet their own needs" [11 , p. 160]. It is crucial to highlight that substantial changes in the concept of green employment were introduced by the Kyoto Protocol to the United Nations Framework Convention on Climate Change in 1997. In this protocol, leading nations committed to reducing greenhouse gas emissions. Russia ratified the provisions of Sections I and II, wherein each country undertakes commitments to reduce emissions [12]. Clearly, the first definition of green employment is directly influenced by the positions of international United Nations documents, the UN Environment Program, the International Labour Conference, and the International Organisation of Employers.

P. V. Trifonov analyzes an alternative interpretation of the concept of green economy [13]. It is considered as a process of fair and just transition. From the perspective of this concept, it envisions providing necessary support to individuals affected by various climate changes and creating new opportunities. These opportunities are collaboratively utilized by employers, employees, their corporate organizations, and other entities and structures operating in the field of social services [13 , p. 22].

P. V. Trifonov's definition aligns most comprehensively with the issue of cultivating green professionals in contemporary cooperation. In our view, governmental organizations, enterprises, and business communities will be the ones providing the opportunity to train future specialists in the field of green economy through institutional foundations and sustainable development strategies.

It is important to observe that several factors pose the main challenges in the training of young professionals in the field of environmental education in Russia. First, there is an excess of general educational knowledge at the higher education level. In other words, students shape their worldview but do not acquire sufficient information about their future activities. Second, there are disparities between the economy and university competencies, where the first problem becomes relevant, with a majority of universities producing ecologists in the field of natural sciences, while the country has been experiencing a crisis in industrial ecologists for several years. Third, compiling current reporting on environmental documentation in enterprises and organizations does not make it possible to focus on direct responsibilities. Fourth, there is a problem with the formal approach of higher education, where students do not have the opportunity to shape their educational trajectory due to a lack of elective disciplines. Fifth, there is the misallocation of funds for student internships, as well as unprofitable material and technical equipment. Sixth, there is a gap between the production attitude of teachers. Seventh, there is a lack of internship systems for regular faculty members.

O. A. Lantsev believes that a significant portion of higher and secondary vocational education programs has already transitioned to the new generation of Federal State Educational Standards (FGOS). This transition is expected to overcome the crisis in preparing personnel for the green economy in cooperation. Currently, Russian educational institutions are taking various actions to develop educational programs to equip students with a set of skills and competencies in the field of green technologies [14]. However, it is worth noting that the Federal State Educational Standards for secondary vocational and higher education currently do not include specific indications or descriptions of green competencies. This signifies a noticeable gap between the developed educational programs and the current education standards.

3 Reevaluation of the Training of Young Green Professionals in Higher Education Institutions to Develop Cooperation

It is believed that trade unions, employers' associations, and political structures, including the Russian Ecological Party "The Greens," the All-Russian Public Organization "Business Russia," and the Russian Union of Industrialists and Entrepreneurs, should play a crucial role in creating green vacancies in cooperation. This thesis supports our notion of a profound crisis in preparing young specialists in the field of cooperation for the green economy. It is essential to emphasize that the previously described problems require a reevaluation of the training of young professionals to develop cooperation (Fig. 1).

In our opinion, the educational process should incorporate interactive learning and activities related to information and communication technologies. Additionally, it is crucial to address the issue of the number of lecturers from the real economic sector to actively integrate project modeling into the learning process for solving regional environmental challenges. Moreover, introducing requirements for professional internships for teachers in real enterprises and organizations is also essential.

Effective support for developing the green economy in terms of resource conservation is possible through innovative methods and ideas. From this perspective, the following recommendations proposed by UNEP [15 , p. 1265] seem timely and effective:

1. Establishing a regulatory framework for transitioning to a green economy involves purposeful development and subsequent implementation of legislative and regulatory acts. The primary aim is to ensure a thoughtful transition to an economy characterized by a commitment to resource conservation and environmental sustainability.
2. Prioritizing state investments in environmentally sustainable sectors involves directing financial resources to economic industries and sectors that contribute to environmentally responsible production, active support of emerging environmental innovations, and developing green technologies.

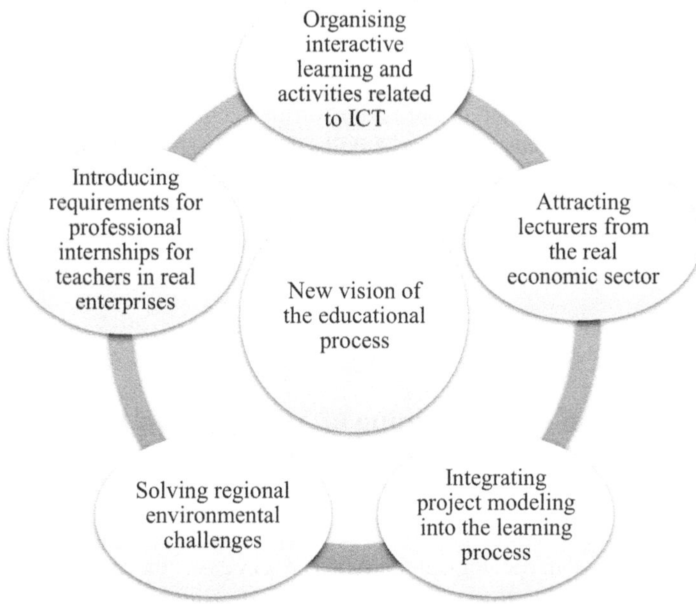

Fig. 1 The new vision of the training of young green professionals in higher education institutions to develop cooperation. Source: Compiled by the authors

3. Limiting state expenditures in resource-intensive areas entails imposing certain restrictive measures on state expenditures that apply to areas with a high degree of natural resource consumption.
4. Strengthening international cooperation for sustainable development means organizing extensive exchanges—resources, knowledge, and experience—based on common environmental interests among organizations or countries.
5. Using tax mechanisms to stimulate green investments implies implementing a tax policy characterized by providing significant incentives to innovative and environmentally sustainable projects.
6. Investing in education in sustainable development areas is particularly interesting in the context of the research topic. This recommendation involves investing in education and emphasizing the development of environmental awareness and continuous improvement. Particularly, there is a thorough review of existing educational programs and the development of more innovative ones.

In the context of such educational formation, the foreground in the education and upbringing of bachelor's students is the creation of conditions where students can realize their potential and demonstrate individual qualities. There are several main perspectives on the issue. The first belongs to E. A. Aleshugina, who advocates for a format of group activities using various scenarios. According to the researcher, e-learning maintains a high quality of teaching, comparable to traditional face-to-face forms, due to the organization of information systemically, providing links to additional informational resources, diversity of educational tasks, and ways of interacting with educators [16 , p. 187].

L. N. Knyazkova highlights the following important aspects of organizing individual work for bachelor's students: creating and implementing projects of individual educational activities for students, their training and upbringing, and supporting students in the development and implementation of their individual professional and life paths prove a high interest from the educational institution and society in shaping competent specialists [17]. The main focus shifts towards independent student work and a well-thought-out strategy for developing an educational path, which will serve as the basis for individual work. In this context, the student's personal motivation gains importance, and a certain degree of freedom in choosing the course and its practical application and deepening is developed. Additionally, we must note the crucial moment of the student's integration into the educational model. By the learning process, we mean not self-education according to one's scenario as the ultimate goal of individual work but a purposeful process systemically managed by the teacher. As a result, individual work during distance learning becomes a fixed idea for the bachelor. One of the primary tasks of individualized student work includes two main functions: orienting and stimulating (the learning process is accelerated and motivated) and informational-educational (students' educational activities, mediated by independent work, become more effective).

4 Conclusion

Preparing green professionals in the context of sustainable development and existing environmental problems has become an integral part of higher education strategy. In this research, the authors discussed the importance of the issue from the perspectives of pedagogy and psychology, emphasizing the feasibility of integrating these disciplines for the effective training of professionals capable of addressing environmental challenges and contributing to the development of the green economy.

One of the key conclusions of this research is the need to reconsider educational programs and teaching methods in light of contemporary challenges. The preparation of green professionals should focus on humanitarian aspects, including the development of ethical values, responsibility towards the environment, and the ability to collaborate within cooperatives. Teachers and psychologists must closely collaborate to create favorable learning environments that contribute to the formation of civic engagement and a conscious relationship with the surrounding reality.

Information and communication technologies provide extensive opportunities for actively involving students in the learning process and promoting their environmental development. Thoughtful use of advanced technologies can make the educational process more interactive and dynamic, facilitating the development of skills and competencies essential for green professionals.

In this regard, the interdisciplinary nature of the issue should be acknowledged. Collaborative efforts between educators and psychologists, along with the integration of knowledge from various disciplines, make it possible to create more comprehensive educational programs.

References

1. President of the Russian Federation. (2012). *Fundamentals of state policy in the field of ecological development of Russia until 2030* (approved on April 30, 2012). Moscow. Consultant Plus. Retrieved from https://www.consultant.ru/document/cons_doc_LAW_129117/ (Accessed 9 October 2023).
2. Uchitelskaya Gazeta. (2017, December 13). *Dmitry Medvedev proposed to launch a priority project "Digital school."*. Retrieved from https://ug.ru/dmitrij-medvedev-predlozhil-zapustit-prioritetnyj-proekt-czifrovaya-shkola/ (Accessed 11 October 2023).
3. Ministry of Education of the Russian Federation. (2022). *The concept of environmental education in the system of general education* (approved by the decision of the Federal Educational and Methodological Association for General Education, Minutes of April 29, 2022 No. 2/22). Moscow. Retrieved from https://docs.edu.gov.ru/document/3da3f2dbd81de632a44729cf4fc40ea9/ (Accessed 10 October 2023).
4. Khan, S. A., Magd, H., & Batra, M. (2023). An empirical investigation of higher education students' intention towards green products usage/purchase: What can educators do? *International Journal of Pluralism and Economics Education, 14*(3-4), 297–314. https://doi.org/10.1504/IJPEE.2023.138576
5. Saad, Z. A., Fauzi, M. A., Zulkepeli, L., Moshood, T. D., & Hussain, S. (2024). Green human resources in higher education institutions: A systematic literature review. *Journal of Applied Research in Higher Education*. https://doi.org/10.1108/JARHE-01-2024-0033
6. Sachdeva, C., & Singh, T. (2024). The impact of green human resource management on employee green behaviour in higher education institutions: The mediating role of green self-efficacy. *International Journal of Education Economics and Development, 15*(3), 337–358. https://doi.org/10.1504/IJEED.2024.139305
7. Wong, L. J., Ling, P. S., & Ling, T. H. Y. (2023). A conceptual framework for higher education student loyalty from the green marketing perspective. *Higher Education, Skills and Work-based Learning, 13*(2), 387–402. https://doi.org/10.1108/HESWBL-08-2022-0165
8. Zheng, K., Zheng, X., Yang, Y., & Chang, J. (2023). Advancing higher education and its implication towards sustainable development: Moderate role of green innovation in BRI economies. *Heliyon, 9*(9), e19519. https://doi.org/10.1016/j.heliyon.2023.e19519
9. Nenastev, N. A., & Yashalova, N. N. (2023). "Green" employment: The essence and genesis of the concept. *Vestnik Universiteta, 5*, 96–104. https://doi.org/10.26425/1816-4277-2023-5-96-104
10. United Nations. (1972). *Declaration of the United Nations Conference on the Human Environment* (adopted June 16, 1972 at the 21st plenary meeting). Stockholm. Retrieved from https://documents-dds-ny.un.org/doc/UNDOC/GEN/NL7/300/05/PDF/NL730005.pdf?OpenElement (Accessed 11 October 2023).
11. Begun, T. V. (2012). Sustainable development: Definition, concept, and factors in the context of single-industry towns. In *Economics, Management, Finance: Proceedings of the II International Scientific Conference* (pp. 158–163). Mercury.
12. Russian Federation. (2004). *Kyoto Protocol to the UN Framework Convention on Climate Change* (ratified by Federal law No. 128-FZ of November 4, 2004). Moscow. Retrieved from https://docs.cntd.ru/document/901880645 (Accessed 10 October 2023).
13. Trifonov, P. V. (2015). Green work places – The basis of ecological-oriented economy in the XXI century. *Business Strategies, 3*(11), 21–27.

14. Lantsev, A. O. (2016). Training of workforce in Russia for the transition to a green economy: Problems and perspectives. In *Russian Regions in the Focus of Change: Collection of Papers of the XI International Conference, November 17–19, 2016; in 2 vols.* (pp. 701-706) (Vol. 2). Yekaterinburg: UMC-UPI.
15. Egorova, M. S. (2014). Economic mechanisms and transition conditions to green economy. *Fundamental Research, 6-6*, 1262–1266.
16. Aleshugina, E. A., Vaganova, O. I., & Kutepov, M. M. (2019). Designing of individual trajectories of training with the use of online courses. *Baltic Humanitarian Journal, 8*(1), 185–187.
17. Knyazkova, L. N. (2017). Designing individual educational activity of students as a means of individualization of educational process in higher education institution. In E. V. Karpova (Ed.), *Realization of the Standards of the Second Generation in School: Problems and Prospects: Collection of Scientific Articles of the VII All-Russian Scientific-Practical Extramural Internet Conference.* Yaroslavl: Yaroslavl State Pedagogical University named after K.D. Ushinsky. Retrieved from https://yspu.org/conferences/Realizaciya_standartov_2-go_pokoleniya_v_shkole/proektirovanie_individual'noi_obrazovatel'noi_deyatel'nosti_studentov.pdf (Accessed 11 October 2023).

Impact of Neural Networks on the Labor Market

Tatiana M. Lebedintseva, Andrey V. Nikitin, Anna M. Vasilieva, Olga Yu Dmitrieva, and Valery A. Filippov

Abstract The emergence of artificial intelligence and neural networks creates a significant potential for automating an increasing number of business processes of large, small, and medium-sized businesses in various sectors of the economy. The financial sector, education, medicine, retail trade, and transport are the most promising from the point of view of the introduction of neural networks. Having undoubted advantages (e.g., saving time and financial resources, improving coordination when working together, etc.), neural networks contribute to the release of jobs, create a problem of rising unemployment, and can increase social tension in society. Additionally, being a tool for replacing human labor with technologies, neural networks can lead to a reduction in the possibilities of a person's social realization, zeroing out the demand for the skills and abilities acquired by people during their lives. The research considers the main opportunities created by neural networks for business. The authors assess the labor market indicators of the Russian Federation and forecast the unemployment rate in the labor market in the conditions of mass withdrawal from the market of "retired professions" due to their replacement by artificial intelligence technologies and neural networks.

Keywords Information and communication technologies · Neural networks · Artificial intelligence · Labor market · Unemployment · Economic security

JEL Code F63

T. M. Lebedintseva (✉) · A. V. Nikitin · A. M. Vasilieva · V. A. Filippov
Cheboksary Cooperative Institute (Branch) of the Russian University of Cooperation, Cheboksary, Russia
e-mail: tlebedinceva@ruc.su; a.v.nikitin@ruc.su; a.m.vasilyeva@ruc.su; v.a.filippov@ruc.su

O. Y. Dmitrieva
I.N. Ulyanov Chuvash State University, Cheboksary, Russia
e-mail: o.yu.dmitrieva@ruc.su

B. S. Sergi et al. (eds.), *University 4.0 and Educational Technology for Sustainable Development*, Education in the Asia-Pacific Region: Issues, Concerns and Prospects 75, https://doi.org/10.1007/978-981-96-7376-6_27

1 Introduction

Scientific and technological progress is a necessary condition for social development. It shapes new trends and regularities and elevates the overall comfort level of society. However, new technologies can create problems, depriving members of society of opportunities for social realization, as is the case with neural networks. The excitement caused by the possibilities of neural networks is primarily explained by their potential to replace a multitude of jobs, which causes certain concerns.

The research aims to study the impact of neural networks on the labor market. To achieve the research goal, the authors solve the following tasks: (1) to consider the role of information and communication technologies for the economy; (2) to assess the demand for ICT in various sectors of the economy; (3) to explore possible prospects for developing neural networks in terms of their use in the economy; (4) to calculate the forecasted unemployment rate in the Russian Federation under conditions of the mass exit of "retired professions" from the market due to their replacement by neural networks.

The scientific validity and degree of reliability of the main provisions are confirmed by the utilization of scientific works by Russian and foreign scholars in the field of information and communication technology development and their impact on the economy and labor market. Additionally, the research relies on the examination of socio-economic characteristics of contemporary reality and the representativeness of the empirical research base.

2 Prospective Directions for Using ICT in the Business Environment

An integral part of any business, regardless of its scale, type of activity, or ownership structure, is information and communication technologies (ICT). ICT is actively used to optimize companies' costs, increase revenue and profits, and enhance labor productivity. Business digitization with attraction of high educated ICT-specialists graduated from universities significantly saves time through the speed and accuracy of executing professional tasks [1–6].

The list of advantages provided by adopting ICT is quite extensive, including the following:

- Automation of business processes within companies;
- Increase in communicative efficiency in solving professional tasks;
- Saving time and financial resources;
- Encouragement for the development of new types of services and products.

A prominent example of using ICT for business process automation is the automation of accounting using the accounting and tax accounting program 1C:Accounting.

This software makes it possible to automate routine operations in accounting, speed up processes, and ensure higher accuracy in execution.

The use of messengers and electronic communication tools (e.g., email and video conferencing) creates broad opportunities for quick resolution and discussion of business issues and tasks in an online mode, facilitating the organization of remote work for companies with their employees and clients.

The application of specialized software provides opportunities for the efficient use of company resources, optimizing the time required to solve professional tasks. To achieve these goals, companies implement various corporate information systems, such as CRM systems (software assisting with customer base management), MRP systems (designed for industrial production planning), ERP systems (enterprise resource planning systems coordinating data flow between corporate business processes), SCM systems (supply chain management), MPS systems (implementing volume-calendar planning methodology), FRP systems (financial resource planning systems), and others.

Additionally, the use of information and communication technologies created new opportunities for companies engaged in traditional types of business, such as organizing services for delivering ready-made meals and groceries, in addition to the main activities.

Nevertheless, despite the advantages of using ICT in the business environment, their demand in various sectors of the economy is not uniform. Based on research conducted by Deloitte CIS and SAP [7] among nearly fifty companies in the top 500 largest companies in Russia, it was concluded that companies and enterprises in the B2C sector have the highest degree of digitization. This is primarily explained by the fact that companies dealing directly with end consumers in the B2C sector are more likely to adapt to rapidly changing customer requirements. The banking sector, retail companies, the telecommunications industry, and higher education are the leaders among the sectors of the Russian economy in terms of digitization. Simultaneously, difficulties in transitioning to digital technologies are observed in companies related to the fuel and energy complex, metallurgy, and mechanical engineering.

Despite the uneven digitization of sectors in the Russian economy and the somewhat passive response of some industries to the emergence of new ICT, progress in information technology continues. As early as 1933, Soviet mathematician A. N. Kolmogorov formulated a widely accepted formal system for mathematically describing probability theory, which became one of the foundations of the work of today's neural networks. In the 1940s, W. McCulloch and W. Pitts created an artificial neural network node based on a simplified version of the natural neuron, as described in their article "A Logical Calculus of Ideas Immanent in Nervous Activity."

The emergence of powerful graphics processors in the 2000s opened access to the use of large volumes of data and laid the foundation for the development of machine learning algorithms called Deep Learning, the development of which initiated the creation of new algorithms and tools for designing and training neural networks.

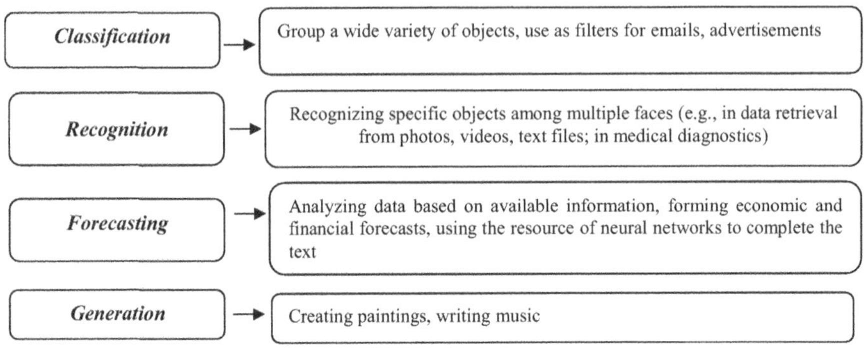

Fig. 1 Tasks that can be solved using neural networks. Source: Compiled by the authors

The process of creating neural networks is complex because it requires knowledge in several scientific areas, including mathematics, statistics, programming, and others. Moreover, this process is constantly evolving. However, despite its complexity, neural networks are currently among the fastest-growing and attracting attention from investors and users worldwide. Figure 1 outlines the main directions of tasks that can be addressed using neural network resources.

3 Case Study of the Neural Networks Impact on the Labor Market in Russia

In 2014, the Skolkovo Foundation's Agency for Strategic Initiatives released the "Atlas of New Professions," which included a section describing professions (around 40 titles) that would exit the labor market in the future (over 10–15 years) [8]. These professions, encompassing intellectual and manual labor, were termed "retired professions." Artificial intelligence was also identified as the factor influencing the departure of these professions from the labor market, alongside the use of robots, outsourcing, and migrant labor.

The authors of the "Atlas of New Professions" noted that if robotic automation would displace workers from heavy and eventually light physical labor industries, artificial intelligence and neural networks would pose serious competition to intellectual labor even in relatively complex fields. Specifically, intellectual operations with routine characteristics would be algorithmized; the management of such operations would be entrusted to computer programs. It is anticipated that the process of "retiring" professions, gaining momentum, will rapidly become widespread. In other words, accelerated progress in digitization and information technologies will lead to the replacement of 70%–75% of "retired professions" with automated intellectual systems or robots within the next 15 years. Experts predict the departure of numerous mass professions from the market in the coming decade. A significant number of individuals engaged in physical and intellectual labor across various economic sectors may be at risk of unemployment.

Research conducted by the Center for Labor Market Studies at the National Research University Higher School of Economics (HSE) revealed that 62% of employed Russians are engaged in intellectual labor [9]. Considering this research and the forecasts of the Skolkovo Foundation's Agency for Strategic Initiatives regarding retired professions and key industries where artificial intelligence and neural networks will revolutionize in the coming years (medicine, retail, finance, education, and transportation), it is possible to identify the main risk group in the labor market and the projected level of unemployment that may arise due to technological progress (implementation of artificial intelligence and neural networks) (Table 1).

The calculation presented in Table 1 is approximate and conditional because the data for the calculation were taken from studies conducted by different research groups at different times. Nevertheless, even these results show that the mass replacement of professions in the labor market by neural networks and artificial intelligence will lead to a colossal increase in social tension due to the rise in the unemployment rate to a mass level (21.1%), with an acceptable level of unemployment being 5.0%–7.9% [11]. The main group of potential unemployed individuals

Table 1 Calculation of the forecast unemployment rate in the Russian Federation under conditions of a mass withdrawal of "retired professions" from the market due to their replacement by artificial intelligence technologies and neural networks

	Indicators	Total	Men	Women
1.	Total population [10]	146,171	67847.8	78323.2
2.	Employed population, thousand people	70,601	36,208	34,393
3.	Employment rate, %	58.4	66.1	52.1
4.	Unemployed population, thousand people	4321	2237	2085
5.	Unemployment rate, % [10]	5.8	5.8	5.7
6.	Including by sectors of the economy, people			
6.1	– Medicine	5499	1130	4369
6.2	– Retail and wholesale trade	10,907	4213	6694
6.3	– Transportation	6198	4797	1401
6.4	– Financial sector	1587	492	1095
6.5	– Education	6676	1193	5483
	Total by industry:	30,867	11,825	19,042
7.	% employed in intellectual labor total, %	62.0		
9.	% of the retirement of professions according to the forecast in the next 15 years	60%		
11.	Forecasted number of employed in the economy considering the percent of intellectual labor employment and retirement of retired professions, persons	59,112	31,811	27,301
13.	Projected number of unemployed population, thousand people	15,810	6634	9176
14.	Forecasted unemployment rate, %	21.1	17.3	25.2

Source: Compiled by the authors

resulting from the replacement of mass professions by neural networks consists of women (58.0% of the total predicted number of unemployed). One of the main reasons for this result is that the majority of women are employed in the sectors where the introduction of artificial intelligence and neural networks is expected (e.g., education, medicine, trade, and the financial sector). Some experts note that more than half of the professions will be automated.

It is also worth noting that the unemployment rate is one of the main indicators of the economic security of a country, representing one of the key supports for maintaining the country's sovereignty. It implies the state of the national economy that ensures the protection of the country's interests, resistance to internal and external pressure and threats, the ability to develop, and the protection of the vital interests of people, society, and the country [12].

Thus, the development of neural networks and their impact on the labor market, which they will undoubtedly exert, can lead to serious consequences for the country's overall macroeconomic stability.

On the other hand, emerging innovative technologies create opportunities for the appearance of new professions not previously known to the labor market. As a result of the increasing pace of ICT implementation, each sector of the economy will offer more new professions (Table 2). In this regard, the issue arises about the readiness

Table 2 Examples of new professions in industries that have emerged as a result of the introduction of ICT into the industry

Industry	Examples of new professions
Biotechnology	Systems biotechnologist
	Living systems architect
	City farmer
	Biopharmacologist
Medicine	IT medical worker
	Genetic counselor
	IT geneticist
	Molecular nutritionist
	Network doctor
Financial sector	Intellectual property appraiser
	Multi-currency translator
	Personal pension plan developer
Education	Developer of educational trajectory
	Organizer of project-based learning
	Startup mentor
	Game educator
Ground transportation	Transportation network security engineer
	Intermodal transportation solutions technician
	Smart road builder
	Architect of intelligent control systems

Source: Compiled by the authors

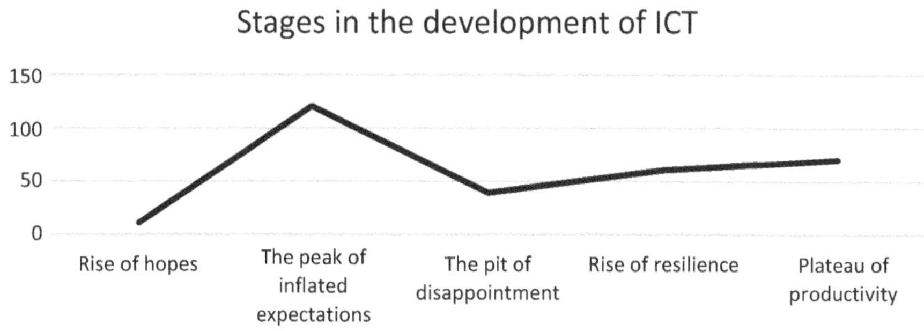

Fig. 2 Stages in the development of ICT. Source: Compiled by the authors

of the professional education system (higher and vocational) to provide training for the workforce that meets the new requirements and standards.

In the context of the discussed issue, it is worth mentioning the opinion of I. S. Ashmanov, the President of the companies "Ashmanov and Partners" and "Kribrum" and a member of the Council under the President of the Russian Federation for the Development of Civil Society and Human Rights. He does not share the excitement regarding the radical changes in the labor market that may occur as a result of using neural networks [13]. In one of his speeches, I. S. Ashmanov noted that, like any new innovative technology, neural networks and artificial intelligence are currently going through a stage of inflated expectations, where their application capabilities can be vastly overestimated.

After this stage, there inevitably comes a stage of disappointment when interest in the new technology significantly decreases. Only after these stages can we talk about the place and significance of innovation for the economy as a whole and the labor market in particular.

Figure 2 presents an approximate graph of the development of any information technology. A similar situation, with the overestimation of the possibilities of a new technology, occurred in the early twenty-first century (2000–2001)—the so-called "dot-com bubble," when many network companies went bankrupt in the USA. A similar path was taken by blockchain technology when, after a surge in activity in terms of investment and scientific interest in 2016–2017, a mass exodus of investors occurred in 2019, especially those who invested in blockchain projects during the peak of high expectations.

4 Conclusion

The emergence of neural network technology and artificial intelligence will undoubtedly contribute to the development of the economy, changing its overall structure and the structure of the labor market. Society should be prepared for the dynamics that will result from the influence of neural networks, paying particular attention to the field of education, which currently struggles to meet the demands of the labor market. For society to benefit from the development and implementation of these technologies and avoid exacerbating issues such as unemployment, it is necessary to consider introducing a certain level of control over the use of neural network technology and artificial intelligence at the state level (e.g., through implementing a Digital Code of the Russian Federation).

References

1. Latorre-Cosculluela, C., Sierra-Sánchez, V., Rivera-Torres, P., & Liesa-Orús, M. (2024). ICT efficacy and response to different needs in university classrooms: Effects on attitudes and active behaviour towards technology. *Journal of Computing in Higher Education, 36*(2), 350–367. https://doi.org/10.1007/s12528-023-09357-2
2. Llorens, A., Trullols, E., Pérez-Poch, A., & Petrović, N. (2025). Soft skills development in ICT students: An evaluation of teaching methods by university educators. *Cogent Education, 12*(1), 2437906. https://doi.org/10.1080/2331186X.2024.2437906
3. Ngcapu, S. R., Simelane-Mnisi, S., & Mji, A. (2024). Aligning preservice teachers' experiences for ICT integration in education in the school of education at the university of technology in South Africa. *International Journal of Learning, Teaching and Educational Research, 23*(2), 203–226. https://doi.org/10.26803/ijlter.23.2.10
4. Román-Graván, P., Fernández-Cerero, J., Montenegro-Rueda, M., & Reyes-Rebollo, M. M. (2024). University teaching skills in ICT and disability. The case of the Autonomous Community of Madrid. *Education and Information Technologies, 29*(10), 12653–12676. https://doi.org/10.1007/s10639-023-12314-2
5. Sanhueza, M., Sandoval, L., Ormazabal, M., & Zúñiga, M. (2025). Effect of a teacher training program with ICT on university students' learning. *Contemporary. Educational Technology, 17*(1), ep556. https://doi.org/10.30935/cedtech/15745
6. Xu, C., Hania, A., & Waqas, M. (2024). Guiding the digital generation: Role of principals' leadership, ICT competence, and teacher professional competence in fostering digital citizenship among university students. *Education and Information Technologies*. https://doi.org/10.1007/s10639-024-13180-2
7. Taizhetinova, M. (2021, March 23). *Experts named the leading and lagging industries in digitalization in Russia*. RBC trends. Retrieved from https://trends.rbc.ru/trends/industry/6054b0a89a7947fd6076994e (Accessed 10 October 2023).
8. Moscow School of Management SKOLKOVO. (2014). *Atlas of new professions*. Moscow. Retrieved from https://www.skolkovo.ru/public/media/documents/research/sedec/SKOLKOVO_SEDeC_Atlas.pdf (Accessed 11 October 2023).
9. HSE Labor Research Center. (2021, October 18). *Index of digitalization of economic and social sectors*. Retrieved from https://issek.hse.ru/news/783750202.html (Accessed 10 October 2023).
10. Federal State Statistics Service of the Russian Federation. (2021). *Russian Statistical Yearbook, 2021*. : Rosstat. Retrieved from https://eng.rosstat.gov.ru/storage/mediabank/Ejegodnik_2021(1).pdf (Accessed 11 October 2023).
11. Lebedintseva, T. M., & Lisitsyna, I. V. (2015). Economic security of Russia and factors affecting on security. *Kazan Science, 9*, 81–83.
12. Lisitsyna, I. V., & Lebedintseva, T. M. (2015). Performance evaluation of economic security of Russia. *Bulletin of the Russian University of Cooperation, 3*(21), 50–55.
13. Ashmanov, I. (2023, March 29). Neural networks will become sophisticated. *Vzglyad*. Retrieved from https://vz.ru/society/2023/3/29/1205108.html (Accessed 14 October 2023).

Improving the Organizational Management Structure of a Research Organization Based on a Project Approach

Ludmila V. Marabaeva, Tatyana E. Shilkina, and Mariya V. Myagkova

Abstract This research justifies the advantages of improving the management of a research organization through the transformation of the organizational structure based on a project approach. The methodological foundation relies on systemic and project approaches, employing methods such as systematization, summarization, critical analysis, retrospective analysis, and SWOT analysis. The chosen research object is a sectoral research institute engaged in the research and production of lighting products for various consumer sectors. To conduct an in-depth analysis of the institute's management system and organizational structure in alignment with development goals and market priorities, the authors analyzed its current state and activities in the lighting product market. The research identified features of research and testing processes, the level of development of the quality management system, the scope of responsibility, and the functionality of top management. Conclusions were drawn from the SWOT analysis regarding the institute's problems and advantages in achieving strategic business tasks in the directions of "LED lighting development" and "Development of halogen and ultraviolet lamps." The authors identified the main issues in the management system and organizational structure and substantiated the benefits of the project approach. The proposals for transforming the institute's organizational structure presented in the research can be utilized by other similar sectoral research organizations to optimize organizational and managerial processes and research activities.

Keywords Research institute · Management · Organizational structure · Change · Project approach

JEL Codes M12 · M14 · M54 · O32

L. V. Marabaeva
National Research Mordovia State University, Saransk, Russia

T. E. Shilkina · M. V. Myagkova (✉)
Saransk Cooperative Institute (Branch) of the Russian University of Cooperation, Saransk, Russia

B. S. Sergi et al. (eds.), *University 4.0 and Educational Technology for Sustainable Development*, Education in the Asia-Pacific Region: Issues, Concerns and Prospects 75, https://doi.org/10.1007/978-981-96-7376-6_28

1 Introduction

Addressing the issues of effective management is one of the key conditions for selecting strategic development priorities for research organizations [1]. An example of developing the system of research and venture organizations is the Institute of Light Sources named after A. N. Lodygin (ILS named after A. N. Lodygin)—one of the leading lighting enterprises in Russia. The ILS named after A. N. Lodygin unites scientists, engineers, and specialists in the field of development, design, and production of light sources and systems based on them.

The research aims to analyze the alignment of the management system and organizational structure with the development goals and market priorities of the research and experimental production activities of the institute and improve its management based on the transformation of the organizational structure in accordance with a project approach. Achieving the stated goal required the formulation of the following research tasks:

- To evaluate the current state and directions of the institute's activities in the lighting product market;
- To identify the features of research and testing processes and the level of development of the quality management system;
- To identify key problems and advantages in achieving strategic business tasks;
- To determine the directions for improving the existing management system and organizational structure based on a project approach.

The methodological framework of this research is based on systemic and project approaches, employing methods of systematization and generalization, critical analysis, retrospective analysis, and SWOT analysis. Several issues of organizational management of a research organization are discovered by [2–7].

The issues of managing research organizations and utilizing a project approach to enhance management efficiency have been explored by numerous scholars, including O. B. Dronova, A. E. Miller, L. M. Davidenko, N. P. Pazdnikova, I. I. Pazdnikov, I. Yu. Pakhomova, A. A. Silaev, S. N. Pilyukova, E. Yu. Pleshakova, E. A. Sorokin, V. V. Tkachev, M. V. Filatova, O. G. Stukalo, L. V. Lebedeva, K. A. Tsukanova, I. Sh. Dzakhmisheva, and others. However, the practical needs of managing research organizations in the search for the most rational approaches to organization and management necessitate the continuation of such research.

2 The Key Features of Organizational Management of ILS Named After A. N. Lodygin

The ILS named after A. N. Lodygin was established on October 1, 2015, in accordance with the state privatization program by transforming the State Unitary Enterprise of the Republic of Mordovia "Lodygin Research Institute of Light Sources."

Nowadays, 54% of the number of employees are under 40 years old, 52% have higher education, 24% have specialized secondary education, and some have a degree of Candidate of Technical Sciences. The staffing level is 97.3%.

The financial results of the ILS named after A. N. Lodygin are presented in Table 1.

The organization's revenue for the studied period increased by 16,427 thousand rubles (20.0%). The growth rate of the cost of sales amounted to 143.3%; in absolute terms, its size increased by 25,731 thousand rubles. Thus, gross profit decreased by 1.7 times. The existing trends in the components of financial results negatively affected the net profit indicators: at the end of 2021, there was a loss of 21,510 thousand rubles.

Figure 1 shows the analysis of the dynamics and composition of the property of the ILS named after A. N. Lodygin.

The aggregate volume of assets increased by 1.0%, comprising non-current and current assets. Throughout the entire research period, non-current assets constituted

Table 1 Financial results of the ILS named after A. N. Lodygin in 2019–2021

Name of indicators	2019, thousand rubles	2020, thousand rubles	2021, thousand rubles	Deviation of 2021 to 2019	
				thousand rubles	%
Revenue	81,943	176,369	98,370	+16,427	120.0
Cost of sales	59,440	108,680	85,171	+25,731	143.3
Gross profit	22,503	67,689	13,199	−9304	58.7
Selling expenses	3746	3608	3223	−523	86.0
Management expenses	32,305	42,226	35,101	+2796	108.7
Profit (loss) on sales	−13,548	21,855	−25,125	−11,577	185.5
Interest receivable	15	6	–	−15	–
Interest payable	123	29	290	+167	235.8
Other income	4461	3580	8542	+4081	191.5
Other expenses	5732	15,567	8550	+2818	149.2
Profit (loss) before income tax	−14,927	9845	−25,423	−10,496	170.3
Current income tax	1138	2356	3913	+2775	343.8
Net income	−13,789	7489	−21,510	−7721	156.0

Source: Developed by the authors

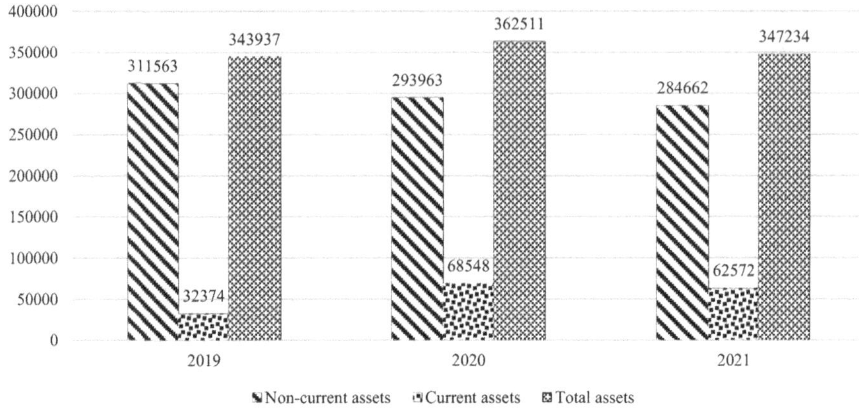

Fig. 1 Composition and dynamics of the assets of the ILS named after A. N. Lodygin in 2019–2021. Source: Developed by the authors

the majority, accounting for over 81.0%. Their size decreased by 93.3% over the period, while the size of current assets, on the contrary, increased by 93.3%.

The institute specializes in the development, production, testing, and research of various sources of optical radiation. The high scientific and technical potential of the institute is realized through the following areas of activity:

- Development and serial production of special-purpose lamps to meet the needs of the defense industry enterprises;
- Development and serial production of ultraviolet gas-discharge lamps designed for air and water purification devices;
- Development and serial production of erythema and bactericidal air irradiators;
- Development and serial production of LED light sources for various purposes and design of lighting systems based on them;
- Provision of services for prototyping printed circuit boards;
- Provision of services for 3D printing and casting into silicone molds;
- Provision of metalworking services, including turning-milling operations, metal cutting and bending, welding, and powder coating;
- Conducting scientific research in developing and photometry of ultraviolet radiation sources and devices with bactericidal action based on them, with increased energy efficiency and reliability [8].

Highly qualified specialists work in the accredited testing laboratory. Its material and technical support meets global standards, making it possible to perform a wide range of metrological and research tasks. The institute's testing laboratory is accredited to conduct tests to confirm the compliance of products with the requirements of the Customs Union "On safety of low-voltage equipment" (CU TR 004/2011). The high qualifications of the staff and advanced testing facilities make it possible to perform the most complex tests and measurements for certification and declaration purposes.

The institute developed and approved a policy that defines quality management objectives, sets tasks to ensure its implementation, complies with legislative requirements, and meets the needs of consumers and customers. The ILS named after A. N. Lodygin continuously works to improve the effectiveness of the quality management system in accordance with the requirements of the ISO 9001:2015 standard related to the research, development, design, production, testing, and sale of electric lamps, lighting products, and their components (certificate No. 31100290 QM 15 issued by the Russian branch of the DQS Holding GmbH Management System Certification).

The institute implemented the quality management system in the Voluntary Certification System "Military Register," which applies to the development and production of products in the quality categories VP in accordance with the requirements of GOST RV 0015–002-2012 (certificate No. VR 22.1.13229–2018 issued by the certification body SMK LLC "MREC").

Work has been organized to assess the quality of products at all stages of the production cycle in accordance with regulatory documentation. Quality control includes incoming inspection of materials and semi-finished products and operational control of products in production with the conduct of acceptance and periodic testing of products in the quality categories OTK and VP.

A distinct area of activity is scientific and technical work, led by the Director of Scientific and Technical Development (DSTD). The DSTD oversees the following divisions:

- Design and technological department;
- Department of LED lighting development;
- Scientific research center for metrology and standardization of optical radiation sources.

Another key focus is market development, overseen by the Director of Market Development, responsible for promoting the institute and its developments and organizing marketing activities. The position of the Chief of Experimental Production stands out in terms of the significance of involvement in R&D projects.

The organization of activities for employees of various categories and their motivation falls under the purview of the Deputy General Director for Personnel. Key structural units directly reporting to the CEO include the following:

- Planning and economic department;
- Sales and marketing department;
- Legal department;
- Quality assurance department;
- Supply and logistics department;
- Mobilization work specialist.

SWOT analysis reveals the organization's strengths, weaknesses, opportunities, and threats (Table 2).

Table 2 Results of SWOT analysis of the ILS named after A. N. Lodygin

Strengths	Weaknesses
• Extensive work experience and a distinguished reputation as an expert in the field of lighting technology; • Well-equipped accredited testing laboratory; • In-house standardization service undertaking IEC-related activities (development of international and national standards in the field of lighting technology); • Capability to offer clients comprehensive lighting solutions (development, testing, production, calculations, and warranty); • Highly qualified personnel; • High profitability of the manufactured products.	• High degree of physical wear and tear on the essential (primary) equipment; • High production cost due to significant fixed (overhead) expenses; • Shortage of working capital for the enterprise.
Opportunities	**Threats**
• Mastering new import-substituting technologies; • Expanding the range of LED products; • Increasing revenue by expanding the LED product line.	• Partial dependence on foreign suppliers for some materials; • High competition with other enterprises producing similar products.

Source: Developed by the authors

In the field of scientific and technical development, the ILS named after A. N. Lodygin is engaged in the design and production of new light sources (LED and traditional incandescent lamps).

Significant attention was given to the organization of school lighting. The ILS named after A. N. Lodygin developed and launched into serial production the luminaires of the DVO07 Standart Opal, DVO09 Standart Prizma, and DVO05 School series, used for general lighting in offices, public spaces, administrative buildings, and other facilities, including educational institutions. An integral part of the school lighting project is the luminaires of the DBO07 School Line series, designed for illuminating classroom boards and wall displays. Luminaires from the DPO12 Sport Opal and DPO13 Sport Prizma series are used for general lighting in sports halls, changing rooms, utility rooms, and other spaces.

The luminaire design of the DVO/UV03 Sunlight Eco series was modernized for general lighting in offices, public spaces, administrative buildings, and other areas with a lack of sunlight and vitamin D deficiency.

The ILS named after A. N. Lodygin developed and launched into serial production the luminaires of the DKU04 Urban series, used for general lighting on roads, streets, squares, parks, yards, production areas, industrial workshops, warehouse areas, and railway platforms.

In collaboration with "Sfera" LLC, the ILS named after A. N. Lodygin developed a range of multifunctional smart lighting poles of the Orion series for landscaping parks and squares, roads, and parking lots, as well as railway stations.

In collaboration with "Konstruktsionnye Kompozity" LLC, the ILS named after A. N. Lodygin developed a fence post based on a composite profile. The production process for a batch of 99 supports for fencing the Depot in Taganrog was launched.

A patent for a utility model was obtained to protect exclusive rights to the support's structure.

Compact halogen quartz lamps KGM 6.6–45 and KGM 6.6–65 were developed and launched into serial production. These lamps are a direct replacement for OSRAM lamps. They are used as high-intensity light sources for illuminating airport runways, taxiways, and landing approach zones.

The ILS named after A. N. Lodygin also developed and launched into serial production compact ultraviolet lamp KL 9E. It is a direct replacement for Philips lamps in medical devices designed to prevent and treat skin diseases.

3 Organizational Management of ILS Named After A. N. Lodygin and Its Improvement with the Use of Project Approach

The quality policy reflects the fundamental principles of the ILS named after A. N. Lodygin, defines goals in the field of quality management, sets tasks to ensure its implementation, complies with legislative requirements, is effective, and meets the needs of consumers and customers. In 2021, four QMS standards of the institute were updated, revised, and put into effect in accordance with the quality management system document list.

The following two external audits were conducted in 2021:

- Certification body DQS Holding GmbH (GOST RV 0015–002). During the recertification, two non-conformities (non-critical) were identified. Corrective actions were developed and implemented; a certificate was issued for 3 years.
- Certification body DECUES (ISO 9001). One non-conformity (non-critical) was identified during the audit. The inspection confirmed the validity of the ISO 9001 certificate issued on October 16, 2020.

Thus, the quality management system is effective, meets the requirements of regulatory documents, policies, and quality goals of the organization, and complies with ISO 9001 and GOST RV 0015–002.

In 2021, the certified laboratory of the Occupational Safety, Environmental Protection, and Industrial Safety Service conducted industrial control of workplaces, air in the working area, washouts, and microclimate at workplaces by the laboratory staff without resorting to the services of external organizations.

The HR department conducts training for employees of experimental production areas in the courses "Gas technician" and "Locksmith for the repair and operation of gas facilities." Instructions on occupational safety are revised. Work is carried out on the certification of the Occupational Safety, Industrial Safety, and Environmental Protection Service. The laboratory received a conclusion on the state of measurements (No. 03–21 dated January 19, 2021), valid until January 19, 2024.

Work is also underway to update the normative-technical environmental documentation of the Production Environmental Control Program, fully aligning with the sustainable development principles.

Employees of the ILS named after A. N. Lodygin participate in the competition "Engineer of the Year of the Republic of Mordovia." The results of the competitions confirm the high professional level of the best scientific and technical workers, whose achievements and developments significantly contribute to the region's overall economic development.

To motivate career growth and stimulate the acquisition of new knowledge and skills among employees, the ILS named after A. N. Lodygin has a talent pool, reducing costs for the search and subsequent adaptation of new employees through the preparation of skilled personnel. This approach makes it possible to resolve personnel tasks, ultimately enhancing the efficiency and competitiveness of the entire organization.

Furthermore, continuous training in related professions is conducted at the enterprise to reduce costs for the search and adaptation of new employees. This helps mitigate risks associated with the temporary incapacity of a worker or during periods of paid leave.

The main clients of the ILS named after A. N. Lodygin include STC "Midis" LLC (procurement of KGS and SM lamps), "Oktyabrskaya" Agrofirm in the village of B. Yelkhovka (a project with the supply of Agro series luminaires), "SEGZ" JSC (procurement of KGS and SM lamps), and "Varyag Techservice" JSC (SG lamps). Some major clients that developed a dealer network are "Promkomplektatsiya" LLC and "TD 'Promoborudovanie'" LLC with the procurement of DB lamps.

Each innovation requires a comprehensive approach to solving economic, technical, personnel, material, and other tasks. This systemic approach has established the practice of project management, particularly through the formation of a separate structural unit responsible for planning, organizing, and monitoring project activities—the project office or project management office (PMO). As a result, there is a need for an integrated approach to organizing production systems, covering processes of digitization, implementing a circular model of resource management, and project management as the foundation for the modernization and improvement of the organization's activities [9] (Fig. 2).

This structural unit, formed on a permanent basis, will make it possible to consolidate key participants in production project management, thereby contributing to the growth of personnel competencies within the emerging project organizational structure. Mandatory project participants in an enterprise should include representatives from the departments of digitalization, production, quality control, and certification. This will make it possible to consider key trends in production management when implementing various measures to optimize the production system [10].

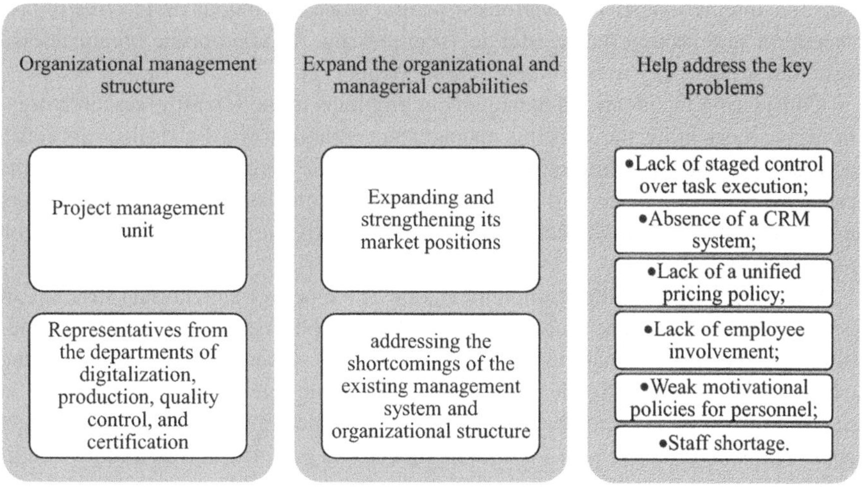

Fig. 2 Project approach to organizational management of a research organization on the example of ILS named after A. N. Lodygin. Source: Developed by the authors

4 Conclusion

As revealed during the research, the ILS named after A. N. Lodygin has an inefficient management system that was established several decades ago and does not meet contemporary requirements. In the author's view, a project approach can be applied to expand the organizational and managerial capabilities of the institute's research and experimental activities, aimed at expanding and strengthening its market positions and addressing the shortcomings of the existing management system and organizational structure [11].

The application of a project approach will help address the key problems of the management system and organizational structure of the ILS named after A. N. Lodygin, including the following:

- Lack of staged control over task execution;
- Absence of a CRM system;
- Lack of a unified pricing policy;
- Lack of employee involvement in solving tasks at various levels of management hierarchy;
- Weak motivational policies for personnel;
- Staff shortage.

Such deficiencies in the organization's functioning significantly impact all areas of its activities and contribute to the worsening of the current situation. To change this situation, it is necessary to take a series of measures to improve the efficiency of the management system of the ILS named after A. N. Lodygin based on the project approach principles. This is justified by the strict requirements for ensuring effective interaction between various structural units, continuous monitoring of task execution, and prompt managerial decision-making. An appropriate organizational structure is necessary to meet these requirements [12].

Considering the identified management problems in the scientific research organization, reorienting the existing management system toward a project approach appears to be the most preferable as the primary mode of operation. In doing so, the proposed transformations will have minimal impact on project groups and the organization's core activities, where processes are already well-established and do not require changes [13].

However, for the systemic implementation of the new organizational structure, it is essential to significantly update and modernize the organizational and methodological support for managing the organization. This task is similarly important and extensive because its effective and timely resolution determines the efficiency of managerial decisions and the control of their implementation [14]. The project approach is an effective tool for enhancing activity performance in today's conditions, acting as a driver for result-oriented transformation and serving as the primary management approach.

References

1. Pleshakova, E., & Yu. (2023). Transformation of the corporate culture during the transition to the project approach in the management of organizations. *Bulletin of the Faculty of Management of Saint Petersburg State University of Economics, 13*, 88–96.
2. Arboleda Muñoz, G. A., & Tenorio, A. P. (2024). Innovation management in university research groups. *Journal of Technology Management and Innovation, 19*(2), 16–27.
3. Forliano, C., De Bernardi, P., Rozsa, Z., & Bertello, A. (2024). Systems dynamics research in management and organization studies: Overview and research agenda. *Journal of Innovation and Knowledge, 9*(3), 100512. https://doi.org/10.1016/j.jik.2024.100512
4. Krlev, G., Hannigan, T., & Spicer, A. (2025). What makes a good review article? Empirical evidence from management and organization research. *Academy of Management Annals, 19*(1), 376–403. https://doi.org/10.5465/annals.2021.0051
5. Robertson, K. (2023). Using a classroom-as-organization method to bridge the research-practice gap in human resource management. *International Journal of Management Education, 21*(3), 100871. https://doi.org/10.1016/j.ijme.2023.100871
6. Subramanian, B., & Bhattacharyya, S. S. (2024). Explicating the factors influencing successful implementation and management of sustainable innovation practices in research intensive organizations. *Journal of Science and Technology Policy Management, 15*(6), 1772–1797. https://doi.org/10.1108/JSTPM-05-2023-0067
7. Zamora, T. A. C., Casanova, A. G. V., Zamora, M. J. C., & Zavala, C. X. N. (2025). A system of tasks to train research skills and information management in university students. *Journal of Educational and Social Research, 15*(1), 352–366. https://doi.org/10.36941/jesr-2025-0027
8. Scientific Research Institute of Light Sources named after A. N. Lodygin. (n.d.). *History of the Institute*. Retrieved from https://vniiis.su/history/ (Accessed 4 October 2023).
9. Pakhomov, I. Y., & Silaev, A. A. (2022). Specifics of the project approach to the organization's HR management. *Business Education and Law, 4*(61), 121–126. https://doi.org/10.25683/VOLBI.2022.61.465

10. Pilyukova, S. N. (2023). Project approach to strategic personnel management. *Bulletin of the Educational Consortium of the Middle Russian University. Series: Economics and Management, 21*, 64–66.
11. Pazdnikova, N. P., & Pazdnikov, I. I. (2020). Project and program management under conditions of strategic planning modernization. *Regional Economy. South of Russia, 8*(2), 56–63.
12. Tkachev, V. V. (2023). The main aspects of the project approach in the management of marketing activities. *International Journal of Humanities and Natural Sciences, 3-3-3*(78), 193–196. https://doi.org/10.24412/2500-1000-2023-3-3-193-196
13. Dronova, O. B. (2023). Project approach to the management of an innovative corporation in constant changes conditions. *News of Higher Educational Institutions. Series: Economics, Finance and Production Management,* 2(56), 78–88.
14. Miller, A. E., & Davidenko, L. M. (2022). Development of a management mechanism for organizing intellectual infrastructure of technological development of industrial enterprises. *Herald of Siberian Institute of Business and Information Technologies, 11*(1), 53–61. https://doi.org/10.24412/2225-8264-2022-1-53-61

Professional Training of Green Personnel in Higher Educational Institutions Through Information and Communication Technologies

Dmitry A. Nazarenko, Maryana V. Arkhipova, Victoria B. Gorbunova, Ekaterina A. Redkina, and Vladimir-Alexander T. Arkhipov

Abstract The research explores a complex of pertinent issues concerning the professional training of green personnel in higher educational institutions by utilizing information and communication technologies (ICT). The relevance of enhancing the professional preparation of ecology experts is substantiated by the necessity to elevate the prestige of environmental sector professionals, ensure the quality of specialists' training, and enhance their competitiveness in the labor market. Furthermore, the use of ICT is advocated for organizing a comprehensive educational process in preparing specialists in the field of ecology. The research is based on a comprehensive examination of the structural components, criteria, and indicators of the readiness of future ecology experts for professional activities through the application of ICT. It is emphasized that incorporating ICT in professional training enhances the motivation for learning and stimulates the cognitive interest of future specialists. In the realm of professional ecological training, the use of ICT is deemed most expedient in addressing current challenges in ecological education. Moreover, integrating naturalistic and humanitarian directions in contemporary ecological education based on ICT learning technologies enables one to maintain interest in and curiosity about the academic discipline. It unleashes the creative abilities of the student body and awakens creative inclinations for further development. The conducted analysis indicates that the professional training of green personnel in higher educational institutions, employing information and communication technologies, makes resolving a range of pertinent issues possible. Specifically, it facilitates the improve-

D. A. Nazarenko · M. V. Arkhipova · V.-A. T. Arkhipov
Kaliningrad Cooperative Institute (Branch) of the Russian University of Cooperation, Kaliningrad, Russia
e-mail: marhipova@ruc.su; v.t.arkhipov@ruc.su

V. B. Gorbunova (✉)
Kaliningrad State Technical University, Kaliningrad, Russia
e-mail: viktoriya.gorbunova@klgtu.ru

E. A. Redkina
Russian State University for the Humanities, Moscow, Russia

B. S. Sergi et al. (eds.), *University 4.0 and Educational Technology for Sustainable Development*, Education in the Asia-Pacific Region: Issues, Concerns and Prospects 75, https://doi.org/10.1007/978-981-96-7376-6_29

ment of students' independent work skills in researching ecological problems and provides opportunities to devise unconventional solutions and draw conclusions about their consequences.

Keywords Green human resources · Professional training · Higher education institutions · Information and communication technologies · Environmental experts · Student · Environmental education

JEL Codes I23 · I25 · O44 · Q01 · Q20 · Q28 · Q30 · Q38 · Q50 · Q57 · Q58

1 Introduction

At the current stage, environmental safety concerns constitute a vital element of the sustainable development strategy. The prevailing theory is that of the pro-ecological uplift of society, encompassing the principles of deep ecology and ecological modernization. Within this context, the professional education of ecology specialists is regarded as preparation for addressing environmental conservation issues.

The primary objective of higher education institutions (HEIs) in cultivating the next generation of competitive ecology experts (green professionals) is the cultivation of ecological competencies essential for their subsequent professional endeavors. Fulfilling this task necessitates exploring avenues for enhancing professional ecological education based on contemporary methodological approaches, progressive concepts, scientific achievements, and cutting-edge methods and technologies.

The analysis of scientific research and an examination of contemporary practices in the organization of professional training for green personnel revealed the existence of contradictions that require effective resolution, including the following:

- The social demand for highly qualified ecology experts conflicting with the mismatch between their competence levels and the requirements of the labor market and employers;
- The complex nature of the professional activities of today's ecology experts contrasting with the absence of educational and methodological tools and practical mechanisms influencing the integrative formation of components of graduates' professional competence;
- The rapid informatization of the environmental sector and sporadic use of information and communication technologies (ICT) in the university's educational process, which fails to ensure the readiness of ecology experts for professional activities;
- The necessity of employing state-of-the-art pedagogical technologies, particularly ICT tools, in preparing green personnel is juxtaposed with the unpreparedness of instructors in ecological and legal disciplines to develop, implement, and utilize them.

2 Readiness of Green Personnel for Professional Activities Through the Utilization of Information and Communication Technologies

This research is grounded in exploring the structural components, criteria, and indicators of readiness of green personnel for professional activities through the utilization of information and communication technologies. The identified components include the following:

1. Motivational-value component (criterion: the formation of motivations and goals for future professional activities) [1];
2. Cognitive component (criterion: possession of the necessary professionally oriented knowledge for the implementation of professional activities) [2];
3. Ecological-legal component (criterion: possession of knowledge, skills, and abilities in regulating ecological-legal relationships, resolving environmental situations) [3, 4];
4. Information and communication component (criterion: the ability to search for and process information using information and communication technologies) [5];
5. Behavioral-activity component (ability to apply knowledge of legislation and environmental law in professional situations, mastery of a set of skills in the field of ecology, environmental protection, and balanced nature management) [6].

3 The Core of Environmental Education

The rapid development of society, the application of new production technologies, and the escalating adverse impact of humanity on the environment lead to significant negative changes in the surrounding milieu. In recent decades, the international community has placed special emphasis on the imperative of providing effective environmental education. In light of such demands, one of the objectives of ecological education in the Russian Federation is the efficient professional ecological preparation of specialists by enhancing the quality of ecological education.

Exceptional ecological situations arise due to various causes. The transformation of this phenomenon into a legal category is ensured by the Federal law "On environmental protection" [7]. Normative documents establishing the legal regime for zones of exceptional ecological situations are found in over 200 legislative acts of various levels and legal force: in the Constitution and laws of the Russian Federation, decrees of the President of the Russian Federation, orders of the Ministry of Natural Resources of the Russian Federation, and acts of central executive authorities [8].

The correlation of the concepts "ecology" and "worldview" becomes paramount in determining the fundamental influences of environmental education on forming the ecological worldview of future ecology experts. Their interplay and interrelation are justified by the fact that "the ecological component of the scientific worldview is most closely linked to a system of limited shared views and principles reflecting the human attitude towards nature, the place and role of humanity (society) in nature" [9].

The ecological future of Russia must rest upon three components: a developed system of ecological law, an effective system of administrative control and regulation, and an ecological culture [10]. Environmental education aims to combine the rational and emotional, influencing the content of the ecological worldview. The functions of environmental education include the following:

- Organizing harmonious relations between society and nature, considering the practical needs of society, and the endeavor of nature to maintain the stability of its normal state, thereby preserving the conditions for the existence and development of humanity;
- Shaping specific behavior patterns regarding nature at the individual and societal levels;
- Creating the ability to predict the negative consequences of human activity and the results of transforming nature;
- Managing society's relationship with nature in the process of economic activities.

Functions influence the selection of components in environmental education. Scholars identify the following components in the content of environmental education:

- Scientific knowledge about the connections in the "human—society—nature" system (a system of ideas, concepts, and regularities reflecting philosophical, technical, legal, and moral aspects of ecology in their development);
- Knowledge about rational ways of individual activity in the natural environment;
- Experience of humanity's emotional and value relationship with nature, including feelings, motives, value orientations, and beliefs, based on which an individual forms awareness of their place in the world and readiness for practical activities;
- Experience in practical ecological activities, including creative endeavors.

Relying on didactic principles enhances the quality of environmental education formation, as the process has a methodological foundation. Consequently, the following is necessary to elevate the quality of environmental education in higher educational institutions preparing future ecology experts:

- To determine priorities for the state and all of humanity among the contemporary tasks of preserving the surrounding natural environment;
- To strengthen the role of ecological-legal education in the development of the younger generation and instill in them a sense of duty in the field of ecology;
- Engage youth in active environmental conservation activities based on the knowledge acquired during training in institutions of higher education for life safety;

- Foster the awareness of a sense of personal responsibility for the state of the environment.

One of the important tasks of higher educational institutions preparing future ecology experts is to create conditions for providing a corresponding level of knowledge about the human, nature, and society. This is a prerequisite for their subsequent ecologically reasonable professional activities. Therefore, introducing disciplines of an ecological-legal nature into the educational process, studying legislation on environmental protection, forms specific behavioral skills for future ecology experts in the environmental conservation sphere.

The professional activities of green personnel should be prepared as a comprehensive, step-by-step system. This system should ensure the transformation of knowledge into ecological practices, with an ecological-professional focus on educational material. It should contribute to developing ecological thinking among future specialists, fostering continuity in environmental training, interdisciplinary connections, and the coherence of theoretical, practical, and scientific work. Moreover, it should encourage the production of exploratory and research works. The outcome of such endeavors will be cultivating creative potential, moral perspectives, individual predispositions, and the development of volitional traits and character [11]. These elements constitute sources of human inclinations and interests and conditions for developing abilities and talents.

Let us delve into a more detailed interpretation of the concepts "preparation" and "professional preparation" as seen in dictionaries and scientific literature.

For instance, the "Explanatory Dictionary of the Russian Language" by Ozhegov defines preparation as "A stock of knowledge acquired by someone" [12]. This notion is directly linked to the term "to prepare," which is interpreted as conducting preliminary work for implementation, execution, etc.

The term "preparation" is scarcely used independently. In most cases, it is combined with other terms related to professional education. Therefore, in this research, the authors employed the concept of "professional preparation."

The Federal law "On education" states that professional training aims at individuals of different ages acquiring professional competence [13]. In the concise dictionary of modern pedagogy, the term "professional preparation" is interpreted as learners acquiring skills necessary for performing specific tasks [14].

Professional education for ecology experts involves preparation in various specializations, the choice of which is determined by the higher educational institution. However, narrow specialization may hinder the mobility of professionals due to changing production needs [15]. Therefore, the Russian system of professional education favors the training of broadly profiled specialists. Future ecology experts acquire basic professional knowledge and skills across various specializations, including ecological safety and balanced nature management; ecological auditing; ecology and life safety; ecology and environmental protection; ecological control; environmental and natural resource economics; information technologies in ecology; environmental safety management; organization of environmental activities; legal aspects of ecology; construction ecology; ecology of forestry, reserves, garden and park management, fishing, resort, and recreational management; production ecology; and urban, hydro, agro, and radioecology.

4 Ecobiocentric Approach to Professional Training of Green Personnel in Higher Educational Institutions Through Information and Communication Technologies

To strengthen the role of ecological-legal education in environmental conservation activities at the state level, teaching methodologies for ecological law and special courses on ecological-legal disciplines are being refined in modern conditions for the training of specialists, including in the field of natural sciences. The formation of an ecological worldview among future ecology experts implies their adherence to future norms of ecologically correct behavior and the execution of practical actions to protect human life and health and the surrounding natural environment. Hence, there arises a need for a sophisticated and qualitative reevaluation by higher educational institutions that prepare future ecology experts for the scientific and methodological approaches to present the fundamental principles of disciplines covering the study of environmental legislation.

Currently, there has arisen a necessity for the professional and practical preparation of green professionals to encompass a comprehensive, balanced integration of disciplines spanning the natural sciences, technology, economics, law, and sociocultural dimensions. Leveraging acquired knowledge, the student should cultivate skills in organizing professional activities, foster ecological awareness and literacy, develop organizational and professional aptitudes, attain the requisite intellectual level, demonstrate an aptitude for systemic creative thinking, and possess the ability to assess historical and contemporary processes and issues in the life of the nation, professionally addressing matters to achieve optimal outcomes.

The curation of the content of ecological education for future specialists in higher education institutions, particularly within the realm of life safety systems, presupposes a specific selection and content of subjects, a structured system, interconnections, and a sequence of delivery. It involves defining the knowledge, skills, and competencies that future ecology experts must possess.

Consequently, ecological education and the preparation of future ecology specialists stand as key considerations for higher education. The ecological education system should be founded on cultivating eco-centric consciousness among students. The readiness of future ecology experts for professional activities is the ability to utilize a set of knowledge and skills within their chosen specialization, foundational knowledge across various aspects of ecological and legal preparation, self-educational skills, and the capability to swiftly acquire new knowledge and master cutting-edge technologies.

Such an approach necessitates the integration of innovative educational technologies into the preparation process for green professionals, including ICT.

The incorporation of state-of-the-art information learning technologies into the educational process contributes to the expansion and deepening of theoretical knowledge, activates the cognitive activities of student communities, unveils their creative potential, and imparts practical significance to the outcomes of their education. The objective of education is to instill in the learner a body of knowledge, skills, and competencies [13]. Nowadays, in light of the global ecological crisis and the determination to prioritize the ecologicalization of education, the combined use of new information technologies in conjunction with traditional methods is increasingly crucial, as experience demonstrates, to develop the ecological competence of specialists.

The majority of the population in society remains captive to an anthropocentric worldview, although there is a current preference for an ecocentric approach. In the philosophical lexicon, anthropocentrism is defined as the inclination to place humans at the center, not of values, as humanism does, but of existence [16].

The ecobiocentric worldview is linked to anthropocentrism. The ecobiocentric worldview demands a higher level of education, comprehension of information volume, and the quality of generalizations. Thus, biocentrism implies an approach where all life holds value, and life is considered the criterion for any activity. Bioecocentrism represents the golden mean between biocentrism and ecocentrism. Bioecocentrism encompasses views that involve moral responsibility for individual living beings, species, and ecosystems.

The introduction of new educational technologies, among which information and communication technologies play a significant role, is one of the factors in addressing ecosystem conservation issues. This is due to their ability to convey a large volume of material over any distance in a short period. Information and communication technologies (ICT) of education are a methodology and technology for the educational process using state-of-the-art electronic learning tools. The search for didactic means and approaches in teaching that could guarantee high effectiveness for any teacher is an ongoing process. Nevertheless, it has a specific characteristic at any given time (Fig. 1).

The integration of ICT into the educational process for preparing green professionals should occur through various avenues, including the following:

- Computer-oriented learning, primarily for disciplines in the economic and legal cycles;
- Application of the latest information technologies in students' scientific work;
- Development of new computer software tools.

The integration of naturalistic and humanitarian directions in contemporary environmental education, based on information and communication technologies of learning, makes it possible to maintain interest and passion for the academic discipline. It unleashes the creative abilities of students, motivating their creative potential for further development.

In training green professionals, ICT provides more opportunities for choosing sources and processing a significant amount of information. It promotes socialization and the development of personal communication qualities, making it possible

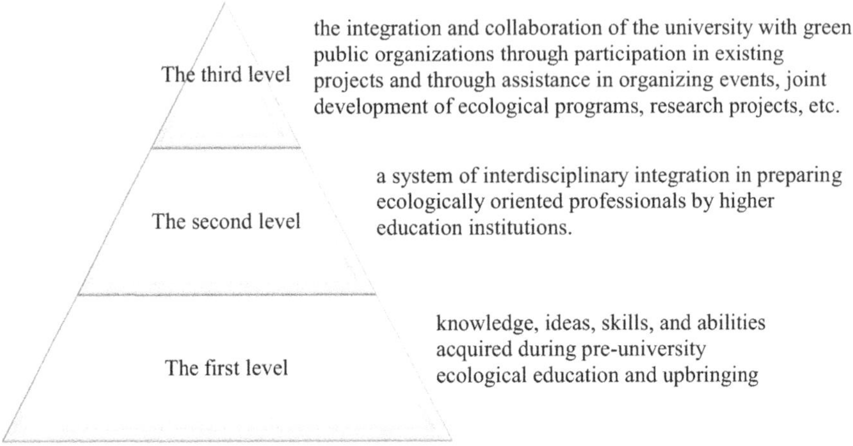

Fig. 1 Ecobiocentric approach to professional training of green personnel in higher educational institutions through information and communication technologies (ICT). Source: Made by the authors

to model various situations that one may encounter in professional activities. Learning with the use of ICT creates an open system of intensive activity with simultaneous adaptation to the individual capabilities of the learner, contributing to the development of their personality.

5 Components of Professional Activities of Environmental Experts

Scientists attribute the tasks of professional activities of environmental experts to information-legal, information-search, information-analytical, and information-communicative components.

Let us delve into these components more thoroughly.

The goal of the information-legal component is to ensure compliance between the provisions of existing environmental legislation and objectively necessary generalizations of specific provisions of the current legislation of the Russian Federation. This requires a revision and restoration. It implements tasks related to analyzing existing regulatory acts on legal support for individuals and legal entities in the field of environmental safety, nature conservation activities of enterprises, interaction with stakeholders, inspections, and reviews.

The information-search component ensures the collection, accumulation, and systematization of various types of documented or material information obtained as a result of legally prescribed actions, examination results, and reports from individuals and legal entities. This information is recorded and registered in accordance with procedural requirements. It is provided in the prescribed manner to the participants in processes affecting the environmental situation at the relevant level, including supervisory and controlling bodies, other government agencies, or individual individuals. The aim is to ensure environmental protection and other functions related to protecting the environment from harmful consequences of human activities, compliance with legal norms, and legislative requirements by all interested individuals and organizations. This is achieved by preventing environmental offenses, stopping them, and taking measures to eliminate conditions and causes contributing to their commission.

The information-analytical component aims to identify and eliminate the causes and conditions contributing to environmental offenses. Its main tasks include knowledge of the forms of statistical reporting, rules of its compilation and maintenance, analysis and consideration of legislation and other regulatory acts, the maintenance of relevant information databases in proper condition, their processing and utilization, analysis of data, and their application to address issues arising during professional activities. It involves analyzing materials of professional relevance presented by adjacent organizations, legal services of executive authorities, institutions, and non-governmental organizations.

The information-communicative component ensures the coordination of interaction in professional activities with all participants in the environmental protection process, including law enforcement and other authorities, in preventing environmental offenses.

For this research, the authors supplemented the considered components with the information-ecological one, which is significant in preparing green professionals. It provides mastery of the system of environmental knowledge in accordance with the level of development of contemporary ecological science.

Usually, information technologies are implemented through computer equipment and software, ensuring high-speed processing of information, its retrieval, data distribution, and access to information sources regardless of location. This contributes to achieving the goals set by the user.

Specialized software for professional ecological training can be classified as follows:

1. Geographic Information Systems (GIS) with various purposes. In the university educational process, the emphasis is primarily on the use of ready-made GIS (e.g., 2GIS);
2. Computer statistical information processing systems, including packages like Statistika, Statgraphics, and SPSS;
3. Software for the analysis and forecasting of environmental actions (e.g., FreeWare and Creative Common);
4. Expert Systems (ES), referring to formally structured knowledge bases obtained from ecology experts;
5. Information-reference systems of specific specialization (e.g., systems like "Consultant").

Information and communication technologies in higher education imply integration and comprehensive use with other educational technologies, applying them in a wide range of educational activities: during lectures, seminars, laboratory and practical sessions, industrial internships, and self-study. The introduction of ICT in higher education contributes to the personalization and individualization of the preparation of future specialists, enhances learning motivation, and forms professional identity.

Multimedia ICT tools make it possible to combine various means of presenting information (text, sound, graphics, animation, video, and modeling), significantly improving the quality and depth of assimilating educational material. In addition to providing an educational environment, they allow teachers to reduce the time spent on information reproduction and significantly increase the time for explaining the material. The high degree of visual clarity enhances learning motivation. Active participation in classes through multimedia educational programs (in contrast to passive observation) contributes to developing students' communicative skills.

Information and communication technologies make it possible to apply various types of knowledge assessment:

- Operational knowledge control (voting and testing);
- Intermediate, final, and ongoing control (hardware and software for various testing forms—closed, open, matching, correct sequence, etc.);
- Distance control (software tools for knowledge control via the Internet).

6 Conclusion

The utilization of software and hardware control tools in the educational process contributes to the development of individual capabilities, all components of the personalization of future specialists' training, and the expansion of connections between teachers and students.

Modeling programs are widely used in preparing green professionals, especially in the context of active learning (during cognitive activities). The use of such programs makes it possible to adapt their operation to the students' developmental characteristics (reaction speed, development of imaginative thinking, etc.). The communicative capabilities of ICT enable students to independently track activities using modeling programs based on databases within the framework of a collective research project. This is particularly crucial for achieving the professional and social competence of future specialists because it provides an opportunity for their active collaboration.

It is necessary to note that the application of information and communication technologies makes it possible to address the following pertinent issues:

- Incorporating the latest information technologies into education;
- Enhancing students' independent work skills in the information environment;
- Shaping individuality considering personal capabilities;
- Providing the opportunity to make unconventional decisions and draw conclusions based on their consequences.

References

1. Aboramadan, M., Barbar, J., Alhabil, W., & Alhalbusi, H. (2024). Green servant leadership and green voice behavior in Qatari higher education: Does climate for green initiative matter? *International Journal of Sustainability in Higher Education, 25*(3), 539–555. https://doi.org/10.1108/IJSHE-03-2023-0112
2. Alimehmeti, G., Ndoka, E., & Paletta, A. (2025). Cultivating green pioneers: Examining the antecedents of sustainable entrepreneurial intent in higher education. *International Journal of Sustainability in Higher Education*. https://doi.org/10.1108/IJSHE-03-2024-0206
3. Fareed El-Tony, Y., & Choo, L. S. (2025). Employee green behavior in laboratories of higher education institutions: A qualitative study. *International Journal of Sustainability in Higher Education, 26*(3), 697–715. https://doi.org/10.1108/IJSHE-03-2024-0168
4. Gomes, S., Lopes, J. M., & Trancoso, T. (2024). The green seed: The influence of pro-sustainable orientation on social entrepreneurship in higher education students. *Industry and Higher Education*. https://doi.org/10.1177/09504222241297550
5. Lakhno, M. (2024). Green or green-washed? Examining sustainability reporting in higher education. *Higher Education Quarterly, 78*(3), 1099–1122. https://doi.org/10.1111/hequ.12513
6. Spada, I., Giordano, V., Chiarello, F., Martini, A., & Fantoni, G. (2025). Text mining on green policies for integrating sustainability in higher education. *International Journal of Management Education, 23*(2), 101126. https://doi.org/10.1016/j.ijme.2024.101126
7. Russian Federation. (2002). *Federal law "On environmental protection"* (January 10, 2002 No. 7-FZ). Moscow: Consultant Plus.
8. Russian Federation. (2012). *Federal law "On education in the Russian Federation"* (December 29, 2012 No. 273-FZ, as amended on August 4, 2023). Moscow: Consultant Plus.
9. Kährik, P. K. (1979). To the characterization of the spiritual element of ecological culture. In *Stimuli of rational nature management in conditions of developed socialism: Theses of reports of the republican scientific conference* (pp. 125–130). Tallinn.
10. Galeeva, G. A. (1982). Environmental education: Problems, solutions. In A. V. Sidorenko (Ed.), *City, nature, people: Problems of environmental education* (pp. 41–74). Mysl.
11. Bolshenko, S. F., Gorbunova, V. B., & Martynenko, O. V. (2021). A methodological approach to determining the competitive positions of the labor potential in regional consumer cooperation. In A. V. Bogoviz, A. E. Suglobov, A. N. Maloletko, O. V. Kaurova, & S. V. Lobova (Eds.), *Frontier information technology and systems research in cooperative economics* (pp. 661–670). Springer. https://doi.org/10.1007/978-3-030-57831-2_71
12. Ozhegov, S. I. (1960). *Explanatory dictionary of the Russian language* (4th ed.). Moscow. Retrieved from https://gufo.me/dict/ozhegov (Accessed 2 October 2023).
13. Batyshev, S. Y. (Ed.). (1999). *Encyclopedia of Vocational Education: In 3 volumes*. Russian Academy of Education.
14. Concise Dictionary of Modern Pedagogy. (n.d.). Retrieved from https://didacts.ru/termin/ (Accessed 8 October 2023).
15. Gorbunova, V. B. (2015). Methodological aspects of management activities in today's economic conditions. *Bulletin of the Kaliningrad Branch of the Saint-Petersburg University of the Ministry of Internal Affairs of Russia, 3*(41), 127–130.
16. Dictionary of Philosophy. (n.d.). Retrieved from https://gufo.me/dict/philosophy_dict (Accessed 3 October 2023).

Preparation of Ecologically Oriented Personnel for Collaboration Through Cooperation with Green Non-Governmental Organizations

Alexander S. Pechatkin, Natalya G. Tarakanova, and Tatyana Yu Pyatkina

Abstract In the current conditions of implementing the cooperative educational process, there is a need to emphasize the development of an ecologically oriented direction in preparing future specialists. In this process, the collaboration of the cooperative system with green non-governmental organizations plays a significant role, exerting a systemic effect on the formation of the ecological consciousness of future specialists. The research aims to analyze the pedagogical and educational possibilities that make it possible to recreate conditions for involving future cooperators in the green movement, shaping their ecological worldview, and developing professional competencies. The research tasks are (1) to summarize the features of implementing environmental education in the context of preparing personnel for cooperation; (2) to identify conditions and pedagogical approaches for the successful realization of students' creative potential in the process of forming ecological consciousness; and (3) to analyze the activities of green non-governmental organizations as essential institutional-organizational entities in preparing ecologically-oriented personnel for cooperation. The research methodology is based on an integrative, interactive, practice-oriented, and systemic approach to education. Special attention is given to the projective method as a leading pedagogical construct in the creative self-realization of students. The authors analyzed the activities of ecologically oriented organizations in the Republic of Mordovia, including the Regional Branch of the Green Movement of Russia "ECA," the Regional Branch of the All-Russian Public Organization for the Protection of Natural Resources "Russian Ecological Society," and the Republican Public Organization "Green World." The experience of the Saransk Cooperative Institute in implementing ecologically-oriented education is examined, including the establishment of the scientific laboratory "Eco Club: Green Science," participation in eco-activist gatherings, and collaboration with the Association of Green Universities in Russia. The

A. S. Pechatkin (✉) · N. G. Tarakanova · T. Y. Pyatkina
Saransk Cooperative Institute (Branch) of the Russian University of Cooperation, Saransk, Russia
e-mail: a.s.pechatkin@ruc.su; n.g.tarakanova@ruc.su; t.yu.pyatkina@ruc.su

B. S. Sergi et al. (eds.), *University 4.0 and Educational Technology for Sustainable Development*, Education in the Asia-Pacific Region: Issues, Concerns and Prospects 75, https://doi.org/10.1007/978-981-96-7376-6_30

research highlights the importance of integration and cooperation between the university and green non-governmental organizations in effectively shaping individuals with an ecological worldview. It is established that ecologically-oriented preparation of specialists in the field of cooperation initiates a process that reflects a qualitative transformation of personal orientations, instilling life-affirming attitudes and a profound understanding of the interconnectedness of humans with the environment.

Keywords Environmental awareness · Cooperation · Personnel training · Environmental organizations · Educational institutions

JEL Codes A12 · A20 · I26 · Q57

1 Introduction

In the conditions of global development and new technologies, ecology emerges not only as a field of scientific knowledge about the environment and nature management but also as a new methodological approach aimed at changing thinking and behavior patterns in the entire process of preparing specialists in any field of activity.

More than half a century has passed since the inception of the sustainable development concept, marking a global shift in the awareness of numerous environmental protection challenges across the world community.

The monograph "Sustainable development: New challenges" notes that "the development of civilization in any foreseeable period is sustainable if it preserves some invariant (i.e., does not change, does not threaten any property, object, relationship, or restriction)" [1 , p. 26]. N. M. Mamedov points out, "As the understanding of the problems and prospects of the transition of modern society to sustainable development has evolved, it has become evident that such a transition is possible with harmonious social relations, moral improvement of people, and a change in human qualities based on eco-humanistic values and ideals" [2 , p. 40].

In the context of comprehending and implementing the sustainable development concept, there is a growing need to realize an integrative approach to the development of the environmental education system and the upbringing of an individual. The goal is to form an ecological worldview. Environmental consciousness is formed through knowledge about global environmental problems and a deep understanding of one's involvement in solving these problems, even at the local level. As the esteemed Russian ecologist and educator V. M. Smirnov rightly asserted, "A person will not strive to restore the purity of a river if it does not flow through his heart." Environmental consciousness is formed on three levels: the knowledge level, the self-awareness level, and the activity level. Therefore, environmental education should primarily aim to shape an active life position based on personal responsibility for making professional decisions. This goal can only be achieved through active participation in diverse environmental activities and applying acquired environmental knowledge in practice during the learning process.

The process of preparing personnel for cooperation serves as a significant catalyst for social transformations because it ensures the creation of a just order in the redistribution of material goods within the framework of the cooperative movement and shapes a worldview that meets the new requirements of the time. Traditional cooperative principles (e.g., justice, democracy, solidarity, equality, and mutual assistance) can possibly be complemented by the principle of ecological sustainability. Analyzing the interrelation between cooperation and ecology, I. M. Hamzin and S. P. Emelyanov identify two positions in understanding this connection. First, the activities of cooperatives influence the environment (environmental pollution and exploitation of natural resources). Second, the environmental activities of cooperatives (waste processing, resource conservation, and land reclamation) [3 , p. 4].

Ecological sustainability refers to economic activities that align with ecosystem laws and are directed towards the non-destructive and co-evolutionary coexistence of humans and nature. In this process, the fundamental importance lies in forming ecological consciousness as a condition for the recreation of an active and creative personality deeply aware of responsibility for the surrounding natural environment.

In this regard, attention should be drawn to the possibility of involving future cooperators in the activities of public youth green organizations, primarily at the regional level, which have significant experience in educational, research, and practical work. This is further emphasized by the fact that collaborative practical or project-research activities develop essential professional competencies such as collaborating, engaging in collective creativity, and working within a team.

Based on scientific methodology, the research aims to analyze the pedagogical and educational possibilities provided by the involvement of future cooperators in the green movement to shape their ecological worldview and professional competencies.

2 Preparation of Ecologically Oriented Personnel at the Green Universities

During the research, the authors employed the projective method as a condition for achieving a high qualitative level of expertise in the field of cooperation. Projecting serves as the process of creating an intellectual product reflecting the novelty of the research, practical significance, and applicability in specific conditions.

The foundational approaches utilized include integrative, interactive, practice-oriented, and systemic approaches in the realm of education.

The integrative approach, outlined notably in A. P. Belyaeva's work "Integrative theory and practice of multilevel continuous professional education" [4], guides subjects toward the comprehensive integration of various components when addressing strategic and tactical educational tasks. The social-practical foundations of such an approach include scientific-cognitive, subject-practical, production-technological, and social aspects (a unified system of activity).

The interactive approach entails organizing the pedagogical process based on engaging learners in collaborative intellectual activities to solve designated tasks, which are significant for each participant.

The practice-oriented or activity-based approach is based on the involvement of learners in practical activities within the framework of developing professional competencies.

When applying the systemic approach, the authors proceed from understanding the educational process as a system – a holistic collection of interrelated elements united by common structural connections.

The authors also employed scientific methods of social forecasting and systemic-structural analysis, along with complexity, specificity, and objectivity principles.

Analyzing the conceptual foundations of cooperation as a social institution [5], the interrelation of cooperation and ecology [3] and green universities [6–11] made it possible to identify the primary components for implementing the educational process in preparing ecologically oriented personnel for cooperation.

3 Case Study of Forming Ecologically Oriented Professionals for Cooperation in Russia

The issue of forming ecologically oriented professionals for cooperation is interdisciplinary because the progressive nature of the cooperative form of social interaction requires the emphasis of various scientific disciplines to achieve a systemic effect. This effect is manifested primarily in constructing the normative, value-based, and institutional-organizational structure of specific cooperative entities. Recreating a systemic effect in ecologically oriented education involves searching for an optimal model that balances the goals of the educational process, the competencies developed in learners, and the approaches to conducting sessions.

The formation of environmental consciousness requires addressing various components of the educational process, including theoretical preparation and a practice-oriented approach. Integrating ecological components into educational programs should be coupled with a practical component, encompassing design work, practice-oriented sessions, and participation in environmental events.

Thus, the ecologically oriented preparation of specialists in cooperation initiates a transformative process that reflects a qualitative shift in personal orientations. It instills life-affirming attitudes and a deep understanding of the interconnectedness between humans and the environment.

Cooperation emerges as a progressive form of social interaction, providing a powerful impetus for the ecological transformation of all sectors of social life and exerting a systemic influence on improving the quality of human life. S. N. Khabakhu notes that cooperation "acts as a powerful factor in group cohesion, stimulates a sense of mutual assistance and collectivism, enhances communication processes among cooperation participants, contributes to the formation of collective representations, norms, and values, creates a specific socio-psychological climate, raises the level of social group incorporation, which lies in its ability to attract new members by providing them with equal rights and opportunities to achieve goals, and enhances the external competitiveness of the group" [5, pp. 48–49].

Therefore, creating conditions for implementing programs for training ecologically oriented professionals for cooperation is the key to the sustainable development of any country. The awareness of this position is gaining practical momentum: environmental public movements and organizations are being established, eco-oriented projects are being integrated into the activities of universities, and increasing attention is being paid to transforming the consumer attitude towards nature.

As a regional example of implementing the idea of ecologization, let us consider the experience of the Saransk Cooperative Institute. The Scientific Laboratory "Eco Club: Green Science" was opened at the Saransk Cooperative Institute on September 22, 2021. The event occurred within the framework of the declared Year of Science and Technology in Russia and the 45th Anniversary of the Saransk Cooperative Institute. The establishment of the scientific laboratory aims to unlock the creative potential of students in the field of ecology, cultivate environmental principles and norms, and prepare them for participation in ecologically oriented competitions and grants.

The team of the Saransk Cooperative Institute (a branch of the Russian University of Cooperation), for the first time, entered the top ten "green universities" in the country under the "Divide with Us 1.0" direction and was invited to the gathering of eco-activists. At the VIII All-Russian gathering of green universities in Russia, the Saransk Cooperative Institute presented the team "Dofamin." An Agreement on Cooperation was signed with the Association of Green Universities of Russia. Collaboration is carried out in the field of joint initiatives and projects for environmental education and the formation of environmental awareness.

Creating conditions for the environmental self-realization of students is a necessary condition for the formation of an environmental education system and the improvement of education quality. This is associated with the fact that the idea of the ecologization of public life, undergoing various metamorphoses, is unfolding in the public consciousness as a scientific concept, ideology, lifestyle, fashion, etc.

4 Environmental Organizations Operating in the Republic of Mordovia (Russia) and Their Activities

The following environmental organizations are actively operating in the Republic of Mordovia:

- The regional branch of the Green Movement of Russia "ECA," founded in 2011. The tasks of this organization include the promotion of environmental awareness, restoration and preservation of forests, and the organization of eco-volunteer movements. Additionally, since 2015, another direction has been developed –

environmental education. The goal of this initiative within the organization's activities is to draw attention to environmental issues among a wide range of the population, from schoolchildren to entrepreneurs. A noteworthy project is the "Green Universities of Russia," a program for greening universities and providing environmental education to students who will integrate environmental technologies into their future professional activities.

- The regional branch of the All-Russian public organization for the protection and conservation of natural resources "Russian Ecological Society."
- The Republican public organization "Green World," established in 1992 as a children's corporation but currently uniting more than five thousand representatives of different generations of students and teachers. It also collaborates with entrepreneurs. Through its constructive activities, this organization became the first public environmental organization in the Republic of Mordovia to receive state support. It implements programs in various directions, all united by one goal: shaping an environmentally oriented worldview in the younger generation and fostering independent creative thinking, initiative, and a responsible civic stance.

These green public organizations have accumulated extensive experience in comprehensive, diverse, but unified-purpose activities.

The main areas of work of these environmental organizations include project research, advocacy, educational activities, spiritual-aesthetic development, and practical initiatives.

Within the framework of the organization's project and research activities, it actively participates in environmental projects with grant support. For instance, the environmental organization "Green World" has implemented 22 projects, with ten of them receiving grants from the Presidential Grants Fund, the Russian Geographical Society, and the Fund for the Support of Children in Difficult Life Situations. One recent project was the online forum "Nature. Knowledge. Unity. Creativity!" held with the support of the Presidential Grants Fund and dedicated to World Nature Conservation Day.

To implement research and enlightenment activities, the Republican public organization "Green World" possesses a unique educational and field center – the Simkin Natural Park for Sustainable Development. It is located 90 km from Saransk, on the territory of the Simkin forestry of the Bolshebereznikovskiy forestry enterprise, covering an area of about 1000 hectares. The southern border of the park is traced by the Sura River. The park's uniqueness lies not only in the diversity of its flora and fauna but also in the variability of ecological conditions. Additionally, more than 30 species of red-listed plants and rare representatives of the animal kingdom inhabit its territory. This provides opportunities for research in various directions, both in studying ecosystems and in nature conservation, developing eco-tourism, and rational nature management.

To achieve this purpose, annual ecological gatherings are held in the park, where participants include schoolchildren and students from various fields of study, united by a common ecological interest. Participants form multi-age research groups based on their chosen themes. The work is conducted under the guidance of experienced educators and mentors, including university professors and school teachers who have undergone relevant training. The geography of participants is not limited to the Republic of Mordovia alone. Over the 30-year history of these gatherings, representatives from various regions, including the Chuvash Republic, the Republic of Tatarstan, the Moscow Region, the Penza Region, the Ulyanovsk Region, the Novgorod Regions, and others, have taken part. This expands the opportunities for social contacts and the exchange of experiences. The gatherings also include educational, creative, and competitive components, featuring visits to educational cycles (e.g., game ecology, floristics, ethnoecology, the ecology of the soul, plant ecology, etc.) and leisure activities (e.g., eco-tournaments, eco-quests, and eco-scenarios), providing each participant in the research group the opportunity to showcase their abilities and talents. This allows the groups to better consolidate and coordinate their work, considering the personal qualities of each member.

Thus, a fusion of ecological and cooperative competencies occurs because cooperation primarily involves being a team player, working towards a collective outcome, and understanding one's role in the common cause.

Research projects undergo testing at regional, all-Russian, and international conferences and competitions. Based on their work, ecologists provide practical recommendations and proposals to representatives of municipal and regional authorities, helping to develop business skills and assert their civic position among young people.

The propagandist direction includes environmental actions, placing posters on public transport, and many other events that attract the attention of residents to environmental issues.

Educational activities are presented through projects involving the inclusion of city dwellers in separate waste collection, as well as activities such as organizing excursions, photo exhibitions, and field expeditions to specially protected areas of the Republic of Mordovia (e.g., "Alatyrskie Dali" and "Zapovednymi Tropamy"). This also involves conducting interactive lessons for schoolchildren by eco-leaders.

Ecologists organize art exhibitions and create eco-theaters within the spiritual and aesthetic education framework. For example, at the initiative of the "Green World," the club "Green Living Room" operates in the A. S. Pushkin National Library of the Republic of Mordovia. It involves writers, scientists, students, journalists, local historians, and specialists from various fields. Meetings take various forms: oral journals, discussions, creative contests, information reviews, and lectures. Based on advanced digital technologies, the eco-club has developed an integrated art project called "ECO-Palette."

The practical activities of organizations are also very diverse, ranging from greening cities and districts and conducting eco-cleanups to participating in environmental expertise and monitoring the environment.

These directions provide a comprehensive, interactive, and practice-oriented approach to shaping an "ecologized" personality. To possess an ecological culture, one must learn to cooperate with nature and know and observe its laws. This can only be achieved through close contact with nature, feeling the inseparable connection between humans and the environment. The activities of youth public environmental organizations are aimed precisely at this. Currently, they have accumulated rich experience expressed in various forms and directions of environmental activities. In conjunction with the educational, scientific, and educational capabilities of universities, this significantly expands the potential for shaping ecologically oriented thinking and behavior among future professionals in the cooperative system.

5 Multi-level Approach to the Preparation of Ecologically Oriented Personnel at the Green Universities

Success in shaping a personality with an ecological worldview can only be achieved through a comprehensive, continuous, and multilevel approach to this process (Fig. 1).

The first level encompasses knowledge, ideas, skills, and abilities acquired during pre-university ecological education and upbringing.

The second level involves a system of interdisciplinary integration in preparing ecologically oriented professionals by higher education institutions. This utilizes the potential of humanitarian and natural sciences, as well as the university's socio-educational work in this direction.

The third level involves the integration and collaboration of the university with green public organizations through participation in existing projects and through assistance in organizing events, joint development of ecological programs, research projects, etc.

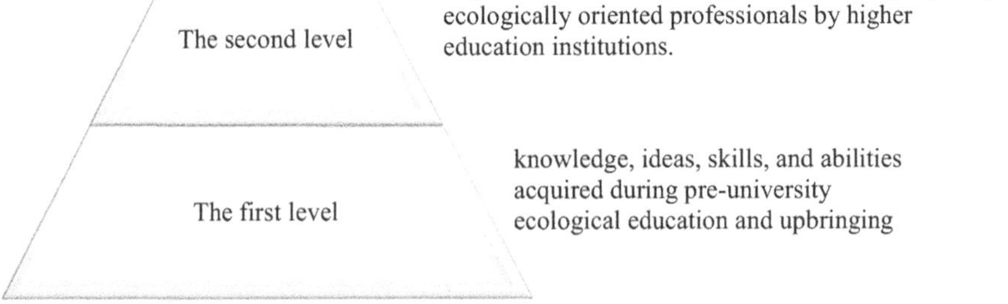

Fig. 1 Multilevel approach to preparation of ecologically oriented personnel at the green universities. Source: Developed by the authors

This ultimately aims to form a system of ecological beliefs and behavioral attitudes that will lead to adopting ecologically sound decisions in practical and professional activities.

6 Conclusion

Contemporary cooperative movements require professionals focused on addressing specific challenges related to the ecological aspects of scientific and technological progress, constructive problem-solving in rational and restorative natural resource management, and active participation in public dialogue on the preservation and multiplication of natural wealth. The success of its development directly depends on this. Therefore, shaping the ecological worldview of graduates is a crucial component of the educational policies of institutions that prepare professionals for the cooperative system.

While the pedagogical support provided within educational institutions plays an unquestionably important role in shaping ecological worldviews, it is contingent upon the teacher's personal commitment to high ecological values and their adept use of ecological materials in the educational process. However, as colleagues rightfully note, "the teacher's readiness for solving problems of ecological education and, in general, humanizing the educational process is not always sufficient" [12 , p. 130].

In this context, to achieve the desired results, it is necessary to engage additional integration resources and collaborate with other civil society institutions to achieve common goals.

Public associations with an ecological focus contribute to the formation of essential personal, professional, and civic qualities necessary for the successful socialization of individuals. Such associations facilitate the establishment of personal development and self-realization guidelines (consciousness, humanity, patriotism, mutual assistance, and justice) and enhance the sustainability of social systems of various scales.

Acknowledgments The authors express their gratitude to Olga E. Ruchina, the Head of the public organization "Green World," for the assistance in preparing this research.

References

1. Danilov-Danilyan, V. I., & Piskulova, N. A. (Eds.). (2015). *Sustainable development: New challenges*. Aspect Press.
2. Mamedov, N. M. (2010). Historical process and the concept of sustainable development. *Journal of Globalization Studies, 2*, 33–46.
3. Hamzin, I. M., & Emelyanov, S. P. (2014). Cooperation and ecology: Organisational and legal aspects. *Vestnik of the Russian University of Cooperation, 2*(16), 4–8.

4. Belyaeva, A. P. (2002). *Integrative theory and practice of multilevel continuing professional education*. Institute of Professional-Technical Education of the Russian Academy of Education.
5. Khabakhu, S. N. (2017). To cooperation research methodology as a social institution: Achievements and prospects. *Theory and Practice of Social Development, 11*, 47–50. https://doi.org/10.24158/tipor.2017.11.9
6. Rathakrishnan, T., Gan, J. L., & Yaacob, A. (2025). Sowing green seeds of mind: Exploring the determinants of green mindfulness among university students. *International Journal of Sustainability in Higher Education, 26*(3), 594–613. https://doi.org/10.1108/IJSHE-03-2024-0158
7. Wahyuningrum, I. F. S., Haumahu, S. A., Puspita, A. S., Budihardjo, M. A., & Humaira, N. G. (2025). From sustainability to superiority: Exploring the impact of green metrics on global university rankings. *International Journal of Sustainable Development and Planning, 20*(1), 133–142. https://doi.org/10.18280/ijsdp.200114
8. Yang, Y. (2024). Impact of institutional support and green knowledge transfer on university students' absorptive capacity and green entrepreneurial behavior: The moderating role of environmental responsibility. *PLoS One, 19*(5), e0304070. https://doi.org/10.1371/journal.pone.0304070
9. Zhang, M., Zhang, R., Li, Y., & Zhou, Y. (2024). Knowing green, buying green: University students green knowledge and green purchase behavior. *Humanities and Social Sciences Communications, 11*(1), 732. https://doi.org/10.1057/s41599-024-03243-7
10. Zhu, K., Meng, W., Jiang, M., & Ma, X. (2024). How to improve green behaviour of employees in Chinese universities? The effects of responsible leadership, organizational ethical climate, and psychological capital. *Eurasian Journal of Educational Research, 2024*(111), 164–180. https://doi.org/10.14689/ejer.2024.111.10
11. Zou, Y., Zhong, N., Chen, Z., & Zhao, W. (2024). Bridging digitalization and sustainability in universities: A Chinese green university initiative in the digital era. *Journal of Cleaner Production, 469*, 143181. https://doi.org/10.1016/j.jclepro.2024.143181
12. Asafova, E. V. (2011). Pedagogical strategies for the development of ecological culture of students at a classical university. *Uchenye Zapiski Kazanskogo Universiteta. Seriya Gumanitarnye Nauki* [Proceedings of Kazan University. Humanities Series], 153(5), 128–135.

Digital Inclusiveness Based on EdTech in the Management Strategies of Universities 4.0 in Russia and Central Asia

Preface to the Fifth Part of the Book

This fifth and final part of this book is devoted to exploring the integration of digital inclusiveness based on EdTech into the management strategies of Universities 4.0 in Russia and Central Asia. The primary thesis of this part is the necessity for universities to adapt flexibly to the new trends in scientific and technological progress.

In this part, it is worth noting the model of digital inclusiveness for Universities 4.0 in Russia and Central Asia developed by Aktam U. Burkhanov, Kudaiberdi G. Kozhobekov, and Yury A. Tikhomirov in the chapter "Flexibility of the University's Digital Inclusion Strategy in a Globalizing World Based on Its Competitiveness." Its advantage lies in the combination of educational content, digital technologies, and pedagogy, all rooted in the principles of inclusiveness in higher education.

The chapter "Organization of Intercultural Dialogue in a Multinational Classical University Based on Its Digitalization: from an Inclusive Environment to Global Openness" by Kobiljon H. Khushvakhtzoda, Tatiana V. Bakhutashvili, Ilyos S. Abdullayev, and Lyudmila M. Lisina further develops the book's topic by identifying new conditions of inclusiveness in higher education in Russia and Central Asia that merge social, local, and global digital inclusiveness in Universities 4.0.

The chapter "Inclusiveness of Universities as the Basis of Their Global Competitiveness: A View from the Standpoint of Rankings ARWU, QS and THE" by Uran N. Busurmankulova, Shakhlo T. Ergasheva, Juliana A. Kitsai, Natalya G. Balibardina, and Maria A. Komissarova supports this part by demonstrating the close link between the inclusiveness in Universities 4.0 in Russia and Central Asia and their competitiveness according to global educational rankings.

In the chapter "Digital Inclusion as a Macroeconomic Mechanism for Solving the Problem of Higher Education in Russia in the Context of the Large-Scale Process of Internationalization of Universities" by Saodat A. Yuldasheva, Munisa J. Khusnuddinova, Tatiana A. Dugina, and Natalya V. Shchukina, it is argued that digital inclusiveness, in the context of internationalizing their activities and alongside trends in globalization and digitalization addresses the microeconomic challenges of Universities 4.0 and enhances the development of intellectual capital in Russia and Central Asia.

The chapter "Prospects for Increasing the Attractiveness of Russian Education based on the Internationalisation of Universities" by Elena G. Popkova and Shakhlo T. Ergasheva reinforces this topic by proposing the development of internationally specialized faculties and an increased online presence of Russian Universities 4.0 in the global educational arena to ensure digital inclusiveness.

The key outcome of the chapter "Reputation Management of a Contemporary University in the Context of the 5-100 Project and the Priority 2030 Program" by Alexandr N. Chumikov and Svetlana Yu. Chumikova is expressed in recommendations to optimize theoretical approaches and practical components of image management for Universities 4.0 in Russia and Central Asia, thereby supporting their digital inclusiveness.

Finally, the chapter "Analysis of the Effectiveness of Channels for Promoting Educational Services in the Tasks of Training Personnel for the Communication Industry: A Case Study of the Educational Program 'Advertising and Public Relations' at the Institute of World Economy and Business of RUDN" by Olga P. Malygina and Ekaterina A. Arkhangelskaya contributes to the discussion by demonstrating the high effectiveness of online channels for promoting educational programs of Universities 4.0 in Russia and Central Asia.

Thus, each of the seven chapters included in this part contributes uniquely to illuminating the topic. Together, they form a compelling body of evidence for the critical importance of EdTech in ensuring the digital inclusiveness and global competitiveness of Universities 4.0 in Russia and Central Asia.

Flexibility of the University's Digital Inclusion Strategy in a Globalizing World Based on Its Competitiveness

Aktam U. Burkhanov, Kudaiberdi G. Kozhobekov, and Yury A. Tikhomirov

Abstract This article elaborates on the theoretical and practical aspects of the formation of a flexible strategy for a university's digital inclusion in a globalising world and the growth of the level of competition in the market of higher education services. The methodological framework of the article is based on the combination of theoretical aspects and practical provisions of pedagogical and technical sciences, social sciences, management of educational services, etc. The concept of e-inclusion, is considered from the application of digital technologies to ensure the inclusiveness of higher education. The methodological apparatus of this article is comprised of the methods of analysis, synthesis, modelling, and observation. This study outlines the important function of universities in the context of global processes and expansion of the concept of inclusion, characterises the parameters of the influence of digital technologies on inclusion in the environment of higher education, discloses the essence and specifics of the concept of e-inclusion in universities, performs a critical assessment of the impact of digital technologies on the quality of higher education and the level of inclusion, offers a structure and complex description of the model of the university's digital inclusion, and describes the key advantages that form due to the combination of the elements of educational content, digital technologies, and pedagogy, which are set onto the principles of inclusion in higher education.

Keywords Digital inclusion · Inclusion · University · Higher education · Globalisation · Competitiveness · Flexibility · Strategy

---

A. U. Burkhanov (✉)
Alfraganus University, Tashkent, Uzbekistan

K. G. Kozhobekov
Osh State University, Osh, Kyrgyzstan
e-mail: kudayberdi.kozhobekov@oshsu.kg

Y. A. Tikhomirov
RUDN University, Moscow, Russia
e-mail: Tikhomirov-yua@rudn.ru

B. S. Sergi et al. (eds.), *University 4.0 and Educational Technology for Sustainable Development*, Education in the Asia-Pacific Region: Issues, Concerns and Prospects 75, https://doi.org/10.1007/978-981-96-7376-6_31

JEL Codes F22 · I21 · I23 · J24 · O15 · O32 · P37

1 Introduction

Globalisation is a complex socioeconomic process that covers a wide range of features connected with the biological integrity of human civilisation, technological and communication development, and value-bases and conceptual provisions that determine the conditions of co-existence and joint activities of humans. The globalising world is a vivid example of the unification of people and their activities, a quick exchange of information, active trade, and technological development. The exchange of thoughts and provision of access to information in such conditions are the key factors in the development of mankind. That is why scientific activities and the sphere of higher education are assigned a special role, which lies in the development of intellectual capital at all levels.

The important trends that define the face of higher education and universities are the comprehensive digitalisation of universities and the development of inclusion. Inclusion here is treated as one of the values of the university, while digitalisation is a tool for solving various problems and a tool for development. One of the directions of using digital technologies in the system of higher education is to support for digital inclusion of people with special needs in social life. The necessity of ensuring the inclusion of education and support for quality education is often considered from the position of the lifelong learning concept and the corresponding employment. This requires constant development and updating of competencies, as well as an increase in the level of the population's digital literacy. This vision allows for effective and targeted use of digital technologies to raise the level of accessibility of higher education and offers better opportunities for success and self-realisation of people regardless of their specific features and needs (European Commission, 2020).

Assessment of digitalisation from the position of its potential, aimed at an increase in the level of the university's inclusion, forms a relevant direction for research, which is connected with its competitiveness [1]. In this context, agreement of global processes, development of inclusion in the world, and consideration of the general dynamics of the environment of the above processes, require from administrations of higher education establishments the preparation of a complex strategy that would take into account not only the factual situation but also the perspective of its development. Therefore, a current task of the educational system in modern conditions is ensuring the flexibility of the strategy of universities' development, aimed at raising the level of their competitiveness, which covers digital inclusion and conforms to global challenges [2].

This study substantiates and describes the key advantages of the model of the university's digital inclusion. It features the combination of higher education elements (educational content, digital technologies, and pedagogy) around the values of inclusion, as well as determination of the necessity of using a flexible strategy for the university's digital inclusion, which is connected with potential risks of using digital technologies to ensure the inclusion of universities and a high level of the dynamics of technological changes.

2 Digital Inclusion of Universities in the Conditions of Globalization

The methodology of this study covers a wide range of scientific theories and approaches. Their combination allows for the comprehensive assessment of the problem of ensuring the digital inclusion of universities in the conditions of globalisation, competitive struggle, and uncertainty. They represent the sphere of educational management, pedagogical and technical sciences, and social sciences and are implemented within the e-inclusion concept. The key methods used in this article are based on the critical generalisation of theoretical and practical developments of experts and scholars in the studied sphere.

The conceptual aspects of this article were researched in detail in scientific works and expert reports, which elaborated on different spheres of the problem posed. This includes the combination of digital and inclusive education [3], university's competitiveness [4–6], institutional substantiation and support for digital inclusion in the sphere of higher education [7–9], theoretical and practical substantiation of the concept of e-inclusion [10], research on the experience of the use of digital technologies in the system of inclusive education [11–13], critical assessment of the influence of digitalisation on inclusive education [14], substantiation of the prospects for the development of digital inclusive education [15], etc.

In this article, the main objective was to study the provisions of the development of a flexible strategy for university's digital inclusion given the globalisation processes and competitiveness of the higher education establishment. The main tasks that were to be achieved include theoretical substantiation of the concept of e-inclusion, assessment of the state of implementing digital inclusion in the system of higher education, and formation of the model of university's digital inclusion.

3 The Model of the University's Digital Inclusion in Accordance with the E-Inclusion Concept

Higher educational establishments perform an important function, which is manifested not only in meeting the labor market's demands for specialists but also in promoting important social values and the development of culture and communication processes. Access to higher education is one of the main human rights, which cannot be limited by biased attitudes of other people or objective barriers, connected with physical or mental problems of individuals. According to this, ensuring the inclusiveness of higher education is among the strategic goals of mankind's development, which is manifested in the creation of fair conditions for education for all people, regardless of the colour of skin, language, religion, health, or other objective needs of the person.

The basis of inclusion in universities is ensuring access to higher education for everyone. This takes place due to adaptation to different needs, improvement of the level of affordability, and creation of barrier-free space for obtaining an education by people with disabilities [13]. The status of inclusion in the environment of higher education offers a sufficient level of the educational environment's adaptation to the special needs of students. Such adaptation became possible due to the corresponding regulatory and institutional changes, as well as the transformation of value orientations, which allows implementing new methodologies and techniques of inclusive education into the system of education [12].

In the modern conditions of a globalised digital world, inclusiveness can be realised in the new information space, which involves the creation of a special design for the maximum use of technological capabilities and the resolution of potential problems faced by people with special needs. Despite the advantages that emerge in the sphere of inclusion due to digital technologies, there are large risks in the mass dissemination of digital technologies in education, the essence of which lies in the exclusion of people who – due to different reasons – do not have skills with digital technologies at the level that is sufficient for mastering educational materials and obtaining education in the form of online learning. In the context of digital development, inclusive education forms a special construct of "e-inclusion," which involves the focus on the processes of involving students with special needs in learning with the help of digital tools and information technologies [10].

The main manifestation of digital education is closely connected with online learning. In the critical conditions of the COVID-19 pandemic, it allowed for ensuring remote accessibility of higher education. The results obtained showed the potential of such a tool from the position of affordability and opportunities for learning. However, unpreparedness and speed of the implementation of the above technologies did not allow implementing this potential in full. That is why further implementation of the elements of digital education must be more systemic. Educators must refrain from the simple transition of traditional educational content to the digital form, but develop real efficient tools of education that would fully use the capabilities of digital technologies and form a better environment for educational inclusion.

The development of the e-inclusion concept has many challenges and requires the use of the system approach. The main problems of dissemination of inclusion in the sphere of digital technologies are connected primarily with the limited level of its implementation in analogue education. As a matter of fact, a low level of inclusion in the traditional space of higher education may be the main barrier to its realisation in the digital form. That is why the development of a strategy for a university's digital inclusion must begin with the preparation and implementation of the system of inclusion in the traditional way – through the development of legal and administrative norms, formation of an inclusive space, and its development according to the needs of vulnerable categories of students and teachers [10]. After this, the developed elements of the university's inclusion can be adapted to the conditions of digital education.

However, the proposed changes are not definitely positive and undisputed. Their effectiveness seems to be obvious only in the conditions of a high level of digital literacy. In these conditions, the proposed tools and communication channels can be used by a wide range of interested parties, which will ensure real accessibility. On the contrary, the low level of digital education or the high cost of the corresponding equipment might create additional barriers to the inclusion of universities. That is why the implementation of this strategy must be flexible and be aimed primarily at the formation of digital competencies and dissemination of the culture of the use of digital technologies [14].

The key aspects to be considered during the formation of inclusive digital higher education are the support for the quality of digitalisation and data protection, training and support for the programmes of digital training of academic staff, and the creation of an educational environment aimed at effective education and educational communications [3]. Thus, the strategy of digital inclusion in universities includes, at the same time, educational and technological aspects. Thus, their successful implementation requires additional attention to the development of inclusion, educational space, and corresponding technologies.

To ensure real inclusion of digital education, its content and tools must be consciously developed for the designated purposes – ensuring access to knowledge for people with impaired vision and hearing, etc. Identification of the special need allows determining the content of educational material, the form of its presentation, and the communication or another digital tool that can solve the tasks posed. The strategy of digital inclusion of universities requires the agreement of different processes, which can be further unified and automatized [15].

University studies in modern conditions is a combination of three main elements – educational contents, technologies (including digital), and pedagogical methodologies. Each of these elements can acquire different forms, but, on the whole, they perform similar functions. Thus, educational content is the main source of information for students. It may have a form of study guides, lectures, methodological materials, presentations, tasks, cases, etc. Assimilation of educational content is one of the main tasks of students seeking higher education. Technologies are a set of devices, tools, and means that help with the assimilation of the content or facilitate the resolution of other tasks of the educational process. In this context, digital technologies often make communication processes easier and facilitate better visualisation, modelling, or immersion in the educational content. The success of education depends to the greatest extent on the third element – pedagogy. It includes teachers and experts, as well as sets of methodologies and tools used by them.

Each of the above elements can ensure the achievement of certain results in the educational process, but only their complex combination allows achieving really good results. In the conditions of an increase in competition and internationalisation of higher education, as well as dissemination of liberal values in the global educational space, inclusion in universities acquires an important role and is often a factor in their competitiveness. The combination of the elements of the educational process and their connection with the conceptual inclusive basis form the foundation of the model of the university's digital inclusion (Fig. 1).

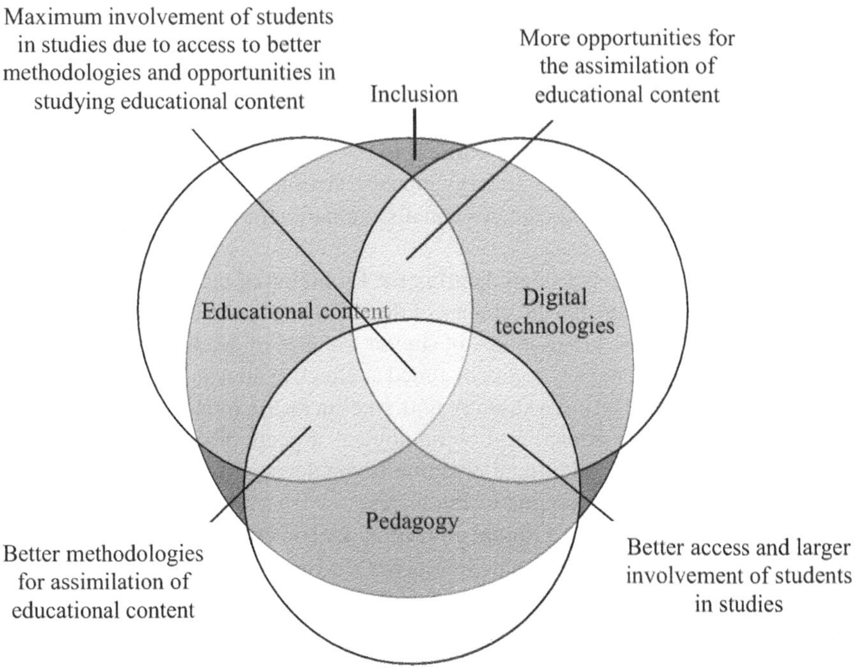

Fig. 1 The model of the university's digital inclusion. *Source* Created by the authors based on [15]

As could be seen, given the formation of an inclusive educational environment, the mix of the educational content with pedagogical components allows offering better methodologies for the assimilation of content. Thanks to this, people with special needs receive the educational material in the form that allows assimilating the volume of information that is required for receiving higher education. If necessary, the form and essence of the content are modified given the needs of the target audience (presented with the help of sign language translation, Braille script, etc.).

The combination of digital technologies with content forms more opportunities for its assimilation without the connection with the place of studies, time of day, weather conditions, etc. According to this, people with impaired motor functions or other problems receive convenient access to materials, educational processes, and social life of universities which takes place in the digital form. The combination of technologies and pedagogy raises the communication capabilities of teachers from the position of the educational process and provides them with additional tools for motivation, education, or control, given the special needs of students. In aggregate, a combination of all three elements of the educational process against the background of educational inclusion offers opportunities for the maximum involvement of students in education due to access to better inclusive methodologies and better opportunities for assimilation of the educational content.

Digital technologies are triggers of development in many spheres. Their smart implementation in the educational process within the proposed model will facilitate an increase in the level of higher education. Given the dynamic changes in the digital environment and implementation of the system of lifelong learning, the given model will be subject to constant reconsideration and correction according to the essence and character of technological changes. This requires the flexibility of the strategy of digital inclusion, which will be developed based on the system of constant forecasting, monitoring, and implementation of new inventions in the system of higher education with the corresponding emphasis on the needs of the vulnerable groups of the population.

4 The Systemic View in University's Digital Inclusion for Its Competitiveness in a Globalising World

Digital inclusion of universities is viewed from the position of online education, which became very popular as a result of restrictions connected with the COVID-19 pandemic. Forced measures allowed for a transition of the educational process to the remote and mixed form, which ensured its continuity and transformation (Fig. 2).

The effectiveness of the measures taken has not been assessed to the full extent yet. Thus, further dissemination of digital technologies in higher education requires additional research and substantiation. Formation of the system of university's digital inclusion in a globalizing world is a complex task, which requires systemic solutions. They must be primarily aimed at the preparation of universities for the corresponding changes.

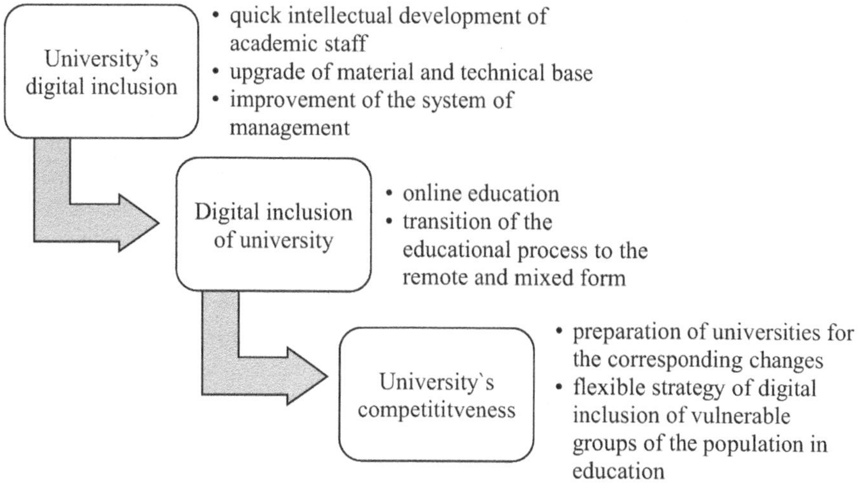

Fig. 2 Formation of the system of university's digital inclusion for its competitiveness in a globalizing world. *Source* Created by the authors

In this context, the main tasks of preparation of educational establishments imply a quick intellectual development of academic staff, upgrade of material and technical base, and improvement of the system of management. All of the above processes have a significant effect on the state of development of universities and directly influence the level of their competitiveness. The possibility of developing a flexible strategy of digital inclusion of vulnerable groups of the population in education based on this requires additional scientific research and discussion to ensure the maximum effect from the implementation of flexible strategic management.

The category of flexibility is not treated ambiguously in the scientific literature. Therefore, the use of this term in the context of strategic management requires specification of the conceptual foundations of flexibility and its influence on the management of the aspects of digital inclusion in the university. Apart from specifying the essence of the very category of flexibility, the criteria of its determination and support require additional discussion. Here it is important to study how flexibility of the strategy of digital inclusion of universities is implemented, which factors influence this, how the level of flexibility is measured and assessed, and whether there are alternatives to the flexible system of digital inclusion in modern conditions.

5 Conclusions

The important role of higher education in the modern globalized and digitalized world draws attention to the gradual domination of social values and justice in the fundamental benchmarks of universities. According to this, inclusion is becoming not only a direction of socially responsible activities of higher education establishments but also a factor that influences the university's competitiveness. Criteria of equality and justice in this context are set onto the moral obligation and legal regulation of the studied process and thus become popular in universities of different levels and profiles.

Active development of digital technologies in modern conditions offers new opportunities not only for quality education but also for an increase in inclusion. In this regard, digital education has a large potential for the involvement of low mobility groups, the creation of special educational content for people with impaired vision or hearing, and other needs and limitations. The concept of the use of digital technologies in the interests of inclusion forms a separate concept of e-inclusion, which distinguishes the advantages and drawbacks of these processes. The key risks are the insufficient level of digital literacy and lack of academic staff with well-developed digital competencies. Therefore, from the position of inclusion, the implementation of digital technologies in universities can be a factor of involvement and a factor of limitation of access to higher education at the same time.

The presented model of university's digital inclusion demonstrates the advantages of complex actions aimed at the creation of quality educational content, the use of effective pedagogical methodologies, and digital technologies. They are combined with the principles and conditions of inclusion in higher education. These advantages serve as clear arguments in favor of the necessity of developing university digital inclusion. However, the dynamic state of scientific & technical and digital environments forms the demand for the development of a separate strategy of digital inclusion of universities and support for its flexibility given the acceleration of the rates of technological progress and an increase in the influence of technologies on the system of higher education.

References

1. Tuan, N. A., & Trang, N. N. (2024). Assessing impacts of university autonomy policies on universities' competitiveness in Vietnam. *WSEAS Transactions on Business and Economics, 21*, 957–966. https://doi.org/10.37394/23207.2024.21.80
2. Wang, C., Zhang, Y., & Rao, C. (2023). Comprehensive evaluation of university competitiveness based on DD-TOPSIS method. *Soft Computing, 27*(24), 18971–18990. https://doi.org/10.1007/s00500-023-08131-x
3. Ahmetova, D. Z., Artjuhina, T. S., Bikbaeva, M. R., Sahnova, I. A., Suchkov, M. A., & Zajceva, J. A. (2020). Digitalization and inclusive education: Common ground. *Vysshee obrazovanie v Rossii [Higher Education in Russia], 2*, 141–150. https://doi.org/10.31992/0869-3617-2020-29-2-141-150
4. Tavoletti, E., Şahin, F., Taras, V., Farrell, W., & Florea, L. (2024). A configurational analysis of university ranking, cognitive ability and cultural intelligence in global virtual teams: The role of academic pedigree. *International Journal of Management Education, 22*(3), 101040. https://doi.org/10.1016/j.ijme.2024.101040
5. Wang, Y., Li, P., Gao, H., & Li, M. (2024). Do the elite university projects promote scientific research competitiveness: Evidence from NSFC grants. *Research Policy, 53*(10), 105074. https://doi.org/10.1016/j.respol.2024.105074
6. Yessimova, S. A., Khassenova, L. A., Mombekova, G. R., & Shadiyeva, A. A. (2023). Academic mobility as a factor of increasing the competitiveness of university graduates. *International Journal of Educational Reform*. https://doi.org/10.1177/10567879231222860
7. European Commission. (2020). Commission staff working document. 30.9.2020. European Commission. Retrieved from https://eur-lex.europa.eu/legal-content/EN/TXT/PDF/?uri=CELEX:52020SC0209&from=EN. Accessed November 18, 2023.
8. European Commission. (2022). *Digital education action plan (2021–2027)*. Retrieved from https://education.ec.europa.eu/focus-topics/digital-education/action-plan. Accessed November 18, 2023
9. UNESCO. (2020). Global education monitoring report 2020: Inclusion and education: All means all. Paris, France. UNESCO. Retrieved from https://unesdoc.unesco.org/ark:/48223/pf0000373718. Accessed November 18, 2023.
10. E-inclusion. (2023). *Background and rationale. e-Inclusion: Building capacity for inclusive education in digital environments*. Retrieved from https://einclusion.net/about/why-e-inclusion/. Accessed November 17, 2023.
11. Ryndak, V. G. (2021). Theory and experience of digitalization in the inclusive education system. *Bulletin of Samara State Technical University Series: Psychological and Pedagogical Sciences, 18*(4), 45–56. Retrieved from https://cyberleninka.ru/article/n/teoriya-i-opyt-tsifrovizatsii-v-sisteme-inklyuzivnogo-obrazovaniya. Accessed November 18, 2023
12. Ryndak, V. G., Allagulov, A. M., & Chelpachenko, T. V. (2021). Digital technologies as a means of developing inclusive education. *Bulletin of Orenburg State University, 3*(231), 70–78. Retrieved from https://cyberleninka.ru/article/n/tsifrovye-tehnologii-kak-sredstvo-razvitiya-inklyuzivnogo-obrazovaniya. Accessed November 18, 2023

13. Troshina, E. P., Baraboshkina, E. A., & Mantulenko, V. V. (2021). Using digital technologies in inclusive education. *Nauka i shkola [Science and School], 1*, 133–142. Retrieved from https://cyberleninka.ru/article/n/ispolzovanie-tsifrovyh-tehnologiy-v-inklyuzivnom-obrazovanii. Accessed November 19, 2023
14. Shemanov, A. Y., & Samsonova, E. V. (2021). Risks and prospects of digitalization for inclusive education. In V. V. Rubtsov, M. G. Sorokova, & N. P. Radchikova (Eds.), *Digital humanities and technology in education (DHTE 2021): Collection of articles of the II All-Russian scientific and practical conference with international participation* (pp. 250–262) Publishing House MSUPE. Retrieved from https://psyjournals.ru/en/nonserialpublications/dhte2021/contents/dhte2021_Shemanov_Samsonova_en.pdf. Accessed November 19, 2023
15. Slootman, M., Korthals Altes, T., Domagała-Zyśk, E., & Nielandt, B. (2023). E-inclusion. Towards inclusive digital learning. *Proceedings of the Erasmus scientific days* 2022 *(ESD 2022)*. Atlantis highlights in social sciences, Education and Humanities. DOI: https://doi.org/10.2991/978-2-38476-036-7_10.

Organization of Intercultural Dialogue in a Multinational Classical University Based on Its Digitalization: From an Inclusive Environment to Global Openness

Kobiljon H. Khushvakhtzoda, Tatiana V. Bakhutashvili, Ilyos S. Abdullayev, and Lyudmila M. Lisina

Abstract In this paper, we considered theoretical and practical aspects of the organization of intercultural dialogue within an inclusive multinational classical university, with the expansion of this experience to a larger number of higher education institutions. The focus of the research was on the transition of higher education institutions from the concept of a local inclusive environment to global openness, formed based on the rational integration of digital technologies into the educational, scientific, and social processes of universities. The methodological framework of these topics comes from the provisions given in the Convention on the Rights of Persons with Disabilities. It combines theoretical aspects of different scientific spheres, including social sciences and psychology, pedagogics and management of higher education institutions, social informatics and digital communications. We determined that the concept of inclusiveness in higher education has a long way of evolution, as a result of which most universities formed a rational educational inclusive environment, which takes into account the needs of persons with disabilities or vulnerable groups of the population. The development of digital technologies against the background of internationalization of higher education allowed creating opportunities for involving people from different countries and cultural environments in higher education. These processes formed new conditions for the inclusiveness of higher education, which combine social, local, and global digital

K. H. Khushvakhtzoda
Tajik National University, Dushanbe, Tajikistan

T. V. Bakhutashvili
North-Caucasus Federal University, Stavropol, Russia

I. S. Abdullayev
Urgench State University, Urgench, Uzbekistan

L. M. Lisina (✉)
Sebryakovskiy Branch of Volgograd State Technical University, Mikhailovka, Russia

B. S. Sergi et al. (eds.), *University 4.0 and Educational Technology for Sustainable Development*, Education in the Asia-Pacific Region: Issues, Concerns and Prospects 75, https://doi.org/10.1007/978-981-96-7376-6_32

inclusiveness. For the transition from the first to the second, it is necessary to pass a range of demands set by digital technologies from the position of technical and competence support.

Keywords Intercultural dialogue · Multinational classical university · Digitalisation · Digital technologies · Higher education · Inclusiveness · Inclusive environment · Globalisation · Global openness

JEL Codes F02 · I21 · I23 · I28 · J15 · O15 · O33 · P48 · Z13

1 Introduction

Trends in the development of higher education in the twenty-first century gradually acquire the features of global unification, which, on the one hand, offers reliable algorithms and methodologies, and, on the other hand, deprives higher education institutions of certain exclusiveness. However, at present, almost all universities work with the principles of inclusiveness and openness, striving towards internationalization and multiculturality [1]. This educational space is supplemented by digital technologies, which penetrate it from all directions, creating additional opportunities for the participants of the educational process, administration of higher education institutions, and government. Apart from vivid advantages, such digital configuration sets demand to the system of education and its participants (in the form of high technical demands and digital competence) and often moves the goals and tasks of higher education in the direction of technology development [2].

This situation is peculiar also to multinational classical universities, which specific feature consists in the presence of principles and norms of multicultural cooperation of a multinational student contingent [3]. Expansion of the ideas of inclusiveness through the creation of the corresponding regulatory environment, its institutionalisation, and development in the direction of an environment of global openness is largely imposed onto the adaptation of administrative, cultural, and educational processes to the capabilities of digital education, which is manifested not only in the form of remote education or online courses but also in the form of creation of digital educational platforms that are open to the entire world [4].

Possibilities of a multinational classical university involving a larger circle of participants in the educational, scientific, and social processes due to the development of digital educational technologies create new risks to intercultural communication and social inclusiveness [5]. Therefore, the close perspective of the development of universities is closely connected with the necessity for the organization of intercultural dialogue, which would cover all participants of the educational process who will be involved in it not only on the university's campus but also through the tools of remote digital education and all available communication channels at the global scale [6].

2 Conceptual Provisions and Values of Inclusiveness in Higher Education

The basis of the methodology of studying the problems of inclusiveness in higher education is formed by the Convention on the Rights of Persons with Disabilities [7], which contains conceptual provisions and values that are manifested in the necessity to ensure the maximum access and creation of the corresponding opportunities for accessibility of higher education to all social groups of the population. It determines efforts to ensure inclusiveness for people who face discrimination or have special educational needs. The general methodology of this research is based on a combination of theoretical and conceptual provisions of a large range of scientific disciplines, including pedagogical and social sciences, spheres of management of higher education institutions, psychology, economics, digital communications, etc. The methods used in this paper include analysis, observation, synthesis, generalisation, grouping, and theoretical and conceptual modelling.

The issue of organization of intercultural dialogue in a multinational classical university in the conditions of digital transition from an inclusive environment to global openness belongs to the relevant interdisciplinary problems, which are widely studied in the scientific and expert circles. The key works that were studied during the preparation of this paper are devoted to the resolution of the following tasks: achievement of social resilience of education in the twenty-first century [8], ensuring inclusive transformation of higher education [9, 10], studying the issues of inter-ethnic and intercultural communication in the sphere of higher education [11, 12], development of inclusive of higher education from the position of formation of the corresponding ecosystem [13] and inclusive culture [14], formation of a digital educational space as a platform for inter-cultural and international communications between universities [15], problems of multinational communication in the inclusive environment based on a multilingual approach [16], and development of communication and didactic competencies in the conditions of digital higher education [10].

The goal of this paper was to consider the conditions of the formation of intercultural dialogue in the conditions of development of an inclusive multinational classical university based on digital technologies and the achievement of global openness. This requires the description of the essence of inclusiveness and components of inclusive higher education, determination of the influence of digital technologies on the development of higher education institutions' inclusiveness, and substantiation of a model of social and digital inclusiveness of higher education and conditions for organization of intercultural dialogue with the combination of social and digital inclusiveness of universities.

3 Model and Directions of Organization of Social and Digital Inclusiveness of Higher Education

The concept of inclusiveness passed a long way from the full inclusion of people with special needs to their maximum involvement and creation of the corresponding inclusive environment. At present, an increase in the range of involved persons who receive access to quality higher education often has a global character, without limitations by the principles of borders, language, or cultural environment. The main factor that ensures the global openness of universities is digital technologies. They reduce spatial limitations and help adapt to a new language's surroundings and cultural specifics.

Certain advantages in these conditions if found with multinational classical universities, which offer a wide range of educational specialities and have a rich experience of internationalization, which is manifested through formed mechanisms of successful involvement, adaptation, teaching, and development of students from different cultural, language, and social environments. The developed tools for inclusiveness of higher education, which exclude any manifestations of discrimination and offer equal conditions of learning for people regardless of their educational, physiological, or mental needs, are set onto a multicultural digital basis and facilitate the global openness of such universities [16].

In the classic understanding, inclusiveness of higher education implies the formation of a favourable educational space at the local level. This involves the creation of legal mechanisms to ensure equal access to educational resources and processes, the formation of an ergonomic inclusive environment (elevators and ramps for movement of students, specially equipped lavatories, gyms, medical offices, etc.) and organizational and methodological support for the educational process [16].

Digital technologies that can reduce the spatial factor do not require the formation of the appropriate physical inclusive space. They are aimed at the resolution of communication and organizational problems of vulnerable groups of the population by supplementing the communication capabilities of devices that are used for remote education for foreign and domestic students. This approach fully implements the principles of accessibility of higher education, which is manifested at many levels through the accessibility of the environment (territorial, information, and architectural accessibility), and educational services (adaptation of educational programs and methods, educational technologies, etc.) [17].

Development of digital technologies and their use for the expansion of universities' inclusiveness led to the division of the notion of inclusiveness of higher education into two separate categories – social inclusiveness, which is connected with involvement of people with special needs or vulnerable groups of population in the educational process, and digital inclusiveness, which is connected with the ability of all participants of the educational space to ensure and participate in the educational, scientific, and social process of the university through digital communication means.

While social inclusiveness usually has a local character and is related to the direct resolution of problems of accessibility and involvement of vulnerable groups of population in the educational activities, digital inclusiveness has a global character and allows offering inclusive opportunities for obtaining higher education for vulnerable groups of population without a connection with the local conditions of residents. However, digital technologies set additional demands to such a form of inclusion, which forms a range of requirements and implies the availability of the corresponding technical devices, communication channels, and a sufficient level of digital competence. Meeting these demands of digital technologies allows the use of the capabilities of digital inclusion.

Thus, a complex social & digital inclusiveness is formed. It has a global character and creates conditions for the accessibility of higher education in any place through communication means. Such inclusiveness is still based on the social factor, which preserves a focus on vulnerable groups of the population. However, it goes beyond the local limitations of such persons and acquires a global dimension. A general model of the interaction of social and global inclusiveness is presented in Fig. 1.

This model is based on the transition from local social inclusiveness to global digital inclusiveness through a range of demands which involve the availability of the corresponding communication tools and competencies to involve people with special educational needs or vulnerable groups of the population in full-fledged receipt of higher education and, in the future, self-realisation in the conditions of the labour market.

The current state of integration of digital technologies into the system of higher education is very dynamic. As of now, the list of partially or completely formed elements of the university's educational environment includes the system of entrance (the possibility to apply online), educational process (educational materials at online platforms, remote lectures in a conference regime), means of control (testing and acceptance of completed tasks via online tools), scientific activity (access to knowledge databases and special software). All the above spheres of activities and processes are moderated by the administration, which determines the cultural policy of interaction, in particular, the availability of materials in different languages, rules of behavior, algorithms for obtaining higher education, etc.

However, the social function in the digital form, which contains the largest number of communication channels, is poorly moderated by the university's administration. Most often, the official channels for such communication are "hotlines" and e-mails, universities' social media pages, and local chats and messenger groups. More communication channels are formed spontaneously and, apart from internal moderation, do not usually have any strict cultural norms of interaction between participants.

Such character of involving the participants of the educational process in the university's social environment gets unclear limits (possibility of participation in the social life of participants who are not connected with the university directly), limited number of contacts (disagreement with other participants leads to exclusion from the communication pool), and weak coordination with cultural and value

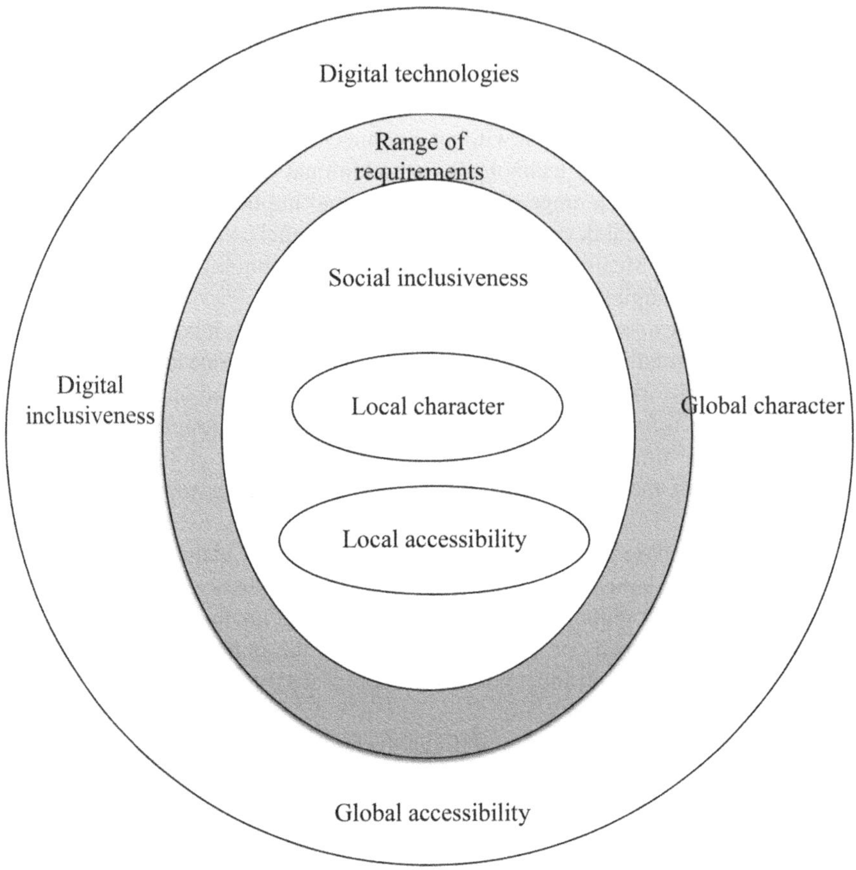

Fig. 1 Model of social and digital inclusiveness of higher education. *Source* Created by the authors based on [10, 12]

systems of the university, including inclusion, striving towards intercultural dialogue, etc. Fragmentation of the communication space in this case is a barrier to the development, and formation of a multicultural environment, and the development of multicultural dialogue [12].

Based on this, there can be assumed the prospective directions of organization of intercultural dialogue in a multinational classical university based on its digitalization (Fig. 2).

The organization of intercultural dialogue in a multinational university, which implements the ideas of inclusiveness in the conditions of digitalization and globalization, requires systemic measures aimed at the development of the corresponding communication platforms. They must cover all components of higher education – learning, scientific activity, and social activity. An equally important aspect in this context is the involvement of stakeholders, who would help prolong the inclusive

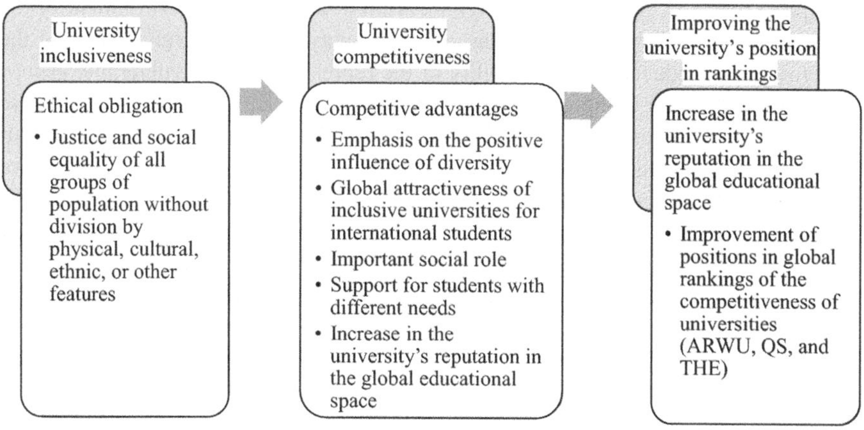

Fig. 2 Directions of organization of intercultural dialogue in a multinational classical university based on its digitalization. *Source* Created by the authors based on [10, 12]

influence of higher education establishment on further professional activity of higher education recipients [13].

Among the current approaches to the development of inclusion to the scales of global openness based on internationalization and digitalization of universities, we can distinguish also suggestions for the development of a specialized educational course "Inclusive culture and digital communications" [14] with the wide involvement of students of different nationalities and lecturers. The result of such training must be the formation of inclusive thinking, the development of tolerance and empathy, and the formation of perceptive skills that are aimed at better understanding representatives of another cultural environment [10].

4 Potential and Risks of Digitalization of an Inclusive Environment of the Multinational Classical University

Many issues revealed in this research have a good potential for further discussion and implementation in the practical spheres of higher education institutions. The most discussed aspects include the necessity to develop unified principles and mechanisms for correct involvement in the inclusive educational environment of representatives of another culture and nationality. These processes are rather complex and require quick reactions. Each action in the process of adaptation, teaching, and inclusion of the representatives of other countries through digital communication means must be coordinated with the general communication strategy of the university and conform to the socio-cultural and other features of each student. Mistakes made at different stages can substantially aggravate the state of people with special educational needs and lead to their losing motivation for learning.

Apart from this, the process of digitalization of an inclusive environment of the multinational classical university has risks that may also negatively influence the general level of education or the ability of the representatives of vulnerable groups of the population for comprehensive integration into the educational process. These risks are mainly connected with the specifics of digital technologies and communication. From the position of cognitive reactions, mistakes during the integration of higher education recipients into the inclusive digital environment of the university can be connected with disorientation or dysfunctions in communication, ethical & regulatory, or cognitive spheres. Also, the prolonged use of digital technologies in the process of learning may contain risks of overload of vision and dysfunctions of the locomotor system. There are also risks of a decrease in the skills of social communication due to the domination of digital remote channels in the person's communicative environment. All the above risks and problems require more detailed discussion with the involvement of the specialists of the corresponding sphere.

5 Conclusions

In global conditions, universities develop according to similar scenarios, which include compliance with the principles of inclusiveness, digitalization, and globalization. For most higher education institutions, these processes were implemented initially through the formation of an inclusive environment with further expansion of the circle of involved people through the tools of internationalization of higher education and its digitalization. At present, the sphere of digitalization largely defines the development of higher education, offering new opportunities not only for remote education but also for solving the problems of people with special educational needs without a connection to a certain country or territory.

These processes are set on the demands of digital communication, with requirements for appropriate competencies and technological equipment of communication channels. These demands form a certain range of limitations, overcoming which will allow a student to receive global access to digital inclusiveness. The presented model of social and digital inclusiveness clearly interprets the transition from an inclusive local environment of universities to their global openness.

Expansion of the circle of people involved in higher education by the university forms additional risks, which are connected with an increase in diversity of the multicultural environment and the formation of new communication channels, most of which have spontaneous character and are not controlled by the university's administration. A necessary condition for avoiding cultural differences and an increase in conflict level in such educational space is an organization of intercultural dialogue with the corresponding acquisition of competencies in multicultural communications and the development of empathy and tolerance.

References

1. Pawar, S. K., & Islam, T. (2025). Hierarchical effects of university social media posts on the enrolment decision-making of international students: Integrating the AIDA and TAM models. *Studies in Higher Education*. https://doi.org/10.1080/03075079.2024.2448530
2. Song, I., & Kim, Y. (2024). Working as a "Token": Different perspectives on the role of international faculty in Korean Universities. *ECNU Review of Education, 7*(4), 926–944. https://doi.org/10.1177/20965311241240482
3. Tao, Y. (2025). Mismatch and detachment: Chinese international students' campus electoral participation in an Australian University. *Asia Pacific Viewpoint*. https://doi.org/10.1111/apv.12437
4. Wang, J., & Xu, W. (2024). Conflictual language ideologies and mismatched language practices of Thai international students at a Chinese-medium instruction university. *Humanities and Social Sciences Communications, 11*(1), 916. https://doi.org/10.1057/s41599-024-03427-1
5. Yu, Q., Brunsting, N. C., Smart, J., & Bingham, W. P. (2025). Language proficiency, second language educational experience, and psychological well-being among international students at U.S. Universities. *Research in Higher Education, 66*(1), 3. https://doi.org/10.1007/s11162-024-09821-7
6. Zhao, K., & Ma, J. (2024). Predicting international student satisfaction in Chinese universities: A comparison between undergraduate and graduate students. *European Journal of Education, 59*(4), e12697. https://doi.org/10.1111/ejed.12697
7. UN. (2006). *Convention on the rights of persons with disabilities resolution adopted by the general assembly on 13 December 2006. 61/106*. Retrieved from https://www.un.org/en/development/desa/population/migration/generalassembly/docs/globalcompact/A_RES_61_106.pdf. Accessed December 12, 2023.
8. Baeva, L. V. (2021). Education for social sustainability in the face of the challenges of the 21st century. *Etnodialogi [Ethnodialogues], 1*(63), 45–60. Retrieved from https://cyberleninka.ru/article/n/obrazovanie-dlya-sotsialnoy-ustoychivosti-pered-litsom-vyzovov-xxi-veka. Accessed December 14, 2023
9. Generalova, L. M., Arzhanovskaia, A. V., & Eltanskaia, E. A. (2021). Pro-inclusive transformation of education—opportunities and prospects. *International Scientific Research Journal, 8-3*(110), 33–36. Retrieved from https://cyberleninka.ru/article/n/proinklyuzivnaya-transformatsiya-obrazovaniya-vozmozhnosti-i-perspektivy. Accessed December 12, 2023
10. Rakhimova, A. E., & Ratner, F. L. (2022). Development of didactic thinking in a digital educational environment. *Problems of Modern Teacher Education, 76*(2), 155–158. Retrieved from https://cyberleninka.ru/article/n/razvitie-didakticheskogo-myshleniya-v-usloviyah-tsifrovoy-obrazovatelnoy-sredy. Accessed December 14, 2023
11. Akinshina, I. B., & Karabutova, E. A. (2021). Interethnic and intercultural interaction of students in the educational environment of Russian universities. *Scientific Notes: Electronic Scientific Journal of Kursk State University, 4*(60), 448–458. Retrieved from https://cyberleninka.ru/article/n/mezhetnicheskoe-i-mezhkulturnoe-vzaimodeystvie-studentov-v-obrazovatelnoy-srede-rossiyskih-universitetov. Accessed December 14, 2023
12. Krivenkaia, M. A., Zakharov, V. F., Omelchenko, E. A., Orlova A. Y., Shevtsova, A. A., Iashina, O. N., ... Dorokhova, M. V. (2022). Education and dialogue in the interests of peace and people. In *Results of the XIII international seminar "Dialogue of Civilizations and Intercultural Cooperation". Etnodialogi* [Ethnodialogues], (Vol. 1, pp. 31–55). Retrieved from https://cyberleninka.ru/article/n/obrazovanie-i-dialog-v-interesah-mira-i-cheloveka-itogi-xiii-mezhdunarodnogo-seminara-dialog-tsivilizatsiy-i-mezhkulturnoe. Accessed December 12, 2023.
13. Bezpalova, A. G. (2022). Algorithm for improving inclusive education as a component of the ecosystem of a higher education institution. *Bulletin of the Rostov State Economic University (RINH), 1*(77), 52–57. Retrieved from https://cyberleninka.ru/article/n/algoritm-

sovershenstvovaniya-inklyuzivnogo-obrazovaniya-kak-komponenta-ekosistemy-vysshego-uchebnogo-zavedeniya. Accessed December 13, 2023

14. Salnikova, O. D. (2022). Analysis of the effectiveness of work to develop an inclusive culture among students of the North Caucasus Federal University (NCFU). *Spetsialnoe obrazovanie [Special Education], 3*(67), 187–200. Retrieved from https://cyberleninka.ru/article/n/analiz-effektivnosti-raboty-po-razvitiyu-inklyuzivnoy-kultury-u-studentov-severo-kavkazskogo-federalnogo-universiteta-skfu. Accessed December 12, 2023
15. Chibisguleva, T. I. (2022). Development of the digital educational environment as an element of international online cooperation between universities. *Bulletin of the Moscow State University of Culture and Arts, 5*(109), 150–158. https://doi.org/10.24412/1997-0803-2022-5109-150-158
16. Egorova, P. A., Prokhorova, A. A., & Sorokoumova, S. N. (2021). Application of a multilingual approach to inclusive education in higher education. *Iazyk i kultura [Language and Culture], 54*, 152–166. Retrieved from https://cyberleninka.ru/article/n/primenenie-multilingvalnogo-podhoda-pri-inklyuzivnom-obuchenii-v-vysshey-shkole. Accessed December 13, 2023
17. Romanenkova, D. F. (2021). Basic approaches to ensuring accessibility of inclusive higher education through distance technologies. *Bulletin of Chelyabinsk State University Education and Healthcare, 1*(13), 27–32. Retrieved from https://cyberleninka.ru/article/n/osnovnye-podhody-k-obespecheniyu-dostupnosti-inklyuzivnogo-vysshego-obrazovaniya-sredstvami-distantsionnyh-tehnologiy. Accessed December 14, 2023

Inclusiveness of Universities as the Basis of Their Global Competitiveness: A View from the Standpoint of Rankings ARWU, QS and THE

Uran N. Busurmankulova, Shakhlo T. Ergasheva, Juliana A. Kitsai, Natalya G. Balibardina, and Maria A. Komissarova

Abstract In this research, the authors focus on the essence and role of inclusiveness, the transformation of its treatment in society, and support for the influence of universities inclusiveness on the level of their competitiveness from the position of the global rankings of higher education institutions. The authors used multidisciplinary and systemic approaches, which combine the theories and concepts of philosophy, law, sociology, pedagogics, and cultural sciences. Due to this, inclusiveness is considered a worldview and moral and legal concept implemented through pedagogical approaches based on cultural diversity and could be used in the interest of higher educational establishments as a factor and indicator of competitiveness. In this research, the authors consider theoretical and conceptual provisions of inclusiveness and identify its role in higher education. They also characterize the main problems and specific features of ensuring inclusiveness in higher education and emphasize the different treatment of this definition in post-Soviet and Western scientific thoughts. According to this, the authors focus on the necessity of expanding the understanding of inclusiveness to other categories of vulnerable groups of society, apart from people with disabilities. The authors determine the connection between the inclusiveness of universities and their competitiveness according to global educational rankings. This is an additional argument in favor of further

---

U. N. Busurmankulova
Kyrgyz National University named after J. Balasagyn, Bishkek, Kyrgyzstan

S. T. Ergasheva
Tashkent State University of Economics, Tashkent, Uzbekistan
e-mail: sh.ergasheva@tsue.uz

J. A. Kitsai (✉)
Pskov State University, Pskov, Russia

N. G. Balibardina
Sebryakovskiy Branch of Volgograd State Technical University, Mikhailovka, Russia

M. A. Komissarova
Platov South-Russian State Polytechnic University (NPI), Novocherkassk, Russia

B. S. Sergi et al. (eds.), *University 4.0 and Educational Technology for Sustainable Development*, Education in the Asia-Pacific Region: Issues, Concerns and Prospects 75, https://doi.org/10.1007/978-981-96-7376-6_33

development of inclusiveness in higher educational establishments in Russia and Central Asia and the necessity for expanding its influence on all categories of vulnerable groups.

Keywords Inclusiveness · Inclusion · University · Higher education · Competitiveness · Ranking

JEL Codes A14 · I23 · I24 · I28 · M12 · O15 · O18 · O49

1 Introduction

The Fourth Industrial Revolution, globalization, and the promotion of liberal values are important factors in the development of humanity. Depending on the interaction with these factors, governments of different countries of the world reach certain success on the path to innovativeness and competitiveness. However, they also consider limitations that are formed under the influence of the dominating trends of global development.

Thus, digital transition, together with substantial advantages, which consist in an increase in labor efficiency and development of human capabilities, actualizes the problem of the lack of talent. Globalization increases the number of contacts between the representatives of different regions, cultures, and nations, but aggravates the problem of intercultural differences. Liberal values popularize the idea of general equality, but they often do not consider traditions or the real situation in separate regions or at the local level.

The leading institutions that represent the above processes are universities. They form intellectual and sociocultural trends, which become popular among the youth and gradually start to dominate society. In this context, the idea of inclusiveness of universities is connected with digitalization, globalization, and liberalization of society.

These circumstances allow considering the concept of competitiveness of higher education establishments through the lens of their ability to include people with different needs and handicaps in the educational process, which ensures higher diversity and wider access to talents and stimulates the improvement of the level of teaching in universities, forming their cultural and social features. Thus, inclusiveness of universities is a current trend in contemporary social development, which characterizes the level of the integration of educational establishments in global processes and forms preconditions for raising the level of their competitiveness in global rankings, especially the ones that consider the criteria of diversity, openness, and inclusiveness.

2 Connection Between University Inclusiveness and the Competitiveness of Higher Education Establishments

The methodology of this paper is based on the cross-disciplinary and systems approaches, which connect the theories and views inherent to philosophy, law, sociology, pedagogics, and economic and cultural sciences. This methodology is based on the equality of people, which is implemented through global and national institutions and is manifested in ensuring equal rights of citizens for education and the use of its results in the professional sector. The key provisions of the philosophical vision of the problem posed focus on the very essence of the notion of equality and its interpretation in the conditions of a right to higher education. This right is considered in detail in the system of legislative norms, which clearly define the principles and limitations of equality in the conditions of obtaining higher education. Sociology supplements these aspects with the research on public view of equal rights for obtaining higher education given different limitations peculiar to certain social groups of the population. Pedagogics offers approaches and means that can ensure the accessibility of education to all citizens. Economics treats the above problems in the context of their influence on the development of universities and support for their competitiveness. Cultural sciences focus on the formation of the culture of inclusive education, intercultural exchange, and the influence of diversity on the educational process and higher education on the whole.

The methodological apparatus of this research is based on the laws of dialectics and involves the use of general scientific methods of cognition, such as analysis and synthesis, observation, generalization, and comparison; expert methods, which determine the level of universities' competitiveness in the form of the corresponding rankings; and methods of mathematical statistics, which allow revealing the connection between the level of inclusiveness of universities and their general rank in the global educational rankings.

The inclusiveness of university education and universities is an important direction of scientific research, which reflects general social processes and the expansion of new values. The key aspects that are studied in the context of the connection between university inclusiveness and the competitiveness of higher education establishments are mainly related to the following aspects:

- The research on the essence and generalization of the experience of implementing inclusion in educational establishments [1, 2];
- Determination of trends and problems of implementing inclusion in higher education [3, 4];
- Study of the problem of training teachers for teaching in the conditions of an inclusive university [5];
- Critical assessment of elite educational establishments from the position of declaration and real implementation of the inclusive education principles [6];

- Development of the research methodology [7] and determination of the level of inclusiveness [8];
- Evaluation of competitiveness of universities at the global rankings [9–13];
- Assessment of the influence of universities' inclusiveness on the level of their competitiveness from the position of the world educational rankings [14].

3 Competitiveness of Universities in Kazakhstan and Uzbekistan by the Criteria of Inclusiveness

Inclusion, or inclusiveness, is a fairly new term for the post-Soviet information and scientific environment. The active use of this term in Russia and countries of Central Asia began in the late 1990s – early 2000s. It was connected primarily with the issue of the creation of special conditions for the access of people with disabilities to the use of public benefits. The basis for the treatment and acceptance of this definition in society was the United Nations Convention on the Rights of Persons with Disabilities, which guarantees the access of people with disabilities to higher education without discrimination and on an equal basis with others [15].

In higher education, inclusiveness is considered an ability to consider diversity from the position of involving all students in the educational process. In this context, the term relates to the quality of the educational system and the effectiveness of teaching, which allows avoiding any limitations in the process of obtaining education and the use of its results in life [4].

In the twenty-first century, the student contingent acquires the features of heterogeneity (diversity) from the position of ethnic, racial, and confessional belonging, as well as health factors. This trend defines advantages and challenges for higher education. Despite this, the higher school still retains high barriers from the position of access to education and achievement of access for many students who belong to socially vulnerable groups [2].

While in the post-Soviet countries, inclusiveness of education is connected only with the rights of people with disabilities [1]. Its treatment in the world science and practice went beyond these limits and implies the involvement of all students in the educational process, regardless of their social state, ethnic or racial origin, and gender and physical abilities [5]. According to this, inclusiveness is treated quite specifically due to its influence on the formation of the corresponding educational space, which covers the culture, practice, and policy of the educational establishment [8]. Thus, as of now, the inclusiveness of higher education is treated from the position of ethical obligation, which is connected with the notions of justice and social equality of all groups of population without division by physical, cultural, ethnic, or other features.

In the conditions of the educational services market, inclusiveness acquires the status of yet another competitive advantage (Fig. 1).

This is proven by a range of factors, particularly the following ones:

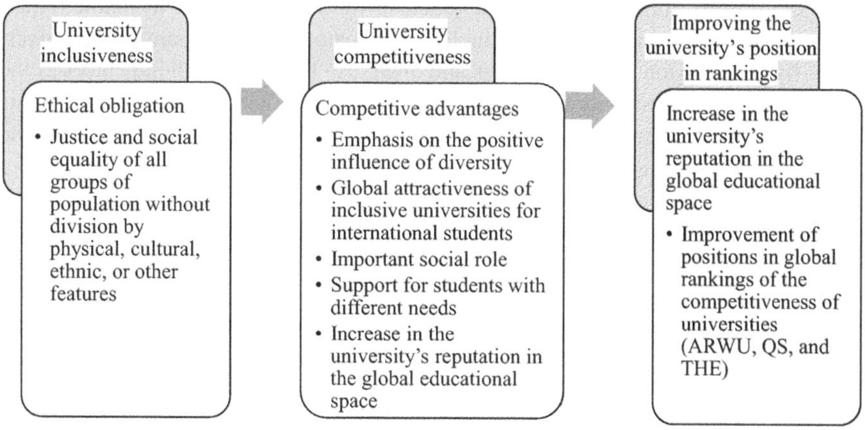

Fig. 1 Competitiveness of university according to its inclusiveness. *Source* Developed by the authors

- Emphasis on the positive influence of diversity, which is ensured by inclusiveness, as to the development of scientific, innovative, and cultural activity;
- Formation of global attractiveness of inclusive universities for international students;
- Execution of an important social role;
- Support for students with different needs;
- Increase in the university's reputation in the global educational space.

The most popular global rankings of the competitiveness of universities include ARWU [16], QS [17], and THE [18]. Each of these rankings has its methodology of ranking universities by different qualities. ARWU focuses on the academic performance of the universities representatives and their scientific indicators. QS and THE evaluate also other aspects, such as international attractiveness, social responsibility, gender proportions, and inclusiveness. These rankings use different sources of information. Thus, while ARWU is based on official data on scientific publications, the rankings QS and THE use information obtained as a result of surveys and from other sources.

To analyze the competitiveness of universities in Russia and countries of Central Asia given the criteria of inclusiveness, the authors studied the above rankings. As a result, the authors revealed different representations of the region's universities in these rankings. Thus, ARWU includes only nine Russian universities, the activities of which were assessed by the academic criteria as to recognition of results of university graduates or lecturers, the level of citation of published works, and the number of published works. Thus, the results of this ranking do not contain representative information that would help the goals of this research; they were not considered.

The ranking THE contains the most representative information on all countries of the region, except for Tajikistan. It contains information on 29 Kazakh, 1 Kyrgyz, 11 Turkmen, and 48 Uzbek universities. The information on 103 Russian higher education establishments that are included in the ranking is not available. Apart from the previous ranking, this one includes indicators that can characterize universities by the criterion of inclusiveness and diversity. These are such indicators as the number of foreign students and gender proportion. A limitation of the results of this ranking for the research is the absence of scores for most of the universities by the criteria of their academic performance.

The information on academic performance, diversity, and inclusiveness is presented most fully and systemically in the ranking QS. Though the representativeness of the sample is limited by universities from Kazakhstan and Uzbekistan (Table 1), accessibility of the data on inclusiveness, as one of the criteria of the ranking, and the possibility of comparing the level of this indicator with the indicators of employment, internationalization, and the level of teaching allowed analyzing the dependence of the total value of the ranking on each separate indicator.

With the help of the correlation analysis of the indicators obtained, the authors formed a correlation matrix (Table 2). The results of the analysis demonstrated a close direct connection between the general level of competitiveness of the region's universities and the level of their inclusiveness ($r(y,x_4) = 0.702$).

As we can see, the connection between the selected indicators is rather high. However, a much larger effect on the general level of the ranking is made by the indicators of the possibility of employment and level of teaching. The influence of the indicator of internationalization on the generalized indicator is somewhat lower. The indicator of inclusiveness within the above list of indicators correlates most closely with the general indicator. This is a proof of the large role of the inclusiveness index in the context of the assessment of universities' competitiveness in the global educational environment.

4 Prospects on Implementation of the Concept of Inclusive Education in Universities of Russia and Central Asian Countries

The problems considered in this research and the conclusions obtained confirm the general theory of the presence of a connection between the level of the inclusiveness of higher education institutions and the level of their global competitiveness. However, many intermediary conclusions and aspects require additional substantiation. Thus, more attention should be paid to the specifics of the policy of inclusive education within each university, which would involve the development and use of a special methodological basis that would allow considering different aspects of inclusiveness in higher education of each educational establishment.

The implementation of the concept of inclusive education in universities of Russia and countries of Central Asia requires additional scientific substantiation. The concept goes beyond the limits of the creation of conditions for teaching people

Table 1 Level of the competitiveness of universities in Kazakhstan and Uzbekistan by the criteria of academic performance, diversity, and inclusiveness

University	University's rank according to the criterion				
	General	Possibility of employment	Internationalzation	Level of teaching	Inclusiveness
	y	x_1	x_2	x_3	x_4
Kazakhstan					
Kazakh State Agrarian Research University	2	1	5	1	7
Abai Kazakh National Pedagogical University	5	4	6	4	1
Narxoz University	6	5	10	5	2
S. Seifullin Kazakh Agro Technical Research University	7	6	11	6	3
L. N. Gumilev Eurasian National University	8	9	3	9	4
Karagandy State University	9	10	4	10	5
South Kazakhstan University named after Mukhtar Auezov (SKU)	10	11	8	11	6
Karaganda University of Kazpotrebsoyuz	11	12	14	12	9
Almaty Technological University	12	13	13	13	10
Kazakh National Medical University	14	7	7	7	8
Uzbekistan					
Tashkent State University of Economics	1	3	1	3	3
Tashkent Institute of Irrigation and Agricultural Mechanization Engineers	2	1	3	1	1
Tashkent University of Information Technologies named after Muhammad al-Khwarizmi	3	2	2	2	2
Tashkent State University of Law	4	4	4	4	4

Source Created by the authors based on [17]

Table 2 Correlation matrix of the indicators of the competitiveness, diversity, and inclusiveness of universities in Kazakhstan and Uzbekistan

		General	Possibility of employment	Internationalization	Level of teaching	Inclusiveness
Indicators		y	x_1	x_2	x_3	x_4
General	y	1.000				
Possibility of employment	x_1	0.892	1.000			
Internationalization	x_2	0.672	0.662	1.000		
Level of teaching	x_3	0.892	1.000	0.662	1.000	
Inclusiveness	x_4	0.702	0.676	0.563	0.676	1.000

Source Created by the authors based on [17]

with disabilities. It also includes consideration of other limitations of vulnerable social groups and the achievement of diversity.

The most significant aspect of inclusive education in universities that requires further disclosure is the determination of the link between implementing the policy of inclusive education and the real level of the university's competitiveness. A specific feature of this task is that the formation of inclusive practices within a university forms conditions for education without limitations, increases the cultural and professional level of the academic staff, promotes the ideas of equality and diversity, and provides better integration with the surrounding world.

The problem of the competitiveness of universities in Russia and countries of Central Asia in the context of their implementation of the concept of inclusiveness requires a more comprehensive approach to studying the constituent and organizational documents of higher education establishments and assessment of their provisions regarding inclusiveness from the position of compliance with them and real implementation in practical activities. Such studies would involve works that include surveys of different groups of participants in the educational process and assessment of the conditions, policy, and culture of inclusiveness in each higher educational establishment.

5 Conclusions

The inclusiveness of universities is an important ethical and institutional criterion of the modern development of society. It determines educational establishments' observance of the key rights of humans and support for equality. In modern conditions, inclusiveness in education has gained popularity in Russia and countries of Central Asia. However, unlike a vision that is popular in Western countries, it is treated only as the universities' ability to create conditions for barrier-free studying of people with disabilities. Contrary to this, an important point is the one that treats inclusiveness in education as a factor of ensuring equal rights for all groups of the population regardless of physical, ethnic, cultural, or other qualities.

The performed correlation analysis confirmed the presence of a strong direct connection between the level of inclusiveness of universities in Kazakhstan and Uzbekistan and their general level of competitiveness. This points to the importance of further development of inclusive policy, practice, and culture in the region's universities with the expansion of emphasis on other categories of vulnerable groups of the population, apart from those with disabilities. In this context, inclusiveness, as a criterion of the university's competitiveness, demonstrates its potential from the position of adaptation and flexibility in ensuring equal rights for studying and self-realisation of different groups of students, including vulnerable categories. This forms a range of positive effects for the university and its students and improves its reputation and the level of attractiveness in the global educational environment.

References

1. Denisova, O. A., & Lekhanova, O. L. (2021). Best practices for interaction in an inclusive educational environment. *Bulletin of A. S. Pushkin Leningrad State University, 4*, 192–205. Retrieved from https://cyberleninka.ru/article/n/luchshie-praktiki-vzaimodeystviya-v-inklyuzivnoy-obrazovatelnoy-srede. Accessed October 24, 2023
2. Volosnikova, L., Zagvyazinskiy, V., Kukuev, E., Fedina, L., & Ogorodnova, O. (2021). The convergence of the concepts of academic and inclusive excellence at research universities. *The Education and Science Journal, 23*, 43–78. https://doi.org/10.17853/1994-5639-2021-4-43-78
3. Detochenko, L. S. (2022). Modern trends in ensuring inclusivity in the higher education paradigm. *Sotsiologiia [Sociology], 4*, 145–152. Retrieved from https://cyberleninka.ru/article/n/sovremennye-tendentsii-obespecheniya-inklyuzivnosti-v-paradigme-vysshego-obrazovaniya. Accessed October 24, 2023
4. Ulybina, O. V., Khakhalkina, U. V., Domnichev, D. I., Goncharova, O. V., & Barinov, S. V. (2023). Inclusiveness as one of the areas of modern practical pedagogy. *Moscow Economic Journal, 1*. Retrieved from https://qje.su/ekonomicheskaya-teoriya/moskovskij-ekonomicheskij-zhurnal-1-2023-45/. Accessed October 25, 2023
5. Kantor, V., & Proekt, Y. (2021). Inclusive educational process at the university: Theoretical and experimental model of teachers' psychological readiness. *The Education and Science Journal, 23*, 156–182. https://doi.org/10.17853/1994-5639-2021-3-156-182
6. Koutsouris, G., Stentiford, L., & Norwich, B. (2022). A critical exploration of inclusion policies of elite UK universities. *British Educational Research Journal, 48*, 878–895. https://doi.org/10.1002/berj.3799
7. Staff, W. (2022). Rating universities on inclusiveness: QS stars. QS Top Universities. Retrieved from https://www.topuniversities.com/qs-stars/qs-stars/rating-universities-access-qs-stars. Accessed October 26, 2023.
8. Booth, T., Ainscow, M., Black-Hawkins, K., Vaughan, M., & Shaw, L. (2002). *Index for inclusion*. Centre for Studies on Inclusive Education. Retrieved from http://csie.org.uk/resources/translations/IndexEnglish.pdf. Accessed October 25, 2023
9. Lee, S., Kim, S. J., Park, S., & Shin, J. C. (2025). Unequal metrics in research publications: The impact of bibliometric databases and faculty size across academic disciplines on university rankings in South Korea. *Scientometrics*. https://doi.org/10.1007/s11192-024-05223-9
10. Pinto-Delacadena, P. A., Liern, V., Acosta-Vargas, P., & Vinueza-Cabezas, A. (2024). A multicriteria approach to ranking Latin-American universities based on region-specific criteria. *Technological Forecasting and Social Change, 208*, 123725. https://doi.org/10.1016/j.techfore.2024.123725

11. Tavoletti, E., Şahin, F., Taras, V., Farrell, W., & Florea, L. (2024). A configurational analysis of university ranking, cognitive ability and cultural intelligence in global virtual teams: The role of academic pedigree. *International Journal of Management Education, 22*(3), 101040. https://doi.org/10.1016/j.ijme.2024.101040
12. Trung, D. D., Dudić, B., Van Duc, D., Son, N. H., & Mittelman, A. (2024). Building a ranking system for lecturers based on student evaluations in teaching a specific course: A case study at a university in Vietnam. *International Journal of Cognitive Research in Science, Engineering and Education, 12*(2), 335–350. https://doi.org/10.23947/2334-8496-2024-12-2-335-350
13. Wang, T., & Guo, C. (2024). Discipline rankings and university admission attractiveness: Quasi-natural experimental evidence from China. *Higher Education Policy*. https://doi.org/10.1057/s41307-024-00380-1
14. Pasupuleti, N. S. (2022). The impact of diversity and inclusivity on university rankings. LinkedIn. Retrieved from https://www.linkedin.com/pulse/impact-diversity-inclusivity-university-rankings-pasupuleti-ph-d-?utm_source=rss&utm_campaign=articles_sitemaps. Accessed October 26, 2023.
15. United Nations. (2006). *United Nations Convention on the Rights of Persons with Disabilities*. Retrieved from https://www.un.org/disabilities/documents/convention/convention_accessible_pdf.pdf. Accessed October 23, 2023
16. Shanghai Ranking. (2023). 2023 Academic Ranking of World Universities. ARWU. Retrieved from https://www.shanghairanking.com/rankings/arwu/2023. Accessed October 26, 2023.
17. QS. (2023). *QS Stars University Ratings*. Retrieved from https://www.topuniversities.com/qs-stars. Accessed October 25, 2023
18. THE. (2023). World university rankings 2024. Times Higher Education. Retrieved from https://www.timeshighereducation.com/world-university-rankings/2024/world-ranking#. Accessed October 25, 2023.

Digital Inclusion as a Macroeconomic Mechanism for Solving the Problem of Higher Education in Russia in the Context of the Large-Scale Process of Internationalization of Universities

Saodat A. Yuldasheva, Munisa J. Khusnuddinova, Tatiana A. Dugina, and Natalya V. Shchukina

Abstract This paper elaborates on the concept of digital inclusion, which is implemented by Russian universities in the conditions of internationalization of higher education. Digital inclusion is considered a factor in an increase in social inclusion, which is combined with digital communications and expands the circle of persons involved in higher education, ensuring the achievement of a range of macroeconomic effects. The methodology of this research is based on the interdisciplinary and systemic approaches, which cover the theoretical basis and practical aspects of pedagogical, sociological, psychological, economic, and management disciplines and are implemented through the use of the methods of critical, retrospective and reflexive analysis, synthesis, and methods of generalisation, comparison, etc. It was revealed that the notion of digital inclusion does not have a unified treatment in scientific literature, but it has a large potential, which goes beyond the limits of working only with vulnerable groups of the population. It was determined that such an approach is combined well with the concept of internationalization of education, in particular, the programme of export of educational services. Aappearance of digital inclusion in the environment of internationalization, combined with the trends of globalisation and digitalisation, ensures not only the resolution of microeconomic problems of higher education establishments but also the achievement of a range of macroeconomic effects, aimed at an increase in the level of intellectual capital of the

S. A. Yuldasheva
Tashkent State Transport University, Tashkent, Uzbekistan

M. J. Khusnuddinova
Tashkent State University of Economics, Tashkent, Uzbekistan
e-mail: m.khusniddinova@tsue.uz

T. A. Dugina (✉)
Volzhsky Institute of Economics, Pedagogy and Law, Volzhsky, Russia

N. V. Shchukina
Sebryakovskiy Branch of Volgograd State Technical University, Mikhailovka, Russia

B. S. Sergi et al. (eds.), *University 4.0 and Educational Technology for Sustainable Development*, Education in the Asia-Pacific Region: Issues, Concerns and Prospects 75, https://doi.org/10.1007/978-981-96-7376-6_34

country, dealing with the situation of social support, and improvement of the level of student's motivation for studying.

Keywords Inclusion · Digital inclusion · Macroeconomic mechanism · Higher education · Internationalization of higher education · Digital technologies

JEL Codes I21 · I23 · I25 · I28 · O15 · O33 · O57 · O15 · F16

1 Introduction

The Russian market of higher education is acquiring the features of a real market every year, which allows applying macroeconomic research tools. In this context, the autonomy of educational establishments is one of the key macroeconomic conditions of the market's functioning, which could be compared to the freedom of entrepreneurial activity in traditional markets. The autonomy of university management is treated as one of the basic principles of higher education. According to this, universities have a sufficient volume of rights and authority in the sphere of decision-making on the management of their development. This is concerned largely with the development of educational courses, number of students and own scientific and international policy.

Internationalization of higher education has been one of the manifestations of such autonomy according to which the management of universities was free to deal with its international relations, invitation of foreign students, and realisation of different international programmes. At present, the internationalization of higher education is not a one-time or local initiative anymore. It is a full element of the national policy, which forms macroeconomic mechanisms and is implemented within the export direction. At that, it solves a range of typical university problems, such as the formation of the number of students, differentiation of the sources of university financing, increase in currency revenues in the Russian economy, expansion of international communication channels, and cooperation. The government initiative of such character allows regulating the processes of interaction of Russian universities with the global educational space at the macro-level, through the corresponding programmes, policies, and financial and socio-cultural processes.

An important element of such policy is adaptation and integration of the regulatory and socio-cultural field with global processes: liberalisation, multiculturalism, openness, and inclusion [1, 2]. In this list, inclusion forms a separate methodological field, which is manifested through the creation of conditions for access to higher education for vulnerable groups of the population, the formation of the space of support and mutual support, and ensuring active constructive dialogue between the participants of the educational process [3]. After the COVID-19 restrictions and the emergence of the necessity for the development of digital education, the notion of inclusion acquired a new meaning [4]. It penetrated the sphere of digital communications, providing universities with new capabilities and creating a concept of digital inclusion as a component of the policy of global accessibility of higher education [5].

2 Digital Inclusiveness in the Context of the Internationalization of Universities

Consideration of the process of internationalization of education from the position of macroeconomics expands the methodological basis of this research due to interpreting the educational service in the sphere of higher education as a product that receives access to the global market of education and becomes its part. According to this, it becomes a part of the global competitive environment, acquires relevant trends and values, reacts to the changes in demand and interests of the target audience, and receives a large stimulus for development. Therefore, the methodology of this research combines the conceptual approaches and theories of pedagogics, social sciences, psychology, economics, management, and other disciplines. This allows considering all studied processes as a mechanism of mutually coordinated actions, which influence each other.

The issue of development and implementation of digital inclusion in the system of Russian higher education, which is manifested in the context of large-scale internationalization of universities and expansion of their influence on macroeconomic processes, is of the complex character and is therefore studied in the context of more specialised topics, which consider it from the position of internationalization of higher education in the conditions of the pandemic [6] and globalisation [1, 7–9]; determination of the specifics and trends of export of educational services by Russian universities [10, 11]; influence of digital inclusion on the development of human capital [12], assessment of the institutional dynamics in the development of higher education in Russia [13]; problems and prospects for the development of inclusive education in Russia [14]; and achievement of the macroeconomic influence on the sustainable development of economy [15].

The goal of this research was to substantiate digital inclusion's ability to influence macroeconomic processes through the tools of internationalization of higher education. To achieve this goal, the following tasks were posed: determining the essence of the key categories of research – digital inclusion and internationalization of education; evaluation of the stages of their development; and preparation of the model of interaction of digital inclusion, internationalization of higher education, and digitalisation of the global educational space from the position of achievement of macroeconomic effects. The value of this work lies in the theoretical substantiation of the macroeconomic model of solving the problems of higher education in the conditions of digital inclusiveness in the context of the internationalization of universities. The proposed model characterises the direction and coordination of the processes of digitalisation, inclusion, and globalisation, which facilitate the resolution of the problems of universities and are manifested through macroeconomic influence.

3 The Macroeconomic Model of Higher Education's Digital Inclusiveness in the Context of the Internationalization of Universities

The notion of inclusion as a tool of involvement in a certain process is well combined with digital technologies, which are inseparable tools of communication. However, digital inclusion, as a scientific or practical category, is a rather new term, and, thus, does not have a unified definition. Moreover, scientific studies have several approaches to the treatment of digital inclusion, which define it in the following way:

- Element of digital education;
- Tool for provision of medical services;
- Parameter of the use of digital technologies in online banking;
- Direction for social inclusion, which is supplemented by digital technologies and includes inclusive education [12].

The latter approach is typical for the Russian scientific environment. However, the potential of digital inclusion cannot be limited only by working with vulnerable groups of the population. In the environment of the world that is connected to the unified information network, inclusion cannot be treated without consideration of the context of access to digital services and possession of digital technologies. In this context, inclusion is a criterion of social accessibility and involvement, and the adjective "digital" characterises the means and tools of this involvement.

In the current conditions, higher education is a complex construct, which combines not only the educational process but also other functions, causes, and processes, among which internationalization acquires an increasingly larger role. Internationalization of higher education institutions is a manifestation of the horizontal interaction of market elements. Horizontal links allow increasing the number of contacts and diversifying them without direct interaction with the higher managerial level [16].

The term "internationalization" regarding higher education was first used in the 1970s–1980s. It gradually evolved from the concept of international education to the full direction of its development, which includes such elements as the presence of an international component in educational programmes of universities, international mobility and cooperation, international technical cooperation and mutual help [7]. However, there is no unified definition of the essence of internationalization of the university. Because of this, many scholars still treat it as a tool for announcing their scientific interests, while other scholars treat it as a mechanism for developing the academic image or the element of prestige of higher education establishments. Another popular thought is that internationalization is opposed to globalisation, similar to the opposition of diversity and multiculturalism to unification [16].

The official participation of Russian universities in the process of internationalization started in 2003 when the Russian Federation joined the Bologna Declaration [9]. Still, this process was better manifested as a result of the adoption of the Concept for the export of educational services, which was aimed at raising the quality and competitiveness of the Russian system of education in the global space, and the development of the federal project "Export of education" [11], when these processes received government support. Stimulation of internationalization of higher education in this context was aimed at raising the competitiveness of not only the system of universities but the entire economic system, through the formation of better conditions for socio-psychological, economic, and socio-cultural adaptation of graduates of higher education establishments to the professional activities [8].

Thus, the development of higher education received a vivid economic basis, which can be manifested at the macro- and micro-levels (Fig. 1).

The microeconomic level of ensuring export and internationalization of higher education is the creation of good conditions for foreign students and their adaptation to the educational process and everyday activities. This vision emphasizes the number of foreign students and the revenues from their paying for their studies [10]. At the macroeconomic level, the internationalization of higher education is assessed from the position of its influence on the competitiveness of the economy, which is manifested in the changes in the structure of external trade and ensuring training of specialists, who are adapted to professional activities in the conditions of the global economy and multicultural environment [11].

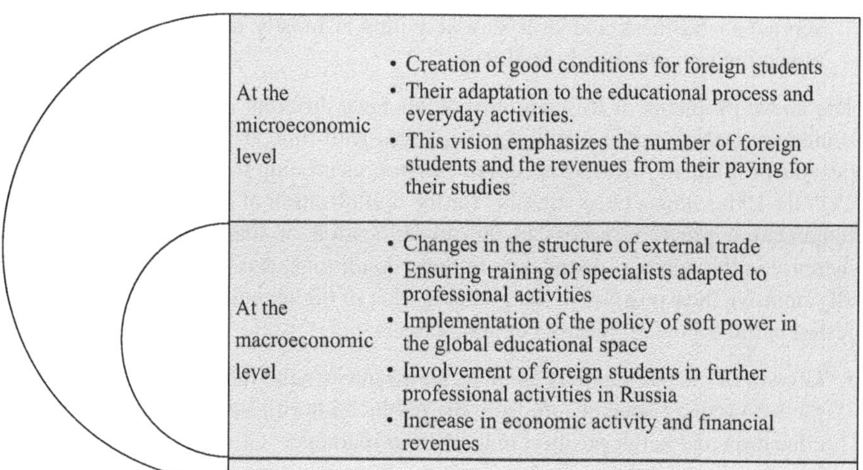

Fig. 1 Benefits of the approach to the export of educational services in modern Russia. *Source* Developed by the authors

Such an approach to the export of educational services is a strict emphasis on the receipt of the macroeconomic effect of socio-political (as a tool of influence on social relations and implementation of the policy of soft power in the global educational space), demographic (as a factor of involvement of foreign students in further professional activities in Russia), and economic (as a source of financial revenues and an increase in economic activity) character.

It should be noted that the above processes took place simultaneously with the constant increase in technological changes in the global and national educational spaces. This results in digitalisation's penetrating all spheres of human life. Higher education was passive in accepting these novelties for a certain period, but restrictions connected with the pandemic became a stimulus for the development of digital and remote education, which, in turn, increased the processes of inclusion and internationalization [6].

All relevant processes allowed for the formation of a stable model, which combines the effects of internationalization, inclusion, digitalisation, and digital inclusiveness, aimed at the improvement of the level of involvement and accessibility of higher education for a wide range of persons. These processes can solve a large part of the problems of higher education establishments in Russia and facilitate an increase in stable macroeconomic effects (Fig. 2).

The main problems that can be solved within such a model could be divided into three groups:

1. Reduction of motivation for learning, lack of talents, and exclusion of certain categories of people from the educational process in universities through the impossibility of creating convenient conditions for inclusive education;
2. Reduction of the number of students, lack of financial resources, and domination of 1–2 sources of financing;
3. Low accessibility of educational materials and their inconvenient form. While service in business and everyday activities is mostly automatized, university education remains mostly analogue.

The above problems of Russian universities have different levels of relevance and acuteness. Besides, their significance changes with time. Thus, for many universities, the problem of inclusive access to classrooms is being solved gradually. In turn, COVID-19 became a clear stimulus for the digitalisation of education, and internationalization provided access to the diversification of financing sources and an increase in the number of students. In these conditions, Russian universities gradually improve their position in the global market of higher education, which is manifested in the following macroeconomic effects:

- Growth of the intellectual capital of the country, reduction in the number of people who require special care from the public authorities due to receipt of higher education and better position in the labour market;
- Increase in the volume of exports in the Russian economy, growth of currency revenues, and ensuring soft cultural influence of Russian culture through foreign students;

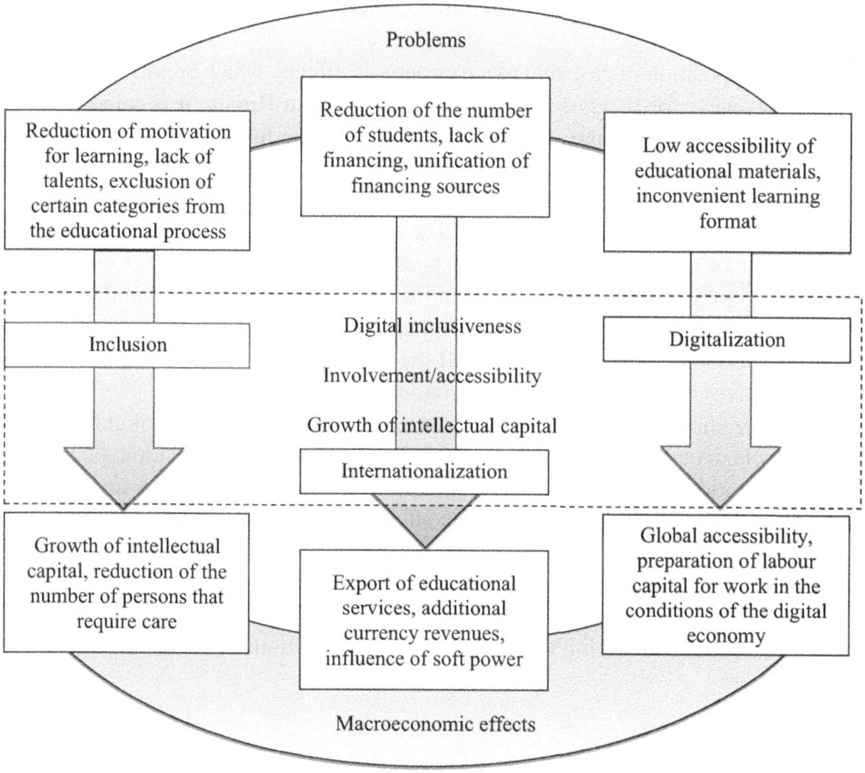

Fig. 2 The macroeconomic model of solving the problems of higher education in the conditions of digital inclusiveness in the context of the internationalization of universities. *Source* Developed by the authors

- Bringing the format of youth communication in accordance with the study format, raising the level of labour potential of the country due to students' preparation for professional activity in the conditions of the digital economy.

4 Quality of the Model of Higher Education's Digital Inclusiveness

The considered scientific problem has many open sources for discussion and further research. Thus, attention should be paid to the necessity of developing a joint conceptual treatment of the category of digital inclusiveness and the formation of the corresponding categorical apparatus, which would allow differentiating the notions of social and digital inclusion, their interaction, and attitude towards the problems of vulnerable groups of the population, as well as ensuring access and conditions for comprehensive use of the abilities of digital communications in the sphere of higher education.

From the position of ensuring macroeconomic effects, which are offered by digital inclusiveness for the system of higher education in Russia, it is important not only to assess the quantitative indicators of an increase in financial currency revenues, growth of the number of foreign students, and an increase in the number of people with special needs involved in higher education but also to develop the tools to measure the quality of these effects. The sign of quality in this context could be the number of really motivated students, who actively participate not only in the educational process but also in the scientific and project activities. Another important aspect is identifying the share of graduates who successfully occupied their niche in the labour market and fulfilled themselves as scholars, entrepreneurs, or representatives of other directions of economic activity.

However, the most debatable issue is an assessment of the long-term influence of digital inclusiveness, which is formed in the conditions of internationalization of higher education, on the further development of the Russian economy and its competitiveness in the global space. In this context, it is important to determine the factor dependencies, offer models and tools that would characterise the mechanism of such influence on macroeconomic processes and allow for measuring and designing of development of the macroeconomic situation through expansion of the abilities of internationalization of higher education in the conditions of an increase in digital inclusiveness.

5 Conclusion

Digital inclusion is an important element of the system of higher education, which has a substantial potential in the sphere of involvement of a much larger number of persons in educational services than can be considered from the traditional position. The influence of digital inclusion on Russian universities has a comprehensive character, for it is set onto the trends of digitalisation and internationalization of the Russian establishments of higher education and is supplemented by the focus on the needs of vulnerable groups of the population. According to this, digital inclusion can stimulate the resolution of many typical problems of Russian universities, which concern the conditions of the number of students, improvement of the financial position, and an increase in the level of students' motivation.

The resolution of these problems has a national character. This is implemented through the corresponding concepts and projects and is manifested in the form of many macroeconomic effects, in particular, in the improvement of the country's intellectual capital and decrease in the number of people requiring special care from the authorities, increase in the volume of currency revenues, improvement of proportions of foreign trade balance of the country, growth of the level of access of the wider group of people to quality higher education, and increase in the level of their motivation for self-fulfilment in the conditions of global digital space.

References

1. Sanz-Benito, I., Lázaro-Cantabrana, J. L., Grimalt-álvaro, C., & Usart-Rodríguez, M. (2023). Training and assessing competences in higher education: An experience on digital inclusion. *RIED-Revista Iberoamericana de Educacion a Distancia, 26*(2), 199–217. https://doi.org/10.5944/ried.26.2.35791
2. Soriano, A. J. Z. G., Montoro, M. A., & Colón, A. M. O. (2024). Digital teaching competence and educational inclusion in higher education. A systematic review. *Electronic Journal of e-Learning, 22*(1), 31–45. https://doi.org/10.34190/ejel.22.1.3139
3. Sudhan, P., & Jahira Parveen, S. (2025). Stress management in higher education: The role of on-line yoga and digital learning. *Journal of Lifestyle and SDG'S Review, 5*(1), e03889. https://doi.org/10.47172/2965-730X.SDGsReview.v5.n01.pe03889
4. Weng, X., Gu, M. M., Xia, Q., & Chiu, T. K. F. (2025). SWOT analysis of AI empowered entrepreneurship education: Insights from digital learners in higher education. *Thinking Skills and Creativity, 56*, 101763. https://doi.org/10.1016/j.tsc.2025.101763
5. Whitmore, C. (2025). Combating student disconnection in higher education using place-based pedagogy: Encouraging its use as a transformative learning practice in digital classrooms. *International Journal of Pedagogy and Curriculum, 32*(1), 1–14. https://doi.org/10.18848/2327-7963/CGP/v32i01/1-14
6. Akopiants, A. M., & Babitova, L. A. (2021). Internationalization of the higher education sector in the era of pandemic. *Servis plus [Service Plus], 15*(2), 137–143. Retrieved from https://cyberleninka.ru/article/n/internatsionalizatsiya-sektora-vysshey-shkoly-v-epohu-pandemii. Accessed December 18, 2023
7. Beregovaya, O. A., & Kudashov, V. I. (2019). Internationalization of higher education in the context of globalization. *Perspektivy nauki i obrazovania [Perspectives of Science and Education], 39*(3), 31–43. https://doi.org/10.32744/pse.2019.3.3
8. Pylneva, T. G. (2014). The role of globalization of socio-economic processes in the formation of international academic mobility. *Scientific Works of the Free Economic Society of Russia, 185*, 133–137. Retrieved from https://cyberleninka.ru/article/n/ol-globalizatsii-sotsialno-ekonomicheskih-protsessov-v-formirovanii-mezhdunarodnoy-akademicheskoy-mobilnosti. Accessed December 18, 2023
9. Riabinina, A. M. (2022). Technologies for internationalization of higher education in modern conditions. *Gosudarstvennoe upravlenie [Public Administration], 94*, 225–246. https://doi.org/10.24412/2070-1381-2022-94-225-246
10. Kostiukov, A. L. (2016). Features of the export of Russian educational services. *Kazachestvo [Cossacks], 7*(20) Retrieved from https://cyberleninka.ru/article/n/osobennosti-eksporta-rossiyskih-obrazovatelnyh-uslug. Accessed December 19, 2023
11. Mironova, S. V., & Timchenko, N. S. (2020). Export of higher education in Russia: A review of theoretical approaches and practical solutions. *Sotsiodinamika [Sociodynamics], 11*, 65–80. Retrieved from https://cyberleninka.ru/article/n/eksport-vysshego-obrazovaniya-v-rossii-obzor-teoreticheskih-podhodov-i-prakticheskih-resheniy. Accessed December 17, 2023
12. Golovina, T. A., Khorolskaia, T. E., & Musostov, Z. R. (2022). Policy of digital inclusion in the development of human capital. *Natural and Humanities Research, 43*(5), 77–83. https://doi.org/10.24412/2309-4788-2022-43-5-77-83
13. Kniazev, E. A., & Drantusova, N. V. (2013). Institutional dynamics in Russian higher education: Mechanisms and trajectories. *University Management: Practice and Analysis, 1*(83), 6–17. Retrieved from https://cyberleninka.ru/article/n/institutsionalnaya-dinamika-v-rossiyskom-vysshem-obrazovanii-mehanizmy-i-traektorii.
14. Kolomiitseva, O. V. (2022). Development of inclusive education: Problems and prospects. *Bulletin of RMAT, 1*, 67–73. Retrieved from https://cyberleninka.ru/article/n/razvitie-inklyuzivnogo-obrazovaniya-problemy-i-perspektivy. Accessed December 18, 2023
15. Trunova, E. Y., & Trunov, V. V. (2011). Macroeconomic mechanism for strategic management of the economy of sustainable development. *Bulletin of Volga University named*

after. V. N. Tatishcheva, 23, 134–142. Retrieved from https://cyberleninka.ru/article/n/makroekonomicheskiy-mehanizm-strategicheskogo-upravleniya-ekonomikoy-ustoychivogo-razvitiya. Accessed December 17, 2023

16. Smoliakov, D. A. (2019). On the genesis of the internationalization of higher education. *Obrazovanie i nauka [Education and Science], 21*(8), 9–28. Retrieved from https://cyberleninka.ru/article/n/o-genezise-internatsionalizatsii-vysshego-obrazovaniya. Accessed December 18, 2023

Prospects for Increasing the Attractiveness of Russian Education Based on the Internationalization of Universities

Elena G. Popkova and Shakhlo T. Ergasheva

Abstract We assessed the level of internationalization of Russian universities by different criteria, determined the influence of the indicators of internationalization on the attractiveness of universities and their competitiveness, identified the problem aspects of the considered process, and substantiated recommendations for their resolution. The study is based on the institutional approach, which includes the aspects of education, international cooperation, and dissemination of knowledge and ideas. The proposed methodological basis is supplemented with the theories and practices of the sphere of organizational change, state regulation, pedagogical sciences, etc. The main methods used in this study represent the general scientific domain, as well as the special methods of statistical research, such as the index method, average values method, method of determination of deviations, etc. This study considered the essence and components of the internationalization of universities, the state of the internationalization of Russian universities according to absolute and relative indicators, and compared with ranking indicators. According to this, we determined the problem aspects of their internationalization and outlined the prospects for their resolution and further development of Russian universities in the conditions of an increase in the globalization of education and science. We identified the problem of imbalance of the indicators of the internationalization в universities with the highest rankings, which involves the use of complex actions for improving the most problematic spheres of the activities. This concerns the development of faculties with international specialization and the improvement of the level of Internet presence of Russian universities in the global educational space.

E. G. Popkova (✉)
The Laboratory of Scientometrics and International Rankings,
Armenian State University of Economics, Yerevan, Armenia

S. T. Ergasheva
Tashkent State University of Economics, Tashkent, Uzbekistan
e-mail: sh.ergasheva@tsue.uz

B. S. Sergi et al. (eds.), *University 4.0 and Educational Technology for Sustainable Development*, Education in the Asia-Pacific Region: Issues, Concerns and Prospects 75, https://doi.org/10.1007/978-981-96-7376-6_35

Keywords Successfulness · Competitiveness · University · Higher education · Rankings · Internationalization · Russia

JEL Codes F02 · I23 · I25 · O15 · O19 · O31 · O38 · Z10 · Z18

1 Introduction

Globalization is one of the determining factors of development in the modern world. The freedom of movement, and international and cross-cultural exchange of thoughts and ideas, together with new forms of international cooperation with the use of digital technologies, form very favorable conditions for an increase in the internationalization of education and science in universities. Universities' ability to independently join international projects and participate in student and teacher exchange and joint scientific projects and other educational and scientific events is an important factor in the universities' development.

By its nature, the internationalization of universities is a bilateral process. It involves the integration of scientific, public, educational, and other initiatives and processes, inherent to foreign or global educational institutions, in the educational and scientific space of the educational establishment and the expansion of own experience outside of the country. Thus, symbiosis is the main factor of higher education's development. It allows accumulating different information and ideas, approbating and evaluating the experience of their application, coming up with new solutions, and, thus, receiving diverse alternatives for actions and creativity in the complex environment of education and science.

Russian universities traditionally belong to highly internationalized educational establishments. This is due to the traditions of Soviet education, in which citizens of Soviet republics studied at Russian universities. This approach was peculiar to citizens of the former Soviet bloc. A large number of such directions for internationalization remained. Its specific feature is the offer of Russian educational products. However, in the conditions of globalization, English-speaking programs and connections, which allow educational establishments to be better integrated into the global educational space, become more popular.

2 The Concept of Internationalization of Universities

The concept of internationalization of universities in the context of improvement of their attractiveness and competitiveness is within the range of the theory of institutionalization [1–3]. However, its use involves the application of a complex approach, which would combine theoretical provisions and practical aspects from the spheres of change management, state regulation, pedagogical sciences, etc. [4–6]. The information framework of this research is formed by ranking assessments and indices which reflect the achievements of universities by different categories. That is why the methodology of this research is comprised of general scientific methods of observation, analysis, synthesis, and generalization; and special methods of statistical analysis, which involve the determination of average indicators and standard deviation index, as well as comparison of the results obtained.

The key works used in this research have a direct relation to Russian universities and deal with a range of relevant issues, which include the following: assessment of the level of internationalization of Russian universities [7, 8]; search for ways to improve their competitiveness in the world [9]; integration of Russian universities in the system of international educational space [10, 11], including in new crisis conditions [12, 13]; consideration of internationalization of Russian universities from the position of international educational rankings [13]; study of the specifics of export of educational services by Russian universities in new conditions [14].

The goal of this research was as follows: identifying the trends and specific features of the internationalization of Russian universities in the context of an increase in the level of their attractiveness and competitiveness in the global educational space. The tasks to be solved were as follows: determining the essence and directions for internationalization of universities, identifying the state of the internationalization of Russian universities, and substantiating the prospects for further internationalization of universities of the Russian Federation.

3 Internationalization of Higher Education in Russia

Higher education belongs to the top priority spheres of state policy. According to this, the issue of its internationalization is also considered at the highest level. The Government of the Russian Federation sees the internationalization of higher education and science as one of the factors in non-raw exports. For this, there are programs for involving foreign students in studying in Russia, which involves the formation of the corresponding educational programs, improvement of the conditions of living and staying for foreign students, improvement of the brand of Russian higher educational institutions, and improvement of the level of the information policy on the activities of Russian universities [11].

Internationalization of higher education is a complex and multilevel phenomenon, which covers the processes of international interaction between universities, and countries – primarily within the Bologna educational process, and at the global level through the institutions of UNESCO, etc. (Fig. 1).

Internationalization of universities has many manifestations and forms. The most typical ones include the following:

- International academic mobility;
- Implementation of international standards in the educational process;
- Internationalization of educational courses;
- Programs of intercultural exchange, etc. [10].

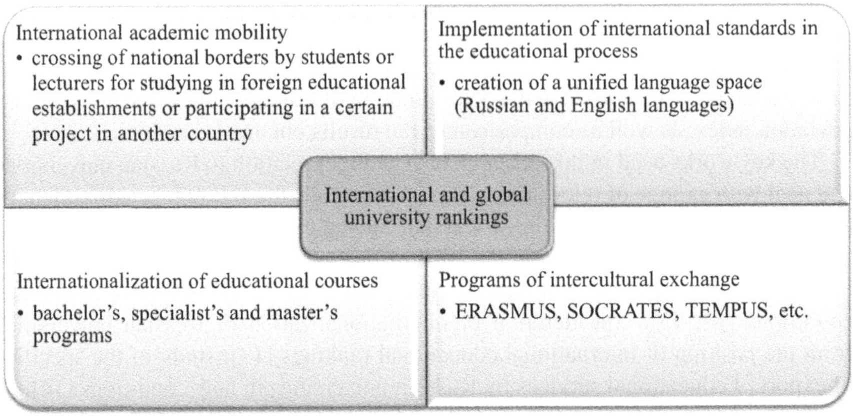

Fig. 1 Manifestations of internationalization of universities. *Source* Developed by the authors

The most popular practice of internationalization is academic mobility – crossing of national borders by students or lecturers for studying in foreign educational establishments or participating in a certain project in another country. From the position of academic mobility, a situation when students from countries with a lower level of income study in the universities of countries with a higher level of income is typical. Thus, Russian universities are peculiar for educational migration from the countries of Central and South-East Asia, as well as African countries. The number of Russian students who study in the USA and the EU is very large as well.

As of the 2021/2022 academic year, the share of foreign students in Russian universities (bachelor's, specialist's and master's programs) for all specialities was 7.9% (104,100 people), which is 3.6 times higher than in 2010/2011 academic year (37,300 people). As for regions, most foreign students were from the CIS countries, Baltic countries, and Georgia (74.8% in 2010/2011 and 57.8% in 2021/2022). Students from Asia accounted for 19.6% and 22.4%, accordingly, and students from Africa accounted for 3.7% and 8.2%, accordingly. Thus, the share of students from post-Soviet countries is decreasing, and the share of students from Africa and Asia is growing [15].

An important element of the internationalization of education is the creation of a unified language space. English is the language of international scientific communication in the world. The position of the Russian language as a language of international scientific communication is rather strong in post-Soviet territories of post-Soviet countries, with the increasing role of the English language in the educational and scientific circles of Russia.

International mobility rarely forms based on own initiative of the student or teacher. In many cases, migration takes place within government or international programs, which have the corresponding financing. The most well-known programs of this character are ERASMUS, SOCRATES, TEMPUS, etc. Due to targeted financing and strict bureaucratization of participation in such exchange projects, the necessary components of the international projects' participants are formed: knowledge of language, mastering of skills of international communication, experience in project management, etc.

The most popular methodology that is used to determine the level of internationalization and competitiveness of universities is international and global rankings. They allow assessing the successfulness of educational establishments by different criteria [13]. The methodology of the ranking QS on the formation of rankings of universities is based on the use of 10 factors, each of which has its weight coefficient for inclusion in the general ranking. The indicators of the internationalization of universities that are taken into account in the ranking are International Students Ratio, International Research Network, International Faculty Ratio, and Web Impact [16].

The first indicator characterizes the level of activity of the university in attracting foreign students. The second indicator characterizes international research activity and points to the diversity of research cooperation with other educational establishments from different countries. The third indicator takes into account the presence of faculties that offer international programs of studies and research. The indicator of web presence uses the data of Webometrics and characterizes the Internet presence of the university, which, in the current conditions, is one of the factors of international activity [17]. The top ten Russian universities from the ranking QS EECA by the criterion of the share of foreign students are presented in Table 1.

Despite the stereotype of the total domination of Moscow universities, there is no strict centralization of the indicator from the position of internationalization. In the ranking of the top ten universities by the criterion of the share of foreign students, only four universities are from Moscow. Other universities in the ranking are located in Tomsk (2), St. Petersburg, Kazan, Voronezh, and Vladivostok. It should be noted that these ten universities are also among the top 30 universities according to generalized indicators.

However, from this list, only Tomsk State University and Peter the Great St. Petersburg Polytechnic University have well-balanced indicators as to all criteria in the table. The indicator of standard deviation (σ) for these universities is 14.3 and 17.9. Most of the educational establishment of the sample is peculiar for the values of standard deviation within 21–32.

From the position of determination of the strongest and weakest positions by different indices of internationalization, we calculated average values for each of them. Given the criterion of ranking (International Students Ratio), the highest value is the value by this indicator ($\overline{x_1}$ =96.0). A rather high ($\overline{x_2}$ =76.5) value is observed in the average level of participation in the international research network. The rest of the average values are rather low and require improvement.

Apart from student mobility, an important factor of internationalization and improvement of the competitiveness of universities is inviting foreign lecturers for the implementation of international projects, support for the preparation of educational programs, etc. However, given the fact that high-quality foreign lecturers require the creation of special conditions and the corresponding wages, not all Russian universities can afford to invite them constantly [9]. However, such

Table 1 Ranking of the top 10 Russian universities by the criterion of the share of foreign students in 2021

Position in the general ranking of the country	University	City	Overall score	International students ratio	International research network	International faculty ratio	Web impact	Average deviation σ
				x_1	x_2	x_3	x_4	(x_1-x_4)
9	Tomsk State University	Tomsk	88.1	99.5	90.8	65.1	85.0	14.3
10	RUDN University	Moscow	71.8	99.4	97.6	36.2	80.5	26.7
12	National Research Nuclear University MEPhI (Moscow Engineering Physics Institute)	Moscow	70.6	98.7	48.7	94.6	84.6	21.3
13	Peter the Great St. Petersburg Polytechnic University	St. Petersburg	63.0	98.5	70.3	51.7	79.9	17.9
14	Kazan (Volga region) Federal University	Kazan	72.8	98.0	96.4	37.7	74.1	25.7
17	National Research Tomsk Polytechnic University	Tomsk	67.7	97.8	93.7	39	77.9	24.6
19	Voronezh State University of Forestry and Technologies named after G.F. Morozov	Voronezh	n/a	96.1	n/a	2.3	3.3	44.5
21	Synergy University	Moscow	n/a	94.6	n/a	25.8	5.0	41.0
28	Far Eastern Federal University	Vladivostok	38.2	89.7	65.0	53.3	17.4	31.7
30	The National University of Science and Technology MISIS	Moscow	60.8	87.6	49.4	44.7	34.7	21.9
Average value ($\overline{x_i}$)			66.6	96.0	76.5	45.0	54.2	–

Source Compiled by the author based on [17]

specialists could be invited to work in Russian universities on joint projects. That is why it is necessary to have a complex approach to the development of universities' internationalization [11].

From the position of the connection between internationalization and competitiveness of universities, it is possible to see a direct link. Not so much quantitative indicators as reputational factors have an important value for competitiveness [13]. In this context, main attention should be paid to the formation of the corresponding infrastructure and organizational and HR support for achieving and retaining high results.

4 Internationalization of Universities from the Position of the Prospects for Their Development

Conclusions that were obtained as a result of the conducted research have significant scientific value, but they need certain clarification, in particular, in the aspects lying in the sphere of a wide discussion. The main topic for discussion and research is the problem of creating a model of competitiveness and attractiveness of Russian universities based on a factor model, including factors characterizing their internationalization. The creation of such a model will allow revealing the most favorable and restraining aspects of the successfulness of universities in the international educational and scientific space and improve the results of their activities due to a more rational distribution of resources.

Internationalization of universities is one of the factors of access to international funds for financing of R&D, as well as databases, which enriches universities and raises the scientific level of researchers. Given the outlined problems of the research, it is necessary to study of the aspects of financing Russian establishments of higher education from international research funds. It is also necessary to clarify the issue of the impact of such financing on the level of success and internationalization of universities.

A current problem of the research on the internationalization of universities from the position of the prospects for their development is narrowing scientific tasks down to the assessment of the processes of educational knowledge export. A wider problem, which has not been disclosed properly in scientific literature, is the issue of attracting talents to the Russian labor market through the system of the internationalization of education. The solution to this problem lies in the context of a strategically important scientific task on the development of the intellectual capital of the country and support for its competitiveness in the conditions of a dynamic global environment [14]. It would also be necessary to discuss the prospects for improving the attractiveness of Russian universities in the new conditions. Apart from traditional factors, such assessment must include new conditions, limitations, and opportunities, which may consist in the reorientation of the direction for international cooperation, change of its focus, etc.

5 Conclusions

Taking into account globalization, the internationalization of universities is considered in the context of their success and competitiveness. Main attention is paid to academic mobility of students and lecturers, participation in international projects, presence of educational programs and conditions for foreign students. Russian universities have a large potential in the sphere of internationalization, given historical factors. According to this, the number and share of foreign students in Russian universities are growing. The main direction for the academic migration of foreign students in Russia is CIS countries, Baltic countries, and Georgia. However, the share of Asian and African countries grows constantly.

Evaluation of internationalization indices demonstrated that internationalization is observed not only with Moscow universities. Educational establishments from other cities also have some results in this regard. However, most of the universities have a weak balance of different indicators of internationalization. To form favorable prospects in the sphere of internationalization of universities it is important to implement complex programs that involve not only the invitation of foreign students but also the formation of the appropriate reputation of the university. Equally important aspects of successfulness of the university are active participation in international programs and attention to the development of the country's intellectual capital due to the search for talent among foreign students and their integration in the Russian labor market.

References

1. Rodero-Acosta, A. L., Barros-Valencia, J., & Huertas, Y. (2024). Characterization of cooperation and academic internationalization initiatives adopted by public universities from the Colombian Caribbean region in 2023. *Formacion Universitaria, 17*(2), 73–82. https://doi.org/10.4067/S0718-50062024000200073
2. Sisavath, S. (2024). Internationalization of higher education in Laos: Perspectives of university staff on rationales, challenges, and strategies for the post-COVID-19. *Higher Education Forum, 21*, 161–180. https://doi.org/10.15027/55013
3. Véliz, D., & Marshall, P. (2024). Internationalization and disciplinary differences: Tensions in the academic career in Chilean universities. *Higher Education Quarterly, 78*(3), 957–970. https://doi.org/10.1111/hequ.12501
4. Wang, J., Yao, H., & Liu, W. (2024). University presidents' international education experience and the internationalization of higher education institutions: A study of leadership in China. *Asia Pacific Journal of Education.* https://doi.org/10.1080/02188791.2024.2344215
5. Wang, X., & Shen, Y. (2024). Is internationalization abroad a prerequisite for university students' intercultural competence development? *Asian Education and Development Studies, 13*(3), 253–266. https://doi.org/10.1108/AEDS-02-2024-0040
6. Yang, X., & Li, T. (2024). Impacts of different internationalization modes on the global competence of Chinese university students. *ECNU Review of Education, 7*(4), 945–970. https://doi.org/10.1177/20965311241256363

7. Minaeva, E., & Taradina, L. (2022). Internationalization in Russian universities during the pandemic of COVID-19: Lessons for succeeding in the new reality. *Higher Education Quarterly, 76*(2), 293–310. https://doi.org/10.1111/hequ.12387
8. Rozhenkova, V., & Rust, V. (2018). Internationalization of higher education in Russia: Aiming for global recognition. In J. Zajda (Ed.), *Globalization and education reforms* (pp. 93–106). Springer. https://doi.org/10.1007/978-94-024-1204-8_6
9. Grishin, V. I., Shtykhno, D. A., & Shubenkova, E. V. (2019). Increasing the competitiveness of Russian universities on the world stage: Problems and solutions. *Bulletin of the Plekhanov Russian Economic University, 4*(106), 85–95. Retrieved from cyberleninka.ru/article/n/povyshenie-konkurentosposobnosti-rossiyskih-universitetov-na-mirovoy-arene-problemy-i-puti-resheniya. Accessed November 6, 2023.
10. Pankratova, M. E. (2011). Internationalization of Russian higher education as a process of integration of the international educational space. *Scientific Results of the Year: Achievements, Projects, Hypotheses, 1–2*, 295–309. Retrieved from https://cyberleninka.ru/article/n/internatsionalizatsiya-rossiyskogo-vysshego-obrazovaniya-kak-protsess-integratsii-mezhdunarodnogo-obrazovatelnogo-prostranstva. Accessed November 7, 2023
11. Riazantseva, I. V., Soboleva, E. L., & Matveeva, I. V. (2022). University internationalization as a key aspect of training competitive specialists. *Tvorchestvo i sovremennost [Creativity and Modernity], 2*(17), 51–59. Retrieved from https://cyberleninka.ru/article/n/internatsionalizatsiya-vuza-kak-klyuchevoy-aspekt-podgotovki-konkurentosposobnyh-spetsialistov. Accessed November 7, 2023
12. Petriakov, P. A., Pevzner, M. N., & Smertin, I. V. (2022). Internationalization of higher education in times of crisis: Searching for an optimal concept. *Chelovek i obrazovanie [A Human and Education], 2*(71), 40–52. Retrieved from https://cyberleninka.ru/article/n/internatsionalizatsiya-vysshego-obrazovaniya-v-krizisnye-vremena-poisk-optimalnoy-kontseptsii. Accessed November 7, 2023
13. Shakirova, I. A., Shchetnikova, I. L., Gariaeva, A. F., & Shugaepova, A. A. (2016). Analysis of the influence of internationalization indicators on the ratings of Russian universities. *University Management: Practice and Analysis, 6*(106), 111–119. Retrieved from https://cyberleninka.ru/article/n/analiz-vliyaniya-pokazateley-internatsionalizatsii-na-reytingovye-otsenki-rossiyskih-vuzov. Accessed November 6, 2023
14. Nikolaev, V. K. (2022). Exporting Russian higher education in the conditions of a new reality. *Vysshee obrazovanie v Rossii [Higher Education in Russia], 31*(2), 149–166. https://doi.org/10.31992/0869-3617-2022-31-2-149-166
15. Utevskaia, M. V. (2022). Internationalization of Russian education in modern conditions. *News of the St. Petersburg State Economic University, 6*(138), 169–173. Retrieved from https://cyberleninka.ru/article/n/internatsionalizatsiya-rossiyskogo-obrazovaniya-v-sovremennyh-usloviyah. Accessed November 6, 2023
16. QS Top Universities. (2021). *QS EECA rankings methodology*. Retrieved from https://www.topuniversities.com/eeca-rankings/methodology. Accessed November 6, 2023.
17. QS Top Universities. (2022). *QS EECA university rankings 2022*. Retrieved from https://www.topuniversities.com/university-rankings/eeca-rankings/2022. Accessed November 6, 2023.

Reputation Management of a Contemporary University in the Context of the 5-100 Project and the Priority 2030 Program

Alexandr N. Chumikov and Svetlana Yu. Chumikova

Abstract The relevance of studying the theory and practice of building the reputation of higher educational institutions is of an enduring nature due to the fixed cyclical nature of their work process and the high degree of specificity of the results obtained. In the second and third decades of the twenty-first century, the rethinking and development of this activity were significantly influenced by the state project "5-100" and the "Priority 2030" program. However, the formal indicators of their effectiveness included in the project and program do not exhaust the list of positions that make up the content of an attractive image of a higher education institution but are the result of their implementation. To update this list, the research analyzes the conceptual basis for building the image, reputation, and brand of an organization – a subject of socio-economic relations). The authors conduct qualitative research on the reputation management of three leading Russian universities: National Research Tomsk State University (TSU), Peoples' Friendship University of Russia (RUDN), and National Research South Ural State University (SUSU). The performance indicators of their participation in these projects are examined in dynamics. The authors analyzed the activities to develop the identification positions of universities in the public space, changing or correcting their semantic messages as guidelines for meaningful activity. Finally, the authors provide recommendations for optimizing theoretical approaches and applied components of the image direction of work at a university.

Keywords Image · Reputation · Brand · Identification · Semantic content · Media rating

JEL Codes I21 · I23 · I28

A. N. Chumikov (✉)
Moscow State Linguistic University, Moscow, Russia
e-mail: chumikov@pr-club.com

S. Y. Chumikova
RUDN University, Moscow, Russia

B. S. Sergi et al. (eds.), *University 4.0 and Educational Technology for Sustainable Development*, Education in the Asia-Pacific Region: Issues, Concerns and Prospects 75, https://doi.org/10.1007/978-981-96-7376-6_36

1 Introduction

The task of shaping an attractive image for higher education institutions has been present throughout their entire history. However, the strategic initiatives in the field of higher education undertaken by the government over the past 20 years have given a new impetus and character to this endeavor. This research focuses on the project aimed at enhancing the competitiveness of Russian universities among the world's leading scientific and educational centers, "5-100" (2013; 21 universities) and the Strategic Academic Leadership Program "Priority-2030" (2021–106 universities). The successful participation in the "5-100" project was measured by achieving high positions in the leading international university rankings such as Quacquarelli Symonds (QS), Times Higher Education (THE) World University Rankings, and Academic Ranking of World Universities (ARWU). This success was attributed to the implementation of indicators such as academic reputation, the quantity and ratio of faculty to students, citation index of academic publications by university representatives, university reputation among employers, and others.

The goal of having five Russian universities in the top 100 of these rankings by the project's completion in 2020 was not achieved. Nevertheless, some experts viewed the project's results as successful, pointing to the improvement in the positions of Russian universities in subject-specific rankings. Others were more skeptical about the rankings, highlighting the growing bureaucratization and leaders' focus on formally defined indicators [1].

The requirements for participation in the "Priority-2030" program are of a more global and comprehensive nature. As researchers note, the program contains a more substantial humanitarian component compared to the "5-100" project, with a shift in focus from quantitative to qualitative indicators [2]. These indicators include the creation of a motivating scientific and educational environment, an integrated entrepreneurial culture, and efforts to strengthen the role of the university as a driver of regional development [3].

2 University Reputation

There are numerous approaches to constructing and analyzing the reputation of a subject in socio-economic relations, including a university [4, 5]. The authors propose a set of dynamically interconnected positions – image, reputation, and brand – with a clear definition of distinct roles for each. Image is the declared position formulated by the subject in the form of a package of visual solutions and meaningful messages, serving the function of presenting (positioning) the university to target groups in the surrounding space [6]. Reputation is interpreted as a perceived position firmly established in the consciousness of target audiences [7]. Brand is the highest level of perception of the declared position when it is quickly and emotionally grasped by the consciousness and maximally aligns with the promoted image.

In this context, the authors position the image to achieve a high reputation and brand; the list of specified operations is referred to as reputation management ([8], pp. 11–13).

In each cycle of this process, along with the permanently present issues, there are situational problems of perception. Typically, a university presents rational propositions to its audience [9]. However, according to the principles of 4D branding theory, the rational part of the motivating image message is just one component of the brand. T. Gad, the author of this theory, calls it the "functional dimension," describing the product's practical utility, in our case, the education received at the university. However, there are the following three equally important dimensions [10]:

- The social dimension allows the consumer to identify with a specific societal group;
- The mental dimension forms new values;
- The spiritual dimension allows for a complex of personal positive sensations

As the volume of the information field containing similar rational propositions constantly increases, the significance of functional messages for the consumer decreases and is replaced by the other three dimensions of the brand.

The declared and promoted image of a university encompasses various components, as outlined by M. V. Gladkova: the image of consumers of educational services, the image of the university's faculty, the image of the rector, internal image in the form of a favorable emotional-psychological climate within the collective; visual image, social image as measures to support students, and business image, implying the business reputation of the university in the region, country, and internationally [11].

This list can be expanded further while maintaining the situational task of determining the hierarchy of reputational priorities and subsequent proportional attention to each of them. For instance, T. Vokueva, in a study among students and faculty of the Financial University under the Government of the Russian Federation, found that the main motivational assets for choosing a university for them were as follows [12]:

- Fame and prestige (brand) – 77.6%;
- Opportunities for successful employment – 63.1%;
- Availability of budget places and passing score – 61.6%;
- The state status of the university and high quality of education – 50.4%;
- Professionalism of teachers – 50.4%.

Accordingly, such positions should be established and promoted primarily, aligning well with the objectives of the "5-100" project and the "Priority-2030" program.

In the subsequent research, the experience of reputation management is examined in the project dynamics of the National Research Tomsk State University (TSU), the Peoples' Friendship University of Russia (RUDN), and the National Research South Ural State University (SUSU).

3 Image Policy of Russian Universities in Connection with the "5-100" Project

If we talk about the results of the image policy of these universities in connection with the "5-100" project, they look as follows (Table 1).

Since the interim results of the universities' participation in the "Priority-2030" program were not available at the time of writing the research, let us refer to the authoritative Russian university ranking TOP-100 by the analytical agency RAEX. In the 2022 ranking, TSU was in 18th place, RUDN was in 19th place, and SUSU was in 56th place [14].

The presented numerical indicators allow us to conclude that the analyzed universities demonstrate positive dynamics in their development year after year and have a high reputation in the international and Russian scientific and educational community. It is essential to note that all three universities were already established regional and federal brands by 2013. To promote their brands towards international recognition, each of them underwent rebranding – updating the visual identity standards for the university and related semantic components that regulate the rules for its further positioning (Fig. 1).

TSU engaged at least three professional agencies in the implementation of the rebranding program. The company "Mitrofanov & Partners" conducted strategic sessions to develop the identification positions (brand code) of TSU with faculty and students. Based on the materials provided, the agency "Lovemedo" developed a new visual image for the university and presented it to the international council, the rectorate, and the administrative council of TSU. It also conducted open presentations for a wide audience. In 2015, the new visual symbolism was implemented on the TSU website and started operating in the external communication space [15].

The agency "International Press Club. Chumikov PR and Consulting" was responsible for shaping the history-legend of TSU, implementing storytelling, the

Table 1 Dynamics of TSU, RUDN, and SUSU in THE, QS, and ARWU world rankings in 2016–2020 (excluding subject rankings)

	2016	2017	2018	2019	2020
TSU (in the "5-100" project since 2013)	THE 601–800 QS 481–490 ARWU –	THE 601–800 QS 377 ARWU –	THE 601–800 QS 323 ARWU –	THE 601–800 QS 277 ARWU –	THE 601–800 QS 322 ARWU 801–900
RUDN (in the "5-100" project since 2015)	THE – QS 601–650 ARWU –	THE 801+ QS 601–650 ARWU –	THE 1001+ QS 491–500 ARWU –	THE 1001+ QS 412 ARWU –	THE 1001+ QS 392 ARWU –
SUSU (in the "5-100" project since 2015)	THE – QS – ARWU –	THE – QS – ARWU –	THE – QS – ARWU –	THE – QS 801–100 ARWU –	THE – QS 801–100 ARWU –

Source Compiled by the authors based on [13]

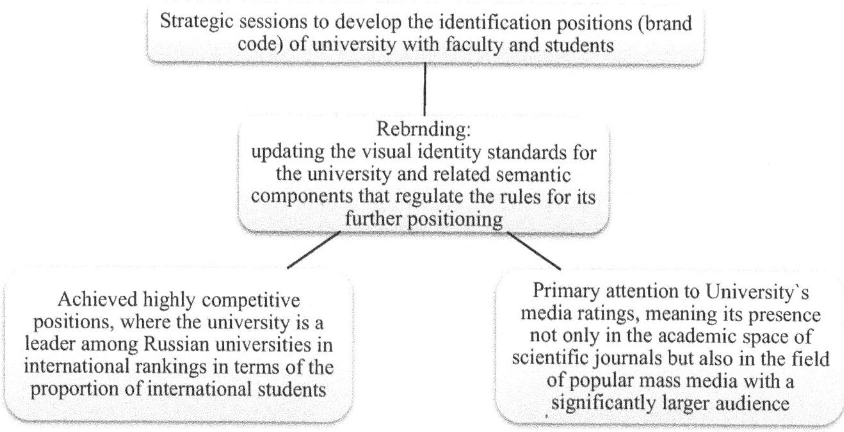

Fig. 1 Image policy of Russian universities in connection with the "5-100" project. *Source* Compiled by the authors

highly demanded media trend – the production of stories of social and economic entities interesting to the mass audience. The agency also developed the TSU brand atlas – a document containing a list of brand-forming thematic directions and associated commodity and personal brand objects for positioning the university in the information space.

The creative efforts of RUDN in the field of rebranding in 2017 were even more extensive. A comprehensive section called "Corporate Culture" appeared on the university's website, containing a full-fledged 117-page brand book. Several other sections were also developed, including the university's history, traditions, and values, the RUDN museum, and more [16]. The company "Oblik" designed and implemented a project for the university's entrance group, which includes seven art objects with sculptures of students from various nationalities and cultures, along with a light sphere featuring motivational phrases in six official languages of the UN [17]. The architectural bureau "Haast" implemented a project for the interior design of the main building of RUDN, showcasing the traditions and atmosphere of the university [18].

Rebranding with a radical change in the components of the corporate style was carried out in 2017 by the SUSU: on the university's logo, graphic images of a griffin symbolizing Europe and a lion as a symbol of Asia replaced the previous student symbol [19].

However, it is much more important to analyze the detailed fixation, change, or correction of semantic messages – the dynamic guides of the substantive activities of the universities. The positioning concept "Eurasian University of World Class" has been approved for TSU. The new slogan for SUSU is "SMART University Uniting Europe and Asia." The aspirations to occupy the Eurasian niche are similar. However, in one case, it is conveyed by a Siberian university. On the other, it stems from a Ural university. The semantic niche of RUDN is more unique and is based on the achieved highly competitive positions, where the university is a leader among Russian universities in international rankings in terms of the proportion of international students; in the second decade of the twenty-first century, young people from 154 countries studied at RUDN. The updated mission of the university is "Uniting people of different cultures by knowledge, RUDN University creates leaders to make the world better," and the slogan is "Discover the World in One University!"

All three universities paid primary attention to their media ratings, meaning their presence not only in the academic space of scientific journals but also in the field of popular mass media with a significantly larger audience. The results of their presence in the media were impressive: RUDN took 8th place, TSU – 18th, and SUSU – 28th in the media rating "TOP-30 Russian Universities" by the company "Medialogia" for 2022 [20].

4 Contradictory Nature of Global Rankings as a Tool for Assessing the Universities Competitiveness

Yu. N. Ebzeeva and Yu. B. Smirnova reflect on the contentious methodology of rankings in the sphere of higher education. The debated aspects include the selection of indicators included in the ranking system, challenges in aggregating indicators into a unified index, and others. When it comes to ranking universities from different countries, the evaluation becomes even more complex [21].

O. Kushneva, I. Rudskaya, and N. Fersman point out the inaccuracy of the sociological tools used in global rankings, preventing claims of the representativeness of the research. Considering that the assessment of academic reputation is based on surveys of scholars from various scientific fields, with over 62,000 scholars worldwide participating in the 2013 survey, an apparent question arises: "Where do respondents learn about conducted scientific research in general and about their level in particular?" [22].

The authors of this research and other researchers note the contradictory nature of global rankings as a tool for assessing the competitiveness of Russian universities and, in our context, as a leading factor in their reputation management. First, the perception of the three international university rankings as the main benchmarks by the scientific and educational community is the subject and outcome of the reputational activity of interested initiators: a specially created ranking company (QS), a journal (THE), and a university (ARWU). Second, as demonstrated by the experience of 2022–2023, the significance and possibility of achieving positions outlined in global rankings depend significantly on the current moment's dynamics. Third, the task of entering the list of leading universities in Russia, the international region, and the world is currently set and implemented by Russian rating structures, the most influential of which are the Moscow International University Ranking "Three University Missions" (MosIUR) (154 universities and 459 universities in the MosIUR ranking family in 2023) and the ranking of the best universities "RAEX-100" (212 universities in 2023) as its integral part.

5 Conclusion

Thus, the reputation management of a university implies a continuous and comprehensive system of actions aimed at creating an image as a declared position for presentation in external and internal environments. Subsequently, it involves promoting this image to gain a positive reputation as a perceived position and building a brand as the highest and unconditional level of positive visibility for the entity on rational and emotional levels.

This universal scheme requires specification and specialization in the form of defining the hierarchy of target groups in society on which reputation impact will be transmitted, as well as the preferred instruments of this impact. Finally, it is crucial to consider the context existing at the time of launching reputation management technologies (i.e., the state of socio-political and socio-economic relations in the world, the country, and the relevant segment of regional and industry activities).

In our case, the initial context was the "5-100" project, which served as the basis for understanding the priority segments of influence and significant updating of visual and meaningful orientations (rebranding) for three universities, the experience of which is analyzed in this article. The "Priority-2030" program and the Moscow International University Ranking "Three University Missions" (MosIUR) became subsequent motivational factors for adjusting the positions previously formed by the universities.

The analysis of reputation activities undertaken by the Russian Peoples' Friendship University, the National Research Tomsk State University, and the South Ural State University in 2013–2023 within the framework of the "5-100" project and the "Priority-2030" program makes it possible to evaluate them as successful and potentially applicable by other universities as instrumental models.

References

1. Dezhina, I. G., & Sorokin, A. N. (2022). The perception of the Project 5-100 by university employees. *Mir Rossii [Universe of Russia], 31*(1), 74–90. https://doi.org/10.17323/1811-038X-2022-31-1-74-90
2. Siraeva, M. N. (2023). Humanitarian potential of the Federal Academic Leadership Programme "Priority 2030." Modern High Technologies, 3, 101–107. DOI: https://doi.org/10.17513/snt.39565.
3. Shtykhno, D. A., Kulapov, M. N., Maslennikov, V. V., Kalinina, I. A., & Karasev, P. A. (2022). Transformation of university structure in the aspect of participation in the program of strategic academic leadership 'Priority-2030'. *Vestnik of the Plekhanov Russian University of Economics, 19*(3), 145–157. https://doi.org/10.21686/2413-2829-2022-3-145-157
4. Rizard, S. R., Waluyo, B., & Jaswir, I. (2023). Impact of brand equity and service quality on the reputation of universities and students' intention to choose them: The case of IIUM and UIN. *F1000Research, 11*, 1412. https://doi.org/10.12688/f1000research.122386.2
5. Shah, S. A., Shoukat, M. H., Ahmad, M. S., & Khan, B. (2024). Role of social media technologies and customer relationship management capabilities 2.0 in creating customer loyalty and university reputation. *Journal of Marketing for Higher Education, 34*(1), 344–367. https://doi.org/10.1080/08841241.2021.1991072

6. Soysal, Y. N., Baltaru, R. D., & Cebolla-Boado, H. (2024). Meritocracy or reputation? The role of rankings in the sorting of international students across universities. *Globalisation, Societies and Education, 22*(2), 252–263. https://doi.org/10.1080/14767724.2022.2070131
7. Sun, Z. (2024). Reactive management of university reputation during the COVID crisis: Governance and the control of public narratives in Chinese higher education. *Higher Education Policy*. https://doi.org/10.1057/s41307-023-00338-9
8. Chumikov, A. N. (2016). *Advertising and public relations: Image, reputation, brand: Textbook* (2nd ed.). Aspect-Press.
9. Zhang, H., Khaskheli, A., Raza, S. A., & Masood, A. (2024). Linkage between students' skills and employability: Moderating influence of university reputation. *Corporate Reputation Review, 27*(4), 229–248. https://doi.org/10.1057/s41299-023-00169-9
10. Gad, T. (2001). *4-D branding: Cracking the corporate code of the network economy*. Pearson Education.
11. Gladkova, M. V. (2022). The image of the university in the views of students. *Economics: Yesterday, Today and Tomorrow, 12*(9A), 443–452.
12. Vokueva, T. S. (2022). Modern Russian university image in the mirror of opinions of lecturers and students (on the example of the Financial University). *Humanities and Social Sciences. Bulletin of the Financial University, 12*(4), 128–132. https://doi.org/10.26794/2226-7867-2022-12-4-128-132
13. Forbes Education. (2020, July 31). *Project 5-100: Program Results*. Retrieved from https://education.forbes.ru/authors/5-100-experts?ysclid=ll1722409v563454058. Accessed September 1, 2023.
14. RAEX Rating Review. (2023, June 7). *RAEX-100 rating of Russia's best universities? 2023*.Retrieved from https://raex-rr.com/education/russian_universities/top-100_universities/2023/?ysclid=ln5om4ugrx100847321. Accessed September 1, 2023.
15. National Research Tomsk State University. (n.d.). *TSU symbolism*. Retrieved from https://www.tsu.ru/help/symbolism.php?sphrase_id=420482. Accessed September 1, 2023.
16. RUDN University. (n.d.). *Corporate culture*. Retrieved from https://www.rudn.ru/about/corporate-culture. Accessed September 1, 2023.
17. Oblik. (2023, April 7). RUDN Project. Retrieved from https://obliks.ru/proekt-rudn/?ysclid=lkvia7u4g6391609849. Accessed September 1, 2023.
18. HAAST. (2023). *Design project of the main building of PFUR*. Retrieved from https://haast.ru/projects/dizajn-proekt-glavnogo-korpusa-rudn/. Accessed September 1, 2023
19. South Ural State University (National Research University) (SUSU). (2023). *SUSU Brand*. Retrieved from https://www.susu.ru/ru/university/our-susu/brend. Accessed September 1, 2023.
20. Medialogia. (2023, January 25). Media rating of Russian higher education institutions for 2022. Retrieved from https://www.mlg.ru/ratings/society/education/11814/?ysclid=lkuxjhtrdt250286929. Accessed September 1, 2023.
21. Ebzeeva, Y. N., & Smirnova, Y. B. (2022). Positions of the Russian universities in the world university rankings in 2022. *RUDN Journal of Sociology, 22*(4), 909–918. https://doi.org/10.22363/2313-2272-2022-22-4-909-918
22. Kushneva, O. A., Rudskaya, I. A., & Fersman, N. G. (2014). World University Ranking and the "5-100-2020" program of the Ministry of Education and Science of The Russian Federation as a way to improve the competitiveness of Russian universities. *Society. Environment. Development (Terra Humana), 2*, 17–26. Retrieved from https://terrahumana.ru/arhiv/14_02/14_02_03.pdf. Accessed September 1, 2023

Analysis of the Effectiveness of Channels for Promoting Educational Services in the Tasks of Training Personnel for the Communication Industry: A Case Study of the Educational Program "Advertising and Public Relations" at the Institute of World Economy and Business of RUDN

Olga P. Malygina and Ekaterina A. Arkhangelskaya

Abstract The research examines the effectiveness of channels for promoting educational services in the tasks of preparing personnel for the communication industry. The high level of professional requirements for graduates of relevant specialties dictates the need to enhance the attractiveness of these educational programs, enabling a more rigorous selection of applicants for these specialties through increased competition. Alongside developing high-quality academic programs and providing a highly professional teaching staff, this approach will facilitate the production of qualified professionals meeting the demands of employers. Active measures are required to address the considered task and promote the university's educational services. As an example, the research explores the promotion practices of the educational program "Advertising and Public Relations" (APR) at the Institute of World Economy and Business (IWEB) of the Peoples' Friendship University of Russia (RUDN). Survey data from first-year students are presented to identify factors influencing the choice of the educational program. Comparing these factors with the statistics on the program's mentions in the Brand Analytics system and analyzing the structure of these factors allows the authors to conclude about the effectiveness of promotional channels for the educational program on the Internet. The analysis of the statistics of the APR program community "Department of Advertising and Business Communications IWEB RUDN" confirms the hypothesis about the significance of resources on the VKontakte (VK) social network for pro-

O. P. Malygina (✉) · E. A. Arkhangelskaya
Department of Advertising and Business Communications, Institute of World Economy and Business, Faculty of Economics, RUDN University, Moscow, Russia

B. S. Sergi et al. (eds.), *University 4.0 and Educational Technology for Sustainable Development*, Education in the Asia-Pacific Region: Issues, Concerns and Prospects 75, https://doi.org/10.1007/978-981-96-7376-6_37

spective students. The research formulates key conclusions and recommendations for promoting the educational services of universities.

Keywords Educational programs · Communication marketplace · Advertising · Public relations · Social media

JEL Codes M31, M37, M53

1 Introduction

The development of the market for goods and services and, consequently, the increase in competition necessitate the thoughtful positioning and communication of key attributes to the target audience [1]. This process is complicated by the constant modernization of communication channels.

The communication market represents economic relationships between companies, clients of communication services, and agencies as executors to achieve the business goals of the former [2].

The market for communication agencies is dynamic and competitive because companies and organizations always require effective communication strategies and solutions to achieve their goals. Communication agencies provide expertise and resources to assist clients in managing their image and audience relationships [1].

As elements of communication, one can identify advertising, public relations, direct marketing, event marketing, and sales promotion [3].

Clients in the communication market can include corporations, non-profit organizations, government institutions, and even individuals [4].

Universities address the preparation of such in-demand professionals for the communication market within the framework of the educational program "Advertising and Public Relations." The guarantee of quality personnel for the industry undoubtedly lies in the curriculum of the specialty, meeting high standards for student preparation, advanced equipment in classrooms, and the professional level of the teaching staff, which includes practitioners from the industry. However, the quality and level of knowledge of applicants entering this field also become crucial factors. To address this issue, it is advisable to enhance the attractiveness of this educational program, thereby increasing competition and enabling the selection of applicants with higher scores on standardized exams. Thus, the task of improving the effectiveness of promoting this educational program becomes relevant.

The research aims to evaluate the qualitative effectiveness of channels for promoting educational services in the tasks of preparing personnel for the communication industry. As an example, the research considers the practice of promoting the educational program "Advertising and Public Relations" (APR) at the Institute of World Economy and Business (IWEB) of the Peoples' Friendship University of Russia (RUDN).

2 The List of Knowledge and Skills Required by Employers from University Graduates in the Field of "Advertising and Public Relations"

The year 2022 was challenging for the industry due to various external factors. For instance, the Internet advertising segment faced the departure of foreign advertisers and a reduction in advertising tools due to blocking certain platforms.

The interests of the communication market in Russia are represented by the Association of Communication Agencies of Russia (AKAR), which annually announces an assessment of market volumes. Despite the challenges of 2022, the association's experts note that the crisis has been overcome. According to data for the first half of 2023, the advertising market is at its peak in its entire history, reaching 315 billion rubles [5]. This figure indicates a growth of +27% compared to the first half of 2022 and +19% compared to the same period in 2021. The quarterly dynamics of +50% in 2023 make it possible to forecast successful results for the entire year for the industry.

With the development of the communication market, there will be an increasing demand for industry professionals. In June 2023, more than 137 thousand vacancies related to the fields of marketing, advertising, and PR were posted on the hh.ru platform [6]. The most sought-after positions were marketing managers and internet marketers (over 10 thousand vacancies), specialists in social media promotion and content managers (3.4 thousand vacancies), copywriters (3 thousand vacancies), and PR managers (1.2 thousand vacancies). Having experience was not mandatory for employers for 29.2% of the job offers. Thus, prospects for starting a career open up for students and entry-level professionals. The presence of vacancies with remote work formats (almost 25 thousand) can also be a positive aspect for students because it provides the opportunity to combine work with education.

According to a study on personnel in the communication industry conducted by the agency "Chief Advisor," "Interium," and the Association of Managers [7], 63% of respondent agencies engage in internships for students and young professionals. For example, during the summer of 2023, 19 students underwent internships at the AMDG agency. Of these, six third-year bachelor students received offers for further part-time work at the agency while continuing their studies.

Communication agencies set diverse requirements for personnel depending on the specifics of their activities, clients, and projects. However, there are common skills, qualities, and characteristics that employers often value and require in the industry, including the following:

1. Professional skills: knowledge and experience in the fields of marketing, PR, advertising, digital marketing, and other related areas. It is important to have skills and experience relevant to the specific position (e.g., writing press releases, managing social media, media planning, etc.);
2. Creativity and innovative thinking: the ability to come up with unconventional ideas to attract attention to clients and their products or services;

3. Analytical abilities: the ability to conduct research, analyze data, and measure the effectiveness of communication strategies;
4. Media and technology skills: knowledge of relevant media platforms, social media, and digital tools, as well as the ability to adapt to a rapidly changing technological environment;
5. Communication skills: excellent written and verbal communication skills, the ability to interact effectively with clients, colleagues, and media representatives;
6. Organizational abilities: the ability to manage multiple tasks and projects, meet deadlines, and be organized;
7. Team orientation: the ability to work in a team, collaborate, and coordinate efforts with colleagues;
8. Client orientation: understanding the needs and goals of clients, the ability to create strategies that meet these needs;
9. Ability to work under pressure: communication agencies often face urgent tasks and crisis management situations, and the ability to work effectively under pressure is crucial;
10. Industry-specific knowledge: for some projects, specialized knowledge in a particular industry or the client's business sector (e.g., pharmaceuticals, automotive, real estate) may be required.

Thus, the list of knowledge and skills required by employers from university graduates in the field of "Advertising and Public Relations" is quite extensive. Bachelor's degree holders should possess fundamental knowledge and practical skills in the professional field.

The chosen research method is regression analysis aimed at identifying the relationship between the dependent variable (mentions through channels) and one or more independent variables (decision-making channels). This approach helps uncover factors influencing the indicator of the dependent variable. For instance, it elucidates which factors influenced the choice of the APR program at the IWEB of RUDN. The outcome of the regression analysis makes it possible to determine the strategy for adjusting the channels for promoting educational services for the selected field of study. In this case, a simple linear regression was employed to address the task.

3 Influence of the RUDN APR Program Mentioning by the Categories of Sites on the Choice of this Educational Program by the Applicants

According to Brand Analytics [8], from March 31 to August 31, 2023, the system analyzed 19,492 messages and 13,035 authors.

The configuration for collecting the statistics on mentions was as follows:

- The program "Advertising and Public Relations" (APR), related fields, specialized disciplines, and professions;
- Extensive mentions of intentions to enroll in the university.

The following data were obtained for the APR program at the IWEB of RUDN (Table 1).

As a result of the admissions campaign in 2023 for the "Advertising and Public Relations" (APR) program at the Institute of World Economy and Business (IWEB) of RUDN, 3523 applications were submitted, of which 8.12% were admitted to the first year.

In September 2023, a survey was conducted among first-year students who entered the APR program at the IWEB. Students were asked a series of questions regarding the choice of educational direction, primary sources of information on the Internet and offline, and the main factors influencing their final choice. A total of 184 surveys were processed.

The analysis of respondent surveys revealed that websites for applicants (aggregators) and pages of educational programs are the primary sources of information about the training program. Additionally, the surveys mentioned communities of educational programs and universities on the VKontakte social network, reviews on the Internet and the university's website, and live broadcasts on aggregator websites. Interestingly, the total number of mentions of reviews is comparable to that of educational program pages. Furthermore, the surveys highlighted factors that influenced the choice of the APR educational program (Table 2).

To assess the impact of the mentions of the APR program at RUDN among individuals expressing an intention to enroll, simple linear regression was employed. The calculation results indicated a weak positive correlation between the number of mentions (X) according to the Brand Analytics system across main platform categories and visits (Y) to these resources based on survey data.

Table 1 Mentions statistics according to the Brand Analytics system

Platform	Number of mentions	Platform's share in the total number of mentions
vk.com	10,602	54%
telegram.org	5122	26%
ok.ru	1073	6%
dzen.ru	763	4%
youtube.com	476	2%
facebook.com	441	2%
tabiturient.ru	321	2%
vuzopedia.ru	279	1%
ucheba.ru	138	1%
Other	277	1%
Total	19,492	

Source: Compiled by the authors based on data from the Brand Analytics system [8] for the period from March 31 to August 31, 2023

Table 2 Statistics on mentions of factors influencing the choice of the educational program based on surveys of first-year students

Factors influencing the choice of the educational program	Share of mentions in the questionnaire
Prestige of the future profession	68%
Opportunity to obtain a translator's diploma	66%
University's ranking	61%
High level of student preparation	57%
Contemporary educational environment	53%
Opportunity to interact with diverse cultures	53%
Profile of standardized exams (Unified State Exam)	45%
Active student life	44%
Academic disciplines	34%
Advice from current students	26%
Proximity to home	18%
Friends or relatives studying in the faculty	15%
Availability of dormitory	14%
Tuition fee	10%
Other	8%

Source: Developed by the authors based on the processing of questionnaires (survey among first-year students enrolled in the APR program at the IWEB of RUDN, September 2023)

The linear regression equation is $\hat{Y} = 41.3392 + 0.0002748X$. The correlation (R) is 0.00564. This implies a very weak positive correlation between X and Y.

Factors influencing the choice of the APR program can be divided into the following two groups based on the category of the information source:

1. Information from aggregator websites for applicants, educational program websites, and social media platforms;
2. Reviews about the university and educational program on the Internet.

Factors falling into the information category include the following:

- RUDN's ranking;
- Opportunity to obtain a translator's diploma;
- Tuition fee;
- Availability of dormitory;
- Profile of Unified State Exam;
- Academic disciplines.

Other factors (high level of student preparation, contemporary educational environment, active student life, the prestige of the future profession, etc.) relate to online reviews and opinions of friends and acquaintances who study or have graduated from the university. The contribution of these factors is influenced by the overall information space formed by online and offline communications. Open Days at RUDN play a significant role in this regard.

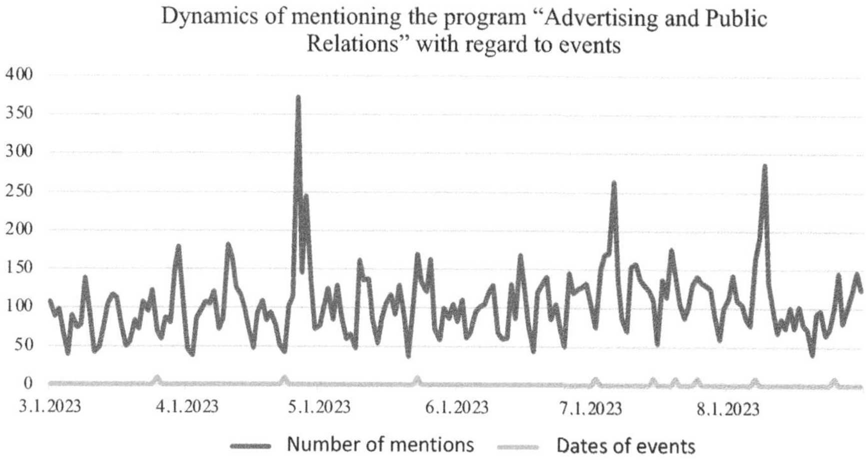

Fig. 1 Dynamics of mentions of the "Advertising and Public Relations" (APR) program by days from March 31 to August 31, 2023, including events. *Source:* Compiled by the authors based on Brand Analytics data [8]

Overlaying the graph of conducted events on the daily mention statistics of the research subject (according to Brand Analytics data) makes this influence more evident. Peak values on the graph coincide with offline events at RUDN that impact the interests of prospective students: open days, document submission deadlines, etc. (Fig. 1).

This approach allows us to change the meaningful side of the interpretation of the assessment of the dependence of the influence of the RUDN APR program mentioning by the categories of sites on the choice of this educational program by the applicants who entered the first year of the university.

4 Significance of VK Resources for Prospective Students of the APR Program at the IWEB of RUDN

The main reason for the zero correlation in calculating the linear regression equation is the high frequency of mentions on the VKontakte (VK) social network. This platform simultaneously serves as a source of information about the educational program and a platform for feedback from students and graduates of the APR program at the IWEB of RUDN.

The structure of the VK community associated with the Advertising and Public Relations program is quite diverse. Communities (e.g., "RUDN Applicants," the "Economics Faculty and IWEB of RUDN," the "Student Committee of IWEB official," "My Advertising and PR at IWEB RUDN," and the "Department of Advertising

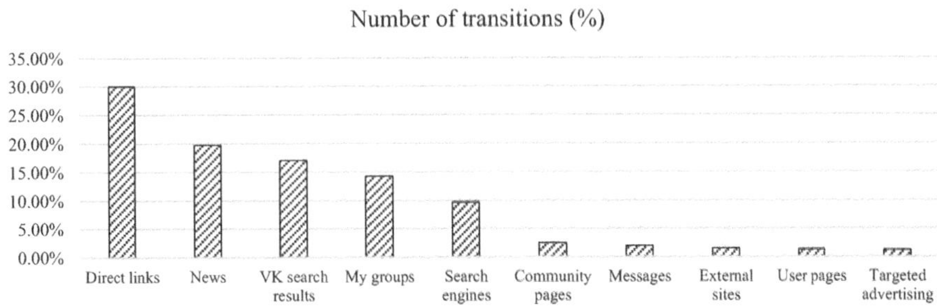

Fig. 2 Sources of traffic over the last year. *Source:* Compiled by the authors based on internal statistics of the VK community "Department of Advertising and Business Communications at the IWEB of RUDN" [9]

and Business Communications at IWEB RUDN") contribute to the overall information space of the APR program. Additionally, there are several publics in the "Overheard at RUDN" category, where there is a constant exchange of opinions among students and prospective students. These platforms contribute to the general information space of the APR program. When obtaining initial information about such an educational program and its main advantages, a prospective student seeks official information on aggregator sites and official RUDN websites and its divisions. This trajectory is optimal for choosing a future specialty and educational program at the university.

The community "Department of Advertising and Business Communications at the IWEB of RUDN" is the official representation of the graduating department in the field of APR at the IWEB of RUDN. The target audience of the public is prospective students. The sources of transitions include targeted advertising (the main share is reflected in the statistics in the form of direct links) during the admission campaign at RUDN, news, VK search results, my groups, and search engines (Fig. 2). This platform has been present on the social network since November 2020.

The analysis of the statistics of the community "Department of Advertising and Business Communications at the IWEB of RUDN" [9] confirms the hypothesis about the significance of VK resources for prospective students.

The main increase in subscribers occurred from May to August of the current year, coinciding with the admission campaign. This figure grows each year and slightly decreases after September. However, a more significant indicator for such a platform is the number of views and reposts of the published materials. The main task is to inform about the APR program, its features, and advantages and publish the opinions of graduates and students' reviews. The format involves video interviews. According to statistics, this type of content gathers the maximum number of views and reposts. Additionally, the community page features a lot of materials about student life: room tours around the campus, interiors of study and leisure areas (lecture halls, computer labs, sports halls, coworking spaces, and cafes with national cuisines), local festivals, KVN competitions, etc.

5 Conclusion

Thus, it is possible to formulate the main conclusions and recommendations for promoting the educational services of universities using the example of the APR program at the IWEB of RUDN. The basis for communication planning should be the analysis of touchpoints with the brand (educational program) throughout the consumer's (prospective student's) journey. Integration of offline and online communications is a significant factor. Nowadays, social media serves as the point of convergence for informational space and is the foundation for the synergy effect of all engaged channels. Any offline event is reflected in the publications of social networks, messengers, forums with reviews, and comments.

For the target audience (i.e., prospective students), creating independent communities to disseminate information crucial for choosing an educational program is necessary. Combining the functions of a structural unit's community with tasks for prospective students is highly inefficient because the platform's content feed format takes valuable information for the target audience out of sight, requiring many screen scrolls. The authors have extensive experience analyzing such combinations with low conversion results for the target audience. Additionally, monitoring mentions of the educational program and its university on the Internet on social media resources, immediate responses from official representatives to negative content, and content managers' comments on neutral posts and reviews are highly significant.

The second important category of platforms for promoting educational programs is aggregator websites for prospective students and official university websites with sections on admission and program descriptions. These are platforms that conclude the decision-making process about choosing a university and an educational program on the client's path.

However, the communication process does not end there because post-communication is equally essential in the digital space. Reviews from first-year students and their parents and content generation on personal social media pages continue the process of forming the informational space of the educational program, drawing in prospective students.

"Consumers have become a factor influencing brands: they advertise, spread, and criticize various products and brands," said Philip Kotler at a seminar on November 12, 2011, in Moscow [10]. The practice of promoting a university's educational program discussed in the research confirms this idea from the well-known American economist and marketer.

References

1. Koekkoek, A., Kleinhans, R., & van Ham, M. (2024). University–community engagement in the Netherlands: Blurring the lines between personal values, societal expectations, and marketing. *Frontiers in Education, 9*, 1005693. https://doi.org/10.3389/feduc.2024.1005693
2. Al-Dmour, R., Al-Dmour, H., & Al-Dmour, A. (2024). The role of marketing mix and social media strategies in influencing international students' university choices in Jordan. *Journal of International Students, 14*(4), 642–663. https://doi.org/10.32674/jis.v14i4.6407
3. Muneja, P. S. (2025). Marketing of electronic resources by universities and colleges in Tanzania. *Global Knowledge, Memory and Communication, 74*(1–2), 564–582. https://doi.org/10.1108/GKMC-08-2022-0186
4. Wahab, A., Aqif, T., & Bint-e-shehzad, Z. (2024). Studying the impact of e-marketing by universities on students' loyalty with the mediation of intention to get enrollment and moderation of eWOM. *Journal of Applied Research in Higher Education, 16*(5), 1486–1499. https://doi.org/10.1108/JARHE-08-2022-0264
5. Association of Communication Agencies of Russia. (2023). *Market size of the media advertising market in January-June 2023*. Retrieved September 30, 2023, from https://www.akarussia.ru/knowledge/market_size/id10462
6. AdIndex. (2023, June 26). *Top 10 high-paying jobs in marketing and advertising – A selection of hh.ru and "More Jobs."* Retrieved September 30, 2023, from https://adindex.ru/publication/opinion/frames/2023/06/23/313468.phtml
7. GlavSovetnik and Interium. (2022). *Research: "Staffing situation in the communications industry."* Retrieved September 30, 2023, from https://glavsovetnik.ru/wp-content/uploads/2022/04/issledovanie-kadrovaya-situacziya-v-kommunikaczionnoj-i-ndustrii.pdf
8. Brand Analytics. (n.d.). *Social media monitoring and analysis system*. Retrieved September 30, 2023, from https://br-analytics.ru/
9. *VK community "Department of Advertising and Business Communications at the IWEB of RUDN."* (n.d.). Retrieved September 30, 2023, from https://vk.com/ad_pr.imeb.rudn
10. SOBOLEV Events. (2011, November 12). *Philip Kotler's seminar "MARKETING 3.0" in Moscow*. Retrieved September 30, 2023, from https://sobolev-events.timepad.ru/event/8648/

Educational Technologies of the Future: Perspective on the Sustainable Development of Universities 4.0 in Russia and Central Asia (Conclusion)

Bruno S. Sergi, Elena G. Popkova, Ainura A. Adieva, and Nurgul K. Atabekova

Educational technologies (EdTech) are dynamically evolving at a rapid pace. Educational technologies have remained unchanged for several centuries since the formation of science and mass higher education. Clear global changes in EdTech began in the late twentieth century and peaked in the first quarter of the twenty-first century. Nowadays, traditional educational technologies are becoming a thing of the past, giving way to multimedia learning tools and distance education.

As shown by the results of the research conducted in this book, Universities 4.0 in Russia and other Central Asian countries are actively implementing advanced educational technologies. This ensures the sustainable development of the system of science and higher education in these countries. Let us summarize the ideas presented in all chapters of the book and formulate its general conclusions:

- The development of EdTech in Universities 4.0 in Russia and Central Asia is not an end in itself but a tool to ensure the inclusiveness of science and higher education in all its aspects—cultural, gender, social, digital, and ecological;
- By leveraging advanced EdTech capabilities, Universities 4.0 enhance management efficiency in science and higher education while fostering entrepreneurial activities that make these institutions increasingly independent from state funding and more autonomous in Russia and Central Asia;
- The use of cutting-edge EdTech ensures the high global competitiveness of Universities 4.0, promotes their international activities, and strengthens their standings in global university rankings, including THE, QS, and ARWU;
- Progressive Universities 4.0 improve the effectiveness of their marketing efforts through EdTech, enhancing their image, reinforcing their reputation, and developing their brands on their websites and social media platforms.

B. S. Sergi et al. (eds.), *University 4.0 and Educational Technology for Sustainable Development*, Education in the Asia-Pacific Region: Issues, Concerns and Prospects 75, https://doi.org/10.1007/978-981-96-7376-6

The comprehensive scientific concept of Universities 4.0 developed in this book enables a comparative analysis of the culture of science and higher education in Central Asia with that of other regions such as China, Europe, and the USA, considering the current status of Universities 4.0. In China, Universities 4.0 are primarily high-tech scientific and educational organizations that utilize EdTech to create advanced technologies and train highly qualified professionals, including IT specialists for high-tech industries that support industrial economic growth.

In Europe, the focus of Universities 4.0 is on training green professionals and creating and commercializing environmental technologies. In the USA, Universities 4.0 are chiefly oriented towards supporting national policies and expanding their global presence, particularly in maintaining leadership in international university rankings. In contrast, Universities 4.0 in Russia and Central Asian countries are primarily dedicated to preserving traditional social values and strengthening integrative ties with neighboring countries, thereby forming and developing a unified and inclusive Central Asian scientific and educational space.

The issue of gender inclusiveness in Universities 4.0 merits special attention because it is interpreted differently in modern cultures. Unlike their counterparts in Europe and the USA, which focus on transgender issues, Universities 4.0 in Russia, Central Asia, and China adhere to traditional values and strive to achieve gender balance by encouraging the participation of women in science and higher education.

The scientific and theoretical framework for Universities 4.0 in Russia and Central Asia developed in this book underscores their broad significance as global centers of science and higher education, opening new avenues for international cooperation. EdTech plays a crucial role in this, expanding opportunities for digital interaction and remote collaboration in science and higher education, which is actievely practiced flourishing in Russia and Central Asia.

In conclusion, educational technologies can represent a challenge and an opportunity. On the one hand, platform-based EdTech solutions have made a significant contribution to crisis management in the global science and higher education systems during the lockdowns amid the COVID-19 pandemic. The experience gained during the pandemic, along with the rapidly established infrastructure supported by the state and the growing attention to EdTech's potential, has greatly expanded the practical application of these technologies.

In the post-pandemic period up to the present, many Universities 4.0 around the world, including those in Russia and Central Asia, have launched large-scale programs to enhance academic mobility and promote lifelong learning through remote work, electronic R&D and conferences, and distance learning. This has increased access to science and higher education for residents of remote areas (e.g., rural regions), strengthened the link between higher education and the labor market, and enabled a balance between educational self-improvement, professional fulfillment, and personal life.

Furthermore, EdTech has automated social interaction processes and decision-making within the science and higher education system, significantly enhancing the ability to ensure social justice and overcome inequality by providing equal opportunities for participation to individuals of varying health conditions, income levels, and social characteristics (e.g., gender, age, and culture), as well as those from different geographic regions—thanks to the neutrality of the digital environment toward these differences.

On the other hand, distance learning has posed a serious challenge to the quality of higher education services, necessitating the development of new educational standards in Universities 4.0. The persistent and widening digital divide hinders the achievement of digital inclusiveness in these institutions, leading to disparities in access to science and higher education for people with differing levels of digital literacy and varying access to telecommunications technologies and infrastructure.

Special attention is warranted for smart EdTech. In considering the impact of artificial intelligence on higher education, the use of generative AI will, in the long run, transform the landscape of Universities 4.0. The automation of data collection and processing in research shifts the functional burden on scientists from conducting field studies to the qualitative interpretation of results and the formulation of scientifically substantiated conclusions and recommendations. Neural networks are being tested as tools for generating scientific texts, and the ethical issues surrounding this process will only grow more pressing over time.

In anticipation of the fifth industrial revolution, it is important to acknowledge that scientific and technological progress continues, and mass implementation of new educational technologies is expected. Robot simulators are one of the promising educational technologies, and widespread adoption of them is expected in the university environment in the near future. It is expected that they will revolutionize the acquisition of practical skills. By using robot simulators in the learning process, university graduates will possess a full set of all necessary practical competencies for professional activities. This will help bridge the gap between education and labor markets.

The promising educational technologies that may gain widespread adoption in the system of science and higher education include virtual and augmented reality. It is expected that this technology will make it possible to simulate various situations, including emergency ones, within university walls, which graduates will later encounter in their workplaces. This will make it possible to flexibly diversify competencies in evolving labor conditions and increase the attractiveness of young professionals to employers.

Another promising educational technology expected to spread among universities in the coming years is artificial intelligence, as well as its derivative, machine learning. The use of smart educational technologies will make the educational process even more flexible and individualized, adapting it to the unique needs of each student. Moreover, the production capacity of universities will increase manifold; their activities will become scalable. The role of university professors will no longer be limited to lecturing or testing students' knowledge. It will involve programming artificial intelligence and launching machine learning algorithms essential for the learning process.

Therefore, in future research and publications, it is advisable to delve deeper into the perspective of sustainable development of universities in Russia and Central Asia with the implementation of future educational technologies. During the peak of the Fifth Industrial Revolution, there will emerge Universities 5.0. It is prudent to begin developing their model for Russia and other countries of Central Asia to reduce uncertainty and risks associated with university modernization and support the sustainable development of science and higher education. These issues are proposed to be further explored in subsequent research following the presented book.

The manufacturer's authorised representative in the EU is Springer Nature Customer Service Centre GmbH, Europaplatz 3, 69115 Heidelberg, Germany. If you have any concerns regarding our products, please contact ProductSafety@springernature.com

Printed and bound by CPI Group (UK) Ltd, Croydon, CR0 4YY
30/09/2026
02225296-0015